SALTERS ADVANCED CHEMISTRY

Chemical
Storylines

Central Team

George Burton

John Holman

John Lazonby

Gwen Pilling

David Waddington

Heinemann

CONTENTS

ACKNOWLEDGEMENTS

The authors and publishers would like to thank the following for permission to reproduce copyright material:

GM and MJ Pilling, 'What is photochemical smog?', Chemistry Review 5:5, 1997, Philip Allan Publishers (Fig. 23, Fig. 24, p.34); AET Technology plc / DETR (Fig. 26, p.35); FS Rowland, 'Light, Chemical Change and Life', © (1982) Open University Press (Fig. 4, p.65); Science, Vol. 228, pp.1309–11, © (1985) American Association for the Advancement of Science (Fig. 7b, p.67); Nature, Vol. 315, no. 6016, pp.207–210, © (1985) Macmillan Magazines Ltd. (Fig. 14, p.73); Environmental Science Technology Vol. 24, no.4, pp.624, © (1991) American Chemical Society (Fig. 17, p.74); NASA Ozone Processing Team at Goddard Space Flight Center (Fig. 18, p.75); AGAGE/CSIRO Atmospheric Research, Australia, UNEP/WMO, Switzerland (Fig. 21, p.78); Paul Brown, © The Guardian (Fig. 22, p.78); Newsman / The Observer (Fig. 23, p.79); Kelter, Carr and Scott, 'A World of Choice', © (1998) The McGraw-Hill Companies (Fig. 28, p.82); The Independent, (Fig. 30, p.83); Dave Keeling and Tim Whorf, Scripps Institute of Oceanography (Fig. 35, p.86); The Independent, (Fig. 39, p.90); Geoff Thompson/The Independent Magazine (Fig. 40, p.90); National Institute of Science Technology (Fig. 6, p.110; Fig. 7, p.267); Bayer AG (Fig. 12, p.114); ABPI (extract, p.117); Zeneca Agrochemicals (Fig. 29, p.192; Fig. 41, Fig. 42, p.199); O. Kaarstafd, Norsk Hydro ASA (Fig. 34, p.195); BASF (Fig. 38, p.197); British Agrochemicals Association, (Fig. 47, p.203); Scientific American, 261(3), pp.80–94, © (1989) Hank Iken (Fig. 23, p.248); New Scientist (Fig. 26, p.250, Fig. 27, p.251); Chemical Industries Association (Fig. 1, p.285); T. Kletz, 'Learning from Accidents in the Chemical Industry', © (1994) Gulf Publishing Company (extract, p.289); Institute of Petroleum (Fig. 18, p.294).

The authors and publishers are grateful to the following for permission to reproduce photographs:

Cover photo J. Paul Getty Museum, Malibu, California/Bridgeman Art Library

EL1, SPL; EL2, Robert Erbe; EL4, Damien Lovegrove/SPL; EL5, John Lazonby, Department of Chemistry, University of York; EL10, Science Museum Library; EL11, Copyright © The Nobel Foundation; EL12, SPL/Royal Observatory Edinburgh; EL14, SPL/NOAA; EL22, SPL/Roger Resmeyer/Starlight; EL23, NASA; EL24, NASA/SPL; DF1, Daimler Chrysler/Mercedes Benz; DF4, Philip Parkhouse; DF5, Roger Scruton; DF6, Robert Erbe; DF7, Martin Bond/SPL; DF11, The Shell Art Collection at the National Motor Museum Beaulieu; DF13, Robert Erbe; DF15, Shell International Petroleum Company Co Ltd; DF18, BP Amoco Chemicals; DF22a&b, Air Resource Specialists Inc.; DF27, Dr Harald Geiger, Physikalische Chemie, Bergische Universitat-Gesamthochschule, Wuppertal, Germany; DF32, Johnson Matthey; DF33, © MSI/Sutton USA; DF34, ICI; DF35, South American Pictures; DF37, A BP Solar, B Roger Scruton, C J Allan Cash, D SPL/Hank Morgan; DF38, Daimler Chrysler/Mercedes Benz; M2, ZEFA; M3, J Allan Cash; M4, Brian Simpson; M5, Brian Simpson; M6, Brian Simpson; M8, Brian Simpson; M9, Brian Simpson; M11, © USDA/APHIS; M12, Brian Simpson; MF15, Kodak; MF16, © Salt Union Ltd: Photo: Weston Point Studios; M17, C M Dixon; M19, left: SPL/Martin Land, right: GSF; M20, Rio Tinto; M24, © Copper Development Association; M27, © Copper Development Association; M28a&b, © Copper Development Association; A1, SPL; A3, Popperfoto; A6, Robert Erbe; A7a, SPL/Phillipe Plailly; A11, Lord Porter/Imperial College; A12, Robert Erbe; A13, Sherry Rowland; A15, SPL/Simon Fraser; A16, © British Antarctic Survey; A20, SPL/Dr Jeremy Burges; A25, Hutchinson Library; A29, SPL/NASA/Goddard Institute for Space Studies; A31, CSIRO Atmosphere Research, Australia; A33, SPL/Graham Ewens; A34, Rex Features/Sipa; A37, SPL/Andrew Syred; PR1, © Pictor International; PR3, Popperphoto; PR4, Science Museum/Science & Society Picture Library; PR5, Archiv for Kunst and Geschichte; PR6, Royal Society of Chemistry; PR7, © The Nobel Foundation; PR9, Shell UK; PR11, © Elenac/BASF; PR12, TEK Image/SPL; PR13a, © London Aerial Photo Library; PR13b, Shell UK; PR14, SPL; PR15, W L Gore & Associates (UK) Ltd; PR16, ENAK Ltd; PR19, Zipperling Kessler and Co; PR20, CDT Ltd; WM1, A–Z Botanical Collection; WM2, A–Z Botanical Collection; WM3, Mary Evans Picture Library; WM4, Mary Evans Picture Library; WM5, Courtesy Bruker UK Ltd; WM8, Courtesy Micromass; WM10, The Garden Picture Library; WM11, © Bayer Pharmaceuticals; WM12, © Bayer Pharmaceuticals; WM13, © Bayer Pharmaceuticals; WM14, © Bayer Pharmaceuticals; WM15, NASA; WM16, © Bayer Pharmaceuticals; WM17, SPL/Prof. P Motta, Dept of Anatomy/University "La Sapienza" Rome; WM18, © Popperfoto; WM21, Zeneca Pharmaceuticals; DP1, The Bridgeman Art Library; DP5, © Hagley Museum and Library; DP6,

Topham Picture Source; DP7, © Hagley Museum and Library; DP8, © ICI; DP9, © Elenac/BASF; DP10, Shell UK; DP12, Shell UK; DP13, David Guyon/SPL; DP14, © Bayer Pharmaceuticals; DP17a, R Guerrin/PPL Ltd; DP17b, Chemical Industry Education Centre, The University of York; DP19, PEEK INC; DP20, © British Aerospace Airbus; DP23, Environmental Images; DP24, British Plastics Federation; DP25, Shell UK; DP26, Shell UK; DP29a, © SPL; DP29b, © Alex Bartel/SPL; EP1, Mary Aitken; EP2, SPL: James King-Holmes; EP3, Diabetic Association; EP5, Corbis; EP10, British Diabetic Association; EP13, Roger Scruton; EP21, SPL/Omikron; EP22, SPL; EP24, European Molecular Biology Laboratory; EP25, BioRad; EP26, DRCT/Custom Medical Stock Photo/SPL; EP27, Mary Evans Picture Library; EP28, Novo Nordisk; EP30, Corbis; EP31, Corbis; EP32, Ace Photo Library; EP33, Dr Apse/Thanh Nguyen; EP35, 36, 37, 39, 43, 51 Prof R E Hubbard/Dept of Chemistry/University of York; EP40, Stuart Priest, Dept of Chemistry/University of York; EP41, Dept of Chemistry/University of York; EP42, Corbis; EP43, Prof R E Hubbard/ Dept of Chemistry/University of York; EP44, Xiao Bing/Dept of Chemistry/University of York; EP50, SPL/Chris Priest and Mark Clarke; EP55, Barnaby's Picture Library; EP56, Barnaby's Picture Library; EP57, Novo Nordisk; SS2, 3, 4, 6, 7, 9, 12, 14, 15, Corus (formerly British Steel); SS18, Corbis; SS19, Mary Evans/Barry Norman Collection; SS20, BASF; SS23, Dr Stuart Thorne/courtesy of The Science Museum; SS24, Avesta, Sheffield; SS25, J Allan Cash; SS26, Ace Photo Library; SS28, John Olive, Department of Chemistry/University of York; SS31, Jim Kershaw/ reproduced by kind permission of the Dean and Chapter at York; AA1, SPL/Jim Gipe/Agstock; AA2, Environmental Images/Amanda Gazidis; AA3, Elenac/BASF; AA8, David Williams Picture Library; AA11a, SPL/Robert de Gugliemo AA11b, Ace Photo Agency; AA14, The Natural History Museum; AA16, GSF; AA20, The Natural History Museum; AA21, Ace Photo Agency; AA24, Ace Photo Agency; AA26, Pete Addis/Environmental Images; AA30, Holt Studios International; AA31, SPL/Nelson Medina; AA33, Rothamsted Experimental Station; AA35, Roger Scruton; AA36, © Hulton Getty; AA37, Elenac/BASF; AA38, Elenac/BASF; AA40, Hydro Media; AA43, Holt Studios International; AA44, British Agrochemicals Association Ltd; AA45, Heather Angel; AA48, Shell; AA51, Zeneca Agrochemicals; AA52, Zeneca Agrochemicals; AA53, Zeneca Agrochemicals; CD1, AKG; CD2, Elenac/BASF; CD3, Mary Evans Picture Library; CD4, Mary Evans Picture Library; CD5, Milepost; CD7, Mary Evans Picture Library; CD8, Philip Parkhouse; CD9, © The National Gallery, London; CD10, The Bridgeman Art Library; CD11, © The National Gallery, London; CD12, © The National Gallery, London; CD13, © The National Gallery, London; CD14, © The National Gallery, London; CD17, © The National Gallery, London; CD18, © The National Gallery, London; CD19, Dr Tony Travis; CD20, adapted from The Rainbow Makers by Tony Travis, Lehigh University Press/Associated University Presses; CD21, Zeneca Pharmaceuticals; CD22, Royal Society of Chemistry; CD23, Dr Tony Travis; CD24, Hoechst; CD27, The Colour Museum, Bradford; CD29, Dr Tony Travis; CD31, Elenac/BASF; CD32, Science & Society Picture Library; CD38, Zeneca Agrochemicals; CD41, Elenac/BASF; O1, J Allan Cash; O2, Robert Harding; O3a&b, Jim Macintosh/Nutrasweet/Kelco; O7, Maldon Crystal Salt Co.; O8, The Skyscan Photo Library: O9, Peter Ryan/SPL; O10, Tom Van Sant/Geosphere Project, Santa Monica/SPL; O12, Mary Evans Picture Library; O16, SPL/Jans Hinsdi; O22, J Allan Cash; O30a&b, SPL/Earth Satellite Corp.; O33, J Allan Cash; O34, J Allan Cash; O35, Scottish National Heritage; O37, © Pictor International; O39, GSF; O40, Planet Earth Pictures; O41, © Skyscan PhotoLibrary/Brian Lee; MD1, SPL/JC Revy; MD2, Popperfoto; MD3, J Sainsbury; MD4, Lion Laboratories Ltd; MD5, Lion Laboratories Ltd; MD6, Lion Laboratories Ltd; MD17, SPL/Simon Fraser; MD19, Teri McDermott/Custom Medical Stock Photo/SPL; MD20, SPL/Chris Priest and Mark Clarke; MD22, Zeneca Pharmaceuticals; MD23, John Olive, Department of Chemistry, University of York; MD24, SPL/Hattie Young; MD25, Chris Mattison; MD27, Steve Mumford, Department of Chemistry, University of York; MD30, SPL/Andrew McClenaghan; MD31, SPL/Andrew McClenaghan; MD32, By kind permission of Smithkline Beecham; MD36a&b, Dr Irene Francois, Chairman of Society for Medicine Research; MD37, SmithKline Beecham, Irvine; VCI2, BP Amoco Chemicals; VCI4, BP Amoco Chemicals; VCI5, BP Amoco Chemicals; VCI6, BP Amoco Chemicals; VCI7, BP Amoco Chemicals; VCI8a&b, © H.M.S.O; VCI9, SPL; VCI10, BP Amoco; VCI11, ICI; VCI12, Westpoint Studios; VCI13, Bayer AG/Leverkusen, Germany; VCI14, Longannet Power Station: Scottish Power; VCI15, BP Amoco; VCI16, BP Amoco; VCI17, BP Amoco Chemicals.

CORE TEAMS

Many people have contributed to the first and second editions of the Salters Advanced Chemistry course, and a full list of contributors is given at the back of this book. They include the following:

Central Team

First Edition

George Burton	Cranleigh School and University of York
John Holman (Project Director)	Watford Grammar School and University of York
Margaret Ferguson (1990–1991)	King Edward VI School, Louth
Gwen Pilling	University of York
David Waddington (Chairman of Steering Committee)	University of York

Second Edition

John Lazonby	University of York
Gwen Pilling (Project Director)	University of York
David Waddington	University of York

Advisory Committee

Dr Peter Doyle	Zeneca Group
Dr Tony Kirby, FRS	University of Cambridge
Professor The Lord Lewis, FRS (Chairman)	University of Cambridge
Sir Richard Norman, FRS	University of Oxford
Mr John Raffan	University of Cambridge

Sponsors

Many industrial companies have contributed time and expertise to the development of the Salters Advanced Chemistry Course. The work has been made possible by generous donations from the following:

The Salters Institute for Industrial Chemistry
The Association of the British Pharmaceutical Industry
BP Chemicals
British Steel
Esso UK
Zeneca Agrochemicals
The Royal Society of Chemistry
Shell UK

Dedication

To Dick Norman

Heinemann Educational Publishers,
Halley Court, Jordan Hill, Oxford, OX2 8EJ
Part of Harcourt Education.

Heinemann is the registered trademark of Harcourt Education Limited.

First published 1994
This edition published 2000

04 03
10 9 8 7 6

ISBN 0 435 63119 5

Edited by Tim Jackson

Designed, illustrated and typeset by Gecko Limited, Bicester, Oxon.

Original illustrations © Heinemann Educational Publishers 2000

Printed and bound in Spain by Edelvives

Tel: 01865 888058 www.heinemann.co.uk

INTRODUCTION FOR STUDENTS

The Salters Advanced Chemistry course is made up of 13 units, together with a structured industrial visit. This book contains the **Chemical Storylines** which form the backbone of each teaching unit. There is a separate book of **Chemical Ideas**, and **Activities**.

Each unit is driven by the Storyline. You work through the Storyline, making 'excursions' to Activities and Chemical Ideas at appropriate points.

The Storylines are broken down into sections and sub-sections. You will find that there are numbered Assignments at intervals. These are designed to help you through the Storyline and to probe understanding, and they are best done as you go along.

Excursions to Activities

As you work through each Storyline, you will find that there are references to Activities. Each Activity is referred to at that point in the Storyline to which it most closely relates. Of course, you may not be able to do the Activity straight away, but it should be done not too long after that part of the Storyline.

Activities are numbered to correspond to the relevant part of the Storyline.

Excursions to Chemical Ideas

As you work through the Storylines, you will also find that there are references to sections in the book of Chemical Ideas. These cover the chemical principles that are needed to understand that particular part of the Storyline, and you will probably need to study that section of the Chemical Ideas book before you can go much further.

As you study the Chemical Ideas you will find Problems to tackle. These are designed to check and consolidate your understanding of the chemical principles involved.

Building up the Chemical Ideas

Salters Advanced Chemistry has been planned so that you build up your understanding of chemical ideas gradually. For example, the idea of chemical equilibrium is introduced in a simple, qualitative way in *The Atmosphere* unit. A more detailed, quantitative treatment is given in *Engineering Proteins* and *Aspects of Agriculture*, and applied to acids and precipitation in *The Oceans*.

It is important to bear in mind that the Chemical Ideas book is not the only place where chemistry is covered! The Chemical Ideas cover chemical principles that are needed in more than one unit of the course. Chemistry that is specific to a particular unit is dealt with in the Storyline itself and in related Activities.

How much do you need to remember?

The AS/A level specifications for Salters Advanced Chemistry define what you have to remember. Each teaching unit concludes with a 'Check your Notes' activity, which you can use to check that you have mastered all the required knowledge, understanding and skills for that unit. 'Check your Notes' tells you whether a topic is to be found in the Chemical Ideas, Chemical Storylines or Activities.

We hope that you will enjoy and learn as much about the fascinating world of chemistry from the Salters Advanced Chemistry books as we have enjoyed and learned writing them.

George Burton   John Holman   John Lazonby   Gwen Pilling   David Waddington

THE ELEMENTS OF LIFE

Why a unit on THE ELEMENTS OF LIFE?

This unit tells the story of the elements of life: what they are, how they originated and how they can be detected and measured. It shows how studying the composition of stars can throw light on the formation of elements which make up our own bodies.

This storyline begins with the elements from which our bodies are formed. You learn how to measure amounts of these elements (in terms of atoms) and, thus, how to calculate chemical formulae. The story then looks in more depth at two elements – iron and calcium – and this leads into learning about patterns in the properties of elements and the Periodic Table.

The second part of the unit looks at the origins of the elements, and introduces you to ideas about the structure of atoms. It concludes with a brief look at how elements combine to form compounds such as the 'molecules of life' which form the body.

Overview of chemical principles

In this unit you will learn more about …

ideas you will probably have come across in your earlier studies

- the Periodic Table
- protons, neutrons and electrons
- radioactivity and ionising radiation
- chemical bonding
- representing reactions
- the wave model of light
- the electromagnetic spectrum

… as well as learning new ideas about

- relative atomic masses and relative formula masses
- amount of substance (moles and the Avogadro constant)
- chemical formulae
- nuclear fusion and nuclear equations
- the photon model of light
- atomic spectra
- the electronic structure of atoms
- ionic and covalent bonding
- shapes of molecules
- metallic bonding.

The chemical ideas about amount of substance, atomic structure and chemical bonding are only *introduced* in this unit. They will be consolidated and developed in later units.

This technique of taking ideas only as far as you need to know them in order to follow the storyline you are studying, and then building on them by repeating the process in later units, is central to the Salters' approach to chemistry at this level.

EL1 *What are we made of?*

Elements and the body

If you asked a number of people the question, 'What are you made of?' you would get a variety of different answers. Some people would use biological terms and talk about organs, bones and so on. Others might answer in more detail and mention proteins, fats and DNA. A chemist would be most likely to talk about atoms and molecules, or elements and compounds.

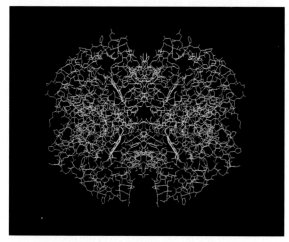

Figure 1 A model of the haemoglobin molecule – a biological molecule essential to life.

As a chemist, you know that you are not really made up of a mixture of elements but rather a mixture of compounds, many of which appear quite complicated. You will be finding out more about some of these compounds later in the unit. To begin with, however, you will look at the elements which are most likely to be in the compounds in your body.

Elements in the body are classified as one of three types:

- **major constituent elements**, which make up 2%–60% of all the atoms present; these are hydrogen, oxygen, carbon and nitrogen
- **trace elements**, which make up 0.01%–1%, eg calcium and phosphorus
- **ultra-trace elements**, which make up less than 0.01%, eg iron and iodine.

Counting atoms of elements

Table 1 gives you the masses and proportions of the major constituent elements in a person of average mass (about 60 kg).

Element	Mass in 60 kg person/g	Percentage of atoms
oxygen	38 800	25.5
carbon	10 900	9.5
hydrogen	5 990	63.0
nitrogen	1 860	1.4

Table 1 The major constituent elements in the human body.

Notice that in the two columns of numbers in Table 1 different elements appear to be more important in the body. For example, there are more atoms of hydrogen in your body than atoms of any other element, but hydrogen contributes far less than carbon or oxygen to the mass of your body. So a list of masses alone does not allow us to decide which of the three categories of 'elements of life' an element belongs to. To do this, we need to determine how many **atoms** of each element there are.

Chemists can convert masses of elements into a measure of the number of atoms they contain by making use of **moles**. When we are dealing with elements, one mole is the amount of an element which contains the same number of atoms as 12 g of carbon.

Because atoms have exceedingly small masses, the number of atoms in one mole of an element is large. In fact it is very, very large – approximately 6×10^{23} atoms per mole. There are almost 1000 moles of carbon in a 60 kg person, so there are an awful lot of carbon atoms in the human body.

Once the number of moles of atoms of each element in the body has been calculated from its mass, these can be added to give the total number of atoms in the body. The percentage of atoms of each element in the body can then be worked out using the total.

Chemical Ideas 1.1 tells you more about moles and how to use them in calculations.

Table 1 shows you that the ratio of hydrogen atoms to oxygen atoms in your body is almost 2:1. That's because 65% of the mass of your body is water and the chemical formula of water is H_2O. You have probably known that formula for a long time. But how did you know it? Did you ever work it out? It's not difficult – provided that you know about moles.

Activity EL1 gives you a chance to practise using moles to work out the formula of a compound.

Putting information in a table like Table 1 is often not the most striking way to present it. Pie charts or bar charts, for example, can be better.

Draw a pie chart to represent the proportions by mass of the four major constituent elements in the body. Label the fifth 'slice' of the chart to represent the contribution of all the trace and ultra-trace elements.

Now draw another pie chart, this time to represent the percentages of atoms.

Do you think that pie charts provide a better way of seeing and comparing the information in this case?

EL2 _Take two elements ..._

A trace is all you need

Section **EL1** shows that about 99% of your body is made up of atoms of only four elements: hydrogen, oxygen, carbon and nitrogen. These elements are obviously vital for life. But the trace and ultra-trace elements, although they make up only the remaining 1% or so of the body, are also essential for good health.

Table 2 lists the proportions of the trace elements in a 60 kg person. The ultra-trace elements – which include cobalt, copper, iodine, iron, manganese, molybdenum, silicon, vanadium and zinc – are not given because their quantities are so small.

Element	Mass in 60 kg person/g	Percentage of atoms
calcium	1200	0.31
phosphorus	650	0.22
potassium	220	0.06
sulphur	150	0.05
chlorine	100	0.03
sodium	70	0.03
magnesium	20	0.01

Table 2 The trace elements in the human body.

Figure 2 Milk, hard cheeses such as Cheddar, cottage cheese and bread are rich sources of calcium.

Table 3 shows just how essential the trace and ultra-trace elements are for proper body function.

Element	Function
calcium	major component of bone; required in some enzymes
phosphorus	essential for the synthesis of chemicals in the body and for energy transfer
sulphur	required in proteins and other compounds
copper	essential in enzymes involved in oxidation processes
iodine	an essential component of thyroid hormones
iron	contained in haemoglobin and many enzymes
zinc	required for the activity of many enzymes

Table 3 Functions of some trace and ultra-trace elements.

The human body is often deficient in iron and calcium. The next section of the unit looks in more detail at the role and chemistry of these elements.

An iron story

Iron carries out a vital role in the body: as part of the substance **haemoglobin**, present in blood, it is responsible for the transport of oxygen.

The body can easily become deficient in iron, particularly young teenagers who are growing fairly rapidly. Iron deficiency, or anaemia, can usually be remedied with a course of iron tablets and an alteration in diet.

There can also be too much iron in the body due to an inherited genetic disorder. Over a lifetime the body can accumulate many times the normal quantity of iron. It gets stored in several organs, including the liver, and can only be treated by taking blood from the patient.

Figure 3 shows part of the haemoglobin molecule. It consists of an iron atom at the centre of a ring structure. The ring is the _haem_ part of the molecule. Also associated with the central iron atom is the protein, _globin_.

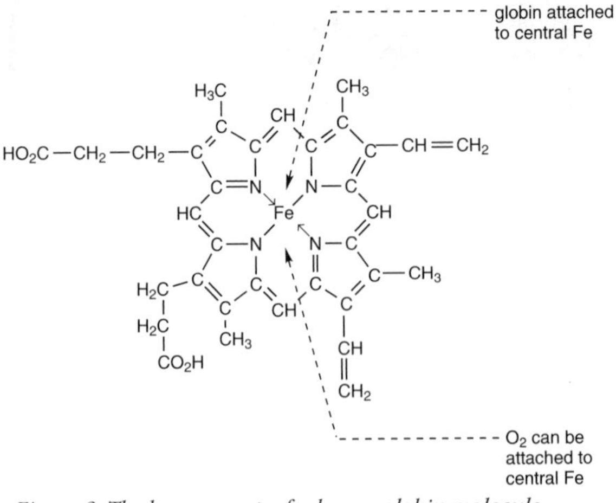

Figure 3 The haem group of a haemoglobin molecule.

Haemoglobin is able to do its job because oxygen molecules become attached to the iron (Fe) atom. Oxygen is not the only substance to behave in this way: carbon monoxide can become similarly attached. Unfortunately, it is attracted more strongly than oxygen. This is why carbon monoxide acts as a poison if it gets into the body: it prevents the haemoglobin from carrying oxygen. One of the reasons why people who smoke are particularly prone to heart disease is that the carbon monoxide they inhale from cigarettes reduces the amount of oxygen their blood can carry. Their hearts therefore have to work much harder to maintain an adequate supply of oxygen to the body.

Once you are 18 years old you can become a blood donor. Because it is inadvisable for people with iron deficiency to give blood, a small sample of blood is always tested first to find the iron concentration. A drop of blood is taken from the donor's thumb and placed in a solution of copper sulphate. The copper sulphate reacts with the haemoglobin to form an insoluble compound, which appears as a white-coloured 'blob' in the solution.

The density of the copper sulphate solution is carefully arranged so that if there is sufficient haemoglobin (and therefore iron) in the blood, the density of the blob is enough to make it sink. If this happens, the person can go on to make a blood donation.

Figure 4 Anyone over 18 years of age can volunteer to become a blood donor.

However, if the blob floats there is insufficient haemoglobin in the blood, and the person is not allowed to make a blood donation. A sample of about $20\,cm^3$ of blood is taken for more detailed analysis from people found to have low haemoglobin levels.

It isn't safe for you to work with samples of blood, but you can find out the concentration of iron in crystals of an iron compound in **Activity EL2.1**.

Joanna's story

There are several places in the storylines where you can read about people who use chemistry in their jobs. All the stories are about *real* people. This is about Joanna who works in a medical laboratory where samples of blood are sent for analysis.

After A-levels in Chemistry, Biology and Physics, Joanna obtained a degree in Applied Biology. This was a sandwich course which included one year's work experience in a hospital laboratory. On leaving university she did work-based training at Leeds General Infirmary and, following a successful viva, she became a State Registered Medical Laboratory Scientific Officer (MLSO). She then continued to work at the hospital and studied, part-time for two years, for an MSc in Pathological Science. This degree enabled her to become a Fellow of the Institute of Biomedical Sciences, which makes her eligible to apply for senior posts in the future. Joanna has been in her current post at York District Hospital for one year. She describes here exactly what happens to blood sent for analysis.

Figure 5 Joanna using an automated chemistry analyser. This instrument can estimate the concentration in blood of ions such as Na^+, K^+ and Ca^{2+}, compounds such as enzymes, urea and bilirubin, as well as iron and a number of drugs.

"Two $10\,cm^3$ blood samples are sent to our laboratories. One sample is allowed to clot. The second has a compound known as EDTA added, which interacts with calcium ions in the blood and prevents the blood from clotting.

"The tests done on the uncoagulated sample are called full blood count tests and involve counting the number of red blood cells, the number of white blood cells and the number of platelets. An electronic cell counter does the job once the sample has been loaded.

"The machine also measures the size of the red blood cells. If the cells are smaller than normal, this indicates that the person has anaemia. If, on the other hand, they are bigger than normal, this suggests that the person might have a vitamin deficiency.

"When the first sample has clotted, the concentration of iron can be measured. First the sample is centrifuged. This separates out the cellular material in the blood from the serum.

"The serum is a pale straw-coloured liquid, and this is the part used for the second set of tests. The iron is dissolved in this serum. The concentration of iron is determined by measuring the colour it produces when it combines with a reagent called ferrozine. The colour is measured by a colorimeter fitted into a multi-analyser. This records the light absorbed when it passes through solutions of different concentrations. The serum from the patient is compared with solutions containing known concentrations of iron.

"These tests are done on the analyser shown in Figure 5. The results are then sent back to the patient's doctor. The doctor will contact the patient it any treatment is required as a result of having the tests carried out."

A calcium story

Calcium is essential for the healthy development of bones, particularly in children and young people who are still growing. It is also important for a pregnant woman to eat a calcium-rich diet to ensure that not only her bones remain strong, but her baby's bones are properly formed.

Figure 6 Diet is important during pregnancy, for both the mother and her baby.

Bones containing insufficient calcium are referred to as **demineralised** bones. One of the consequences of people living longer is that a condition called **osteoporosis** is becoming far more common. People with osteoporosis have demineralised bones and tend to suffer from frequent bone fractures and curvature of the spine. The condition is usually observed in women who have passed through the menopause.

An increased intake of calcium (see Figure 2) can help the condition, and scientists wanted to find out how much of the calcium which is eaten is taken up by the bloodstream and becomes part of the bones. One way is to use a **tracer technique**. Patients were given a *meal* containing a radioactive calcium isotope, and the amount of radioactive calcium which was then absorbed into the bloodstream from the gut was measured.

ASSIGNMENT 2

a What sort of measurements do you think it would be necessary to make on patients if they received treatment involving a radioactive tracer?

b Suggest why there are advantages in using a *short-lived* radioactive isotope of calcium.

c Suggest what might be the disadvantages of using a short-lived isotope.

You can read about the structure of the atom, the occurrence of isotopes and radioactive decay in **Chemical Ideas 2.1** and **2.2**.

However, radioactive isotopes can be dangerous and an alternative method was to use strontium, an element very similar to calcium. The approach developed from observations made in the 1960s that some children had strontium in their bones. The quantities were small, but were still much larger than normal.

The strontium was present as the radioactive isotope strontium-90, a product of nuclear fission. Its uptake into children's bones was thought to result from the increased atmospheric concentration of strontium-90 which accompanied the testing of nuclear weapons in the 1950s.

It revealed a very important fact for scientists, which is that strontium, just like calcium, is taken up into bones during their formation. Normal, non-radioactive strontium can be analysed in the presence of calcium because strontium compounds give out a characteristic red light when they are placed in a flame. The intensity of this red light is a measure of how much strontium is present.

Scientists could therefore use *non-radioactive* strontium to detect the rate at which strontium is absorbed into the blood from the gut. They found that the rate of uptake of strontium was the same as the rate at which calcium was absorbed in tracer experiments. Using strontium is far safer than using a radioactive calcium isotope.

The ability to monitor patterns of absorption of calcium through the use of non-radioactive strontium has led to a greater understanding of calcium deficiencies and improvements in the treatment of bone disorders such as osteoporosis.

The calcium content of bones can now be routinely monitored in hospitals. The procedure is quick and accurate and involves measuring bone density by comparing the transmission of X-rays through bone and soft tissue.

Good study habits, right from the start, are invaluable to your success in this course. **Activity EL2.2** will help you to take better notes – an important aspect of study skills.

EL3 *Looking for patterns in elements*

When the elements were being discovered, and more was being learned about their properties, chemists looked for patterns in the information they had assembled.

You have seen that there are close similarities between calcium and strontium. Magnesium and barium can be added to them to make a 'family' or 'group' of four elements. Your earlier studies probably introduced you to two other 'families' – lithium, sodium and potassium, and fluorine, chlorine, bromine and iodine.

Activity EL3.1 can help you look for patterns in the properties of two of these 'families' of elements.

You will need to write simple balanced equations in **Activity EL3.1**. If you feel in need of revision, **Chemical Ideas 1.2** will help.

Fifty-nine of the 92 naturally occurring elements had been discovered by 1850, so the search for patterns among the elements was particularly fruitful in the mid-19th century.

Much of the work was done by Johann Döbereiner and Lothar Meyer in Germany, John Newlands in England, and Dmitri Mendeleev in Russia. These chemists looked at similarities in the chemical reactions of the elements they knew about, and also patterns in physical properties, such as melting points, boiling points and densities.

Reactions to their suggestions were not always favourable. The March 1866 edition of *Chemical News*, a journal of the Chemical Society, shows you what Professor G. F. Foster had to say about Newlands' 'Law of Octaves' (see Figure 7). Some of Newlands' symbols will look unfamiliar. They are listed below with their modern equivalents:

- G is now Be
- Bo is now B
- Ro is now Rh
- By 1897 it had been established that Di ws not an element, but a mixture of two closely related elements.

Figure 7 An extract from Chemical News, *March 1866.*

PROCEEDINGS OF SOCIETIES

CHEMICAL SOCIETY
Thursday, March 1

*Professor A. W. WILLIAMSON, Ph.D., F.R.S.,
Vice-President, in the Chair*

Mr. JOHN A. R. NEWLANDS read a paper entitled '*The Law of Octaves, and the Causes of Numerical Relations among the Atomic Weights.*' The author claims the discovery of a law according to which the elements analogous in their properties exhibit peculiar relationships, similar to those subsisting in music between a note and its octave. Starting from the atomic weights on Cannizzaro's system, the author arranges the known elements in order of succession, beginning with the lowest atomic weight (hydrogen) and ending with thorium (=231.5); placing, however, nickel and cobalt, platinum and iridium, cerium and lanthanum, &c., in positions of absolute equality or in the same line. The fifty-six elements so arranged are said to form the compass of eight octaves, and the author finds that chlorine, bromine, iodine, and fluorine are thus brought into the same line, or occupy corresponding places in his scale. Nitrogen and phosphorus, oxygen and sulphur, &c.,

are also considered as forming true octaves. The author's supposition will be exemplified in Table II., shown to the meeting, and here subjoined:–

Dr. GLADSTONE made objection on the score of its having been assumed that no elements remain to be discovered. The last few years had brought forth thallium, indium, caesium, and rubidium, and now the finding of one more would throw out the whole system. The speaker believed there was as close an analogy subsisting between the metals named in the last vertical column as in any of the elements standing on the same horizontal line.

Professor G. F. FOSTER humorously inquired of Mr. Newlands whether he had ever examined the elements according to the order of their initial letters? For he believed that any arrangement would present occasional coincidences, but he condemned one which placed so far apart manganese and chromium, or iron from nickel and cobalt.

Mr. NEWLANDS said that he had tried several other schemes before arriving at that now proposed. One founded upon the specific gravity of the elements had altogether failed, and no relation could be worked out of the atomic weights under any other system than that of Cannizzaro.

Table II – Elements arranged in Octaves

H	1	F	8	Cl	15	Co & Ni	22	Br	29	Pd	36	I	42	Pt & Ir	50
Li	2	Na	9	K	16	Cu	23	Rb	30	Ag	37	Ca	44	Os	51
G	3	Mg	10	Ca	17	Zn	24	Sr	31	Cd	38	Ba & V	45	Hg	52
Bo	4	Al	11	Cr	19	Y	25	Ce & La	33	U	40	Ta	46	Tl	53
C	5	Si	12	Ti	18	In	26	Zr	32	Sn	39	W	47	Pb	54
N	6	P	13	Mn	20	As	27	Di & Mo	34	Sb	41	Nb	48	Bi	55
O	7	S	14	Fe	21	Se	28	Ro & Ru	35	Te	43	Au	49	Th	56

	Group I	Group II	Group III	Group IV	Group V	Group VI	Group VII	Group VIII
Period 1	H							
Period 2	Li	Be	B	C	N	O	F	
Period 3	Na	Mg	Al	Si	P	S	Cl	
Period 4	K Cu	Ca Zn	* *	Ti *	V As	Cr Se	Mn Br	Fe, Co, Ni
Period 5	Rb Ag	Sr Cd	Y In	Zr Sn	Nb Sb	Mo Te	* I	Ru, Rh, Pd

Figure 8 A form of Mendeleev's Periodic Table (the asterisks denote elements which he thought were yet to be discovered).

Just 3 years later (1869), however, Mendeleev's groupings (see Figure 8) were seen as much more credible. Elements were arranged in order of increasing atomic masses, so that elements with similar properties came in the same vertical group.

Unlike Newlands, Mendeleev left gaps in his table of elements. These gaps were very important: they allowed for the discovery of new elements. (Look at Dr Gladstone's comments in Figure 7.)

Mendeleev was so confident of the basis upon which he had drawn up his table that he made predictions about elements which had yet to be discovered. In 1871, he predicted the properties of an element he called **eka-silicon**, which he was confident would eventually be discovered to fill the gap between silicon and tin in his Periodic Table. Mendeleev's predictions are shown in Table 4. The element was discovered in 1886 and called **germanium**: its properties are in excellent agreement with Mendeleev's predictions.

Since Mendeleev's death in 1907, 8 elements have been discovered and 24 have been made in the laboratory. The first two elements to be made in the laboratory were neptunium ($Z = 93$) and plutonium ($Z = 94$). They were formed by bombarding uranium with neutrons. By 1961 elements up to $Z = 103$ had been made. By 2000, there were 12 more; the heaviest of these was $Z = 118$.

Figure 9 shows you the historical pattern of the discovery of elements.

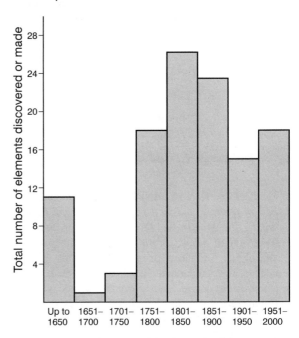

Figure 9 Historical pattern of the discovery of elements – many were discovered in the 19th century. Many 'artificial' elements have been made in the last 50 years.

Property	Prediction
appearance	dark-grey solid
relative atomic mass	72
density	$5.5 \, \mathrm{g\,cm^{-3}}$
reaction with water	none
reaction with acid	very little
reaction with alkali	more than with acid
oxide	basic, reacts with acid
chloride	liquid, boiling point < 100 °C

Table 4 Mendeleev's predictions for the properties of eka-silicon.

ASSIGNMENT 3

Table 4 shows some of the predictions Mendeleev made about the properties of eka-silicon. Use reference books to find out about the properties of silicon and tin, and then make your own predictions about the element which would fill the gap between them.

Find out about some of the properties of germanium and see how well these compare with your predictions and Mendeleev's predictions about eka-silicon.

Figure 10 Dmitri Mendeleev's (1834–1907) ideas form the basis of our modern classification of the elements. In 1934, the centenary of his birth was marked with this 15-m-high version of Mendeleev's Periodic Table, unveiled at the building in St Petersburg where he had worked.

The modern Periodic Table is based on the one originally drawn up by Mendeleev. It is one of the most amazingly compact stores of information ever produced: with a copy of the Periodic Table in front of you, and some knowledge of how it was put together, you have thousands of facts at your fingertips!

Figure 11 Glen Seaborg won the Nobel Prize for Chemistry in 1951 for his discoveries in the chemistry of the 'artificial' elements beyond uranium. He was co-discoverer of plutonium and all the elements from plutonium to nobelium (Z=102). He thought his greatest honour was the naming of element 106 after him (seaborgium, Sg). He was the first living scientist to be so remembered.

Chemical Ideas 11.1 tells you more about the modern Periodic Table.

You can use your IT skills to investigate how the physical properties of elements vary across a period in **Activity EL3.2**.

Chemical Ideas 11.2 looks at Groups 1 and 2 in the Periodic Table and will allow you to check your results from **Activity EL3.1**.

Activity EL3.3 will help you to organise the material you have come across in Sections **EL1** to **EL3**.

EL4 *Where do the chemical elements come from?*

A star is born

In the beginning, there was hydrogen – and a lot of it! There still is: hydrogen is the most common element in the universe. Humans also contain quite a lot of hydrogen, but we also contain other, heavier, elements as well. The theory of the evolution of the stars shows how heavy elements can be formed from lighter ones, and helps to explain the way elements are distributed throughout the universe.

The theory of how stars form is one of the major scientific achievements of the 20th century. It was developed through observation of a range of stars at different stages in their development, studying them as they changed over time.

Although hydrgen is the most common element, its atoms are relatively few and far between. There is about one atom per centimetre cube (cm^3) in the space between the stars, compared with over 1×10^{19} atoms per cm^3 in the air you are breathing now. With a density of hydrogen atoms in space as low as this, there is almost no chance that hydrogen atoms will come together to form hydrogen molecules.

However, there are some regions between the stars where molecules do form. These are called **dense gas clouds** or **molecular gas clouds**, though they are hardly dense by standards on Earth. These regions may contain as few as 100 particles per cm^3 to as many as 1×10^6 particles per cm^3. This sort of density means that there may be distances between the particles many millions of times the size of the particles themselves. The gas clouds are made up of a mixture of atoms and molecules, mainly of hydrogen, together with a **dust** of solid material from the break up of old stars.

The temperatures of the gas clouds vary from 10 K to 100 K (–263 °C to –173 °C). The particles have low kinetic energies and move around relatively slowly, so that gravitational forces between the particles are able to keep them together. Parts of the clouds gradually contract in on themselves and the gases become compressed. 'Clumps' of denser gas are formed.

The densest part of the 'clump' is its centre. Here the gases are most compressed and become very hot, up to 10 million °C. Such temperatures are high enough to trigger nuclear reactions. At these temperatures, atoms cannot retain their electrons and matter becomes a **plasma** of ionised atoms and unbound electrons.

A nuclear reaction is different from a chemical reaction. A chemical reaction involves a *rearrangement* of an atom's outer electrons, while a nuclear reaction involves a change in its nucleus. In a nuclear reaction *one element can change into another element* – something that would be impossible in a chemical reaction.

One nuclear reaction that takes place in the centre of 'clumps' is **fusion**, when lighter nuclei are fused together to form heavier nuclei. The nuclei need to approach one another at high speed – with a large kinetic energy – to overcome the repulsions between the positive charges on the two nuclei.

The nuclei of hydrogen atoms present in the gas cloud join together by nuclear fusion and the hydrogen turns into helium. The process releases vast quantities of energy, which causes the dense gas cloud to glow. The dense gas cloud has become a star (see Figure 12). In the centre of stars, where temperatures can reach hundreds of millions of degrees, fusion is common. Here are two examples of reactions which take place in the Sun:

$$^{1}_{1}H + ^{2}_{1}H \rightarrow ^{3}_{2}He + \gamma$$
$$^{2}_{1}H + ^{3}_{1}H \rightarrow ^{4}_{2}He + ^{1}_{0}n$$

Notice that atomic numbers and mass numbers must balance in a nuclear equation.

After hydrogen, helium is the must abundant element in space.

Chemical Ideas 2.1 and **2.2** provide you with information about the structure of atoms and nuclear reactions. These topics will help you with your study of this section.

Activity EL4.1 will enable you to find out how our current model of the atom developed.

In **Activity EL4.2** you can use data from a mass spectrometer to find the mass numbers of isotopes and work out the relative atomic mass of an element.

The nuclear reactions also generate a hot wind which drives away some of the dust and gas, leaving behind the new star. This is often surrounded by planets which have condensed out of the remaining dust cloud.

Heavyweight stars

What happens next to a star depends on its mass. All stars turn hydrogen into helium by nuclear fusion. This process occurs fastest in the heaviest stars because their centres are the hottest and the most compressed. These **heavyweight** stars have very dramatic lives. The temperatures and pressures at the centre of the star are so great that further fusion reactions take place to produce elements heavier than helium.

Layers of elements form within the star, with the heaviest elements near the centre where it is hottest and where the most advanced fusion can take place.

Figure 13 shows an example of the composition of the core of a typical heavyweight star after a few million years – long enough for extensive fusion to have taken place.

The element at the centre of the core is iron. When iron nuclei fuse they do not release energy but they *absorb* it. When the core of a heavyweight star reaches the stage where it contains mainly iron, it becomes unstable and explodes. These explosions are called **supernovae** – the most violent events in the universe (Figure 14). As a result of the supernova, the elements in the star are dispersed into the universe as clouds of dust and gas, and so the life cycle begins again.

ASSIGNMENT 4

Identify the isotopes which are missing from the following nuclear equations:

a $^{12}_{6}C + ? \rightarrow ^{16}_{8}O$

b $^{14}_{7}N + ^{1}_{1}H \rightarrow ?$

c $^{7}_{3}Li + ? \rightarrow ^{4}_{2}He + ^{4}_{2}He$

---

be no supernova. Once the hydrogen is used up, the Sun will expand into a **red giant**, swallowing up the planets Mercury and Venus. The oceans on Earth will start to boil and eventually it too will be engulfed by the Sun. The good news for Earth is that the Sun still has an estimated 5000 million years' supply of hydrogen left!

As red giants get bigger, they also become unstable, and the outer gases drift off into space, leaving behind a small core called a **white dwarf**, about one-hundredth of the size of the original star.

Figure 12 *The glow of star formation in the Orion nebula.*

The Sun – a lightweight among stars

The Sun is a **lightweight** star: it is not as hot as most other stars and will last longer than heavyweight stars. It will keep on shining until all the hydrogen has been used up and the core stops producing energy: there will

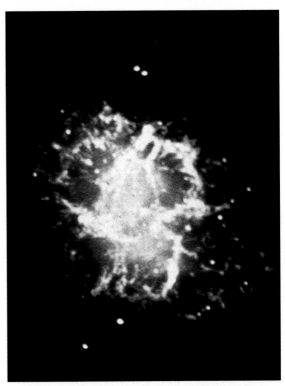

Figure 14 *Violence in the Crab nebula – scientists believe that such pictures are evidence of a supernova.*

Figure 13 *A model of the core of a typical heavyweight star. Heavyweight stars vary in mass, but can be about 8 times the mass of our Sun.*

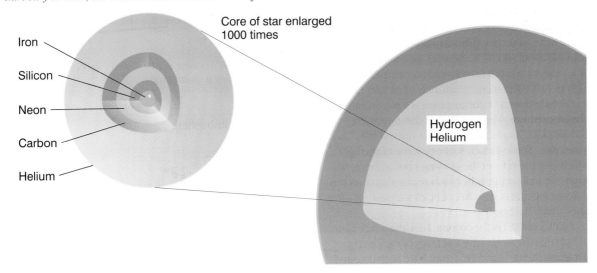

Core of star enlarged 1000 times

Iron
Silicon
Neon
Carbon
Helium

Hydrogen
Helium

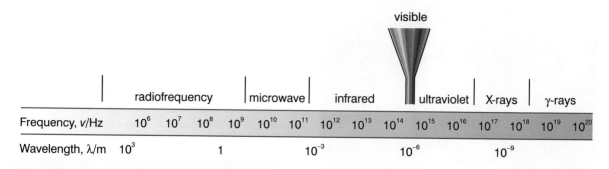

Figure 15 The electromagnetic spectrum.

How do we know so much about outer space?

The work of chemists has made a vital contribution to the understanding of the origin, structure and composition of our universe. To do this, they have used one of the most powerful analytical tools available today – **spectroscopy**.

Many different spectroscopic techniques exist (others are discussed in the **What's in a Medicine?** storyline), but all are based on one very important scientific principle – under the right conditions a substance can be made to **absorb** (take in) or **emit** (give out) **electromagnetic radiation** in a way which is characteristic of that substance. The **electromagnetic spectrum** in Figure 15 shows the different types of electromagnetic radiation.

If we analyse this electromagnetic radiation (such as ultraviolet light, visible light or radio waves) we can learn a lot about a substance. Sometimes we just want to know what it is. At other times we want to find out very detailed information about it, such as its structure and the way its atoms are held together. Figure 16 shows how visible light can be analysed using a spectrograph.

Absorption spectra

The glowing regions of all stars emit light of all frequencies between the ultraviolet and the infrared parts of the electromagnetic spectrum. The Sun emits mainly visible light; its surface (**photosphere**) glows like an object at about 6000 K. Some stars are cooler than the Sun; others are much hotter, reaching temperatures as high as 40 000 K and emitting mainly ultraviolet radiation.

Outside the star's photosphere is a region called the **chromosphere** (Figure 17). The chromosphere contains ions, atoms and, in cooler stars, small molecules. These particles absorb some of the light which is emitted from the glowing photosphere. So when we analyse the light which reaches us from the star, we see that certain frequencies are missing – the ones which have been *absorbed*.

Further out still is the **corona**. Here the temperature is so high that the atoms have lost many of their electrons. For example, Mg^{11+} and Fe^{15+} ions have been detected.

β Centauri is a B-type star (a type of very hot star). The spectrum of the visible light reaching us from β Centauri, in other words the star's visible **absorption spectrum**, is illustrated in Figure 18. You can clearly see the **absorption lines**. Because these correspond to the frequencies which are missing, they appear as black lines on the bright background of light emitted from the star.

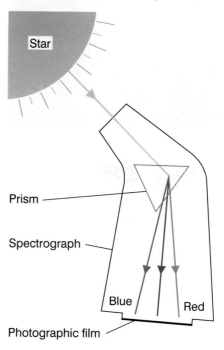

Figure 16 The frequencies in a beam of light can be analysed using a spectrograph.

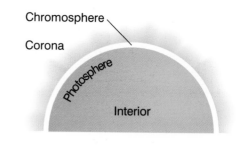

Figure 17 The structure of the Sun, a typical star.

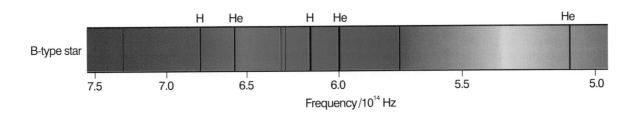

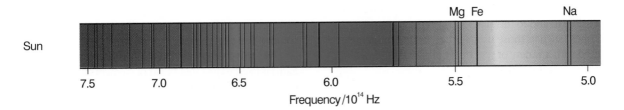

Figure 18 Absorption spectra of a B-type star (eg β Centauri) and the Sun. Black lines occur where frequencies are missing from the otherwise continuous spectra. (Note that here the frequency increases from right to left, which is a common convention for absorption spectra.)

ASSIGNMENT 5

The spectrum of the light received on Earth from Sirius A is shown in Figure 19. Sirius is an A-type star, which is less hot than a B-type star.

Compare this spectrum with those in Figure 18 and name *five* elements which the spectrum shows are present in Sirius A.

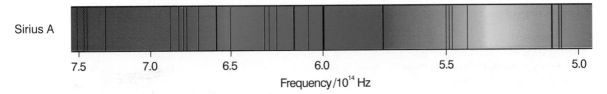

Figure 19 The absorption spectrum of the star Sirius A.

The absorption lines in β Centauri's spectrum arise only from hydrogen atoms and helium atoms. These are the only atoms able to absorb visible light at the very high temperature of β Centauri.

For comparison, the Sun's absorption spectrum is also shown in Figure 18. Because the Sun is at a lower temperature, different particles are able to absorb visible light. For example, lines from sodium, iron and magnesium can be seen. The Sun's chromosphere consists mainly of hydrogen and helium but, at the temperature of the Sun, these do not absorb visible light.

Emission spectra

When the atoms, molecules and ions around stars absorb electromagnetic radiation, they are raised to higher energy states called **excited states**. The particles can lose their extra energy by emitting radiation. The resulting **emission spectra** can also be detected on Earth.

During a total solar eclipse, the glow of the Sun's photosphere is completely blocked out by the Moon. The light being *emitted* by the chromosphere is all that can be seen, and it is then that the presence of

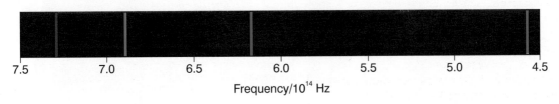

Figure 20 The hydrogen emission spectrum in the visible region.

hydrogen and helium is revealed. Hydrogen atoms dominate the chromosphere's *emission spectrum*, but a helium emission line can also be seen. The hydrogen emission spectrum is shown in Figure 20. Helium gets its name from *helios*, the Greek word for the Sun. The previously unknown element was first detected in the chromosphere during the eclipse of 1868.

Careful and detailed study of all the types of radiation received on Earth from outer space has allowed a picture to be built up of part of the chemical composition of the universe. More recently, it has been possible to add to this picture by sending up space probes (such as those used in the Voyager missions) fitted with a variety of spectroscopic devices.

Chemical Ideas 6.1 will help you to find out more about the information chemists can obtain from spectra. This information includes the arrangements of electrons in shells, which is described in **Chemical Ideas 2.3**.

In **Activity EL4.3** you can look at the emission spectra of some elements for yourself.

Our solar system

Our solar system probably condensed from a huge gas cloud which gradually contracted under the force of gravity. As rings of gas and dust condensed around the

Sun, the planets were formed. This material originated from a supernova and therefore contained a range of elements. The non-volatile elements condensed near to the Sun, where temperatures were greatest, while the more volatile elements condensed further away from the Sun at lower temperatures.

So, our solar system is made up of small, dense, rocky planets near to the Sun, and giant fluid planets further away from the Sun. Conditions on all the planets are very different and some of their chemistry seems very unusual when compared with our experiences on Earth. Figure 21 illustrates how unusual some of the chemistry is.

Does this then mean that the composition of the Earth is fixed to just that blend of elements which condensed around the Sun all those billions of years ago? The answer is no. Some of the atoms which formed the Earth were unstable, and began breaking down into atoms of other elements by **radioactive decay**.

This process is still going on today. Now we are also able to produce our own unstable atoms and usefully apply their radioactive decay.

You can remind yourself about radioactive decay in **Chemical Ideas 2.2**.

Activity EL4.4 introduces you to a current issue concerned with radioactive decay in the UK.

Figure 21 Did you know …?

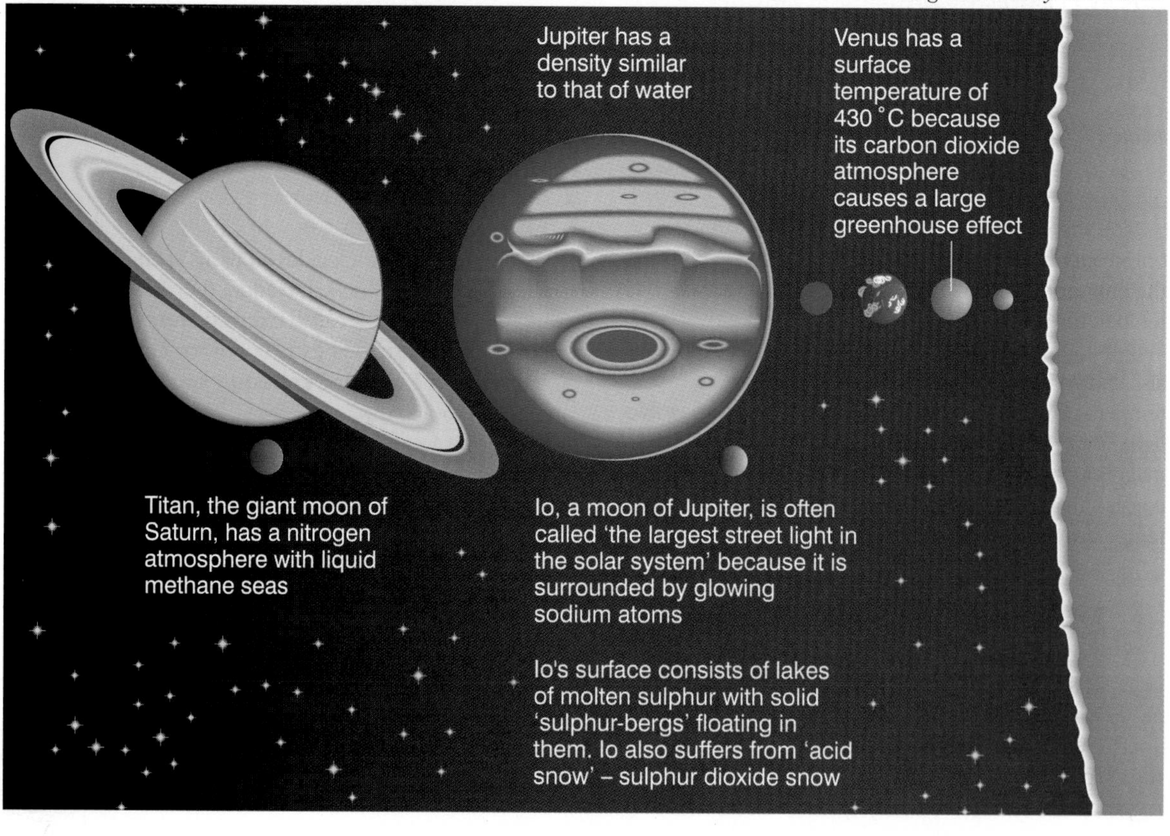

Jupiter has a density similar to that of water

Venus has a surface temperature of 430 °C because its carbon dioxide atmosphere causes a large greenhouse effect

Titan, the giant moon of Saturn, has a nitrogen atmosphere with liquid methane seas

Io, a moon of Jupiter, is often called 'the largest street light in the solar system' because it is surrounded by glowing sodium atoms

Io's surface consists of lakes of molten sulphur with solid 'sulphur-bergs' floating in them. Io also suffers from 'acid snow' – sulphur dioxide snow

Monatomics	Diatomics	Triatomics	Tetra-atomics	Penta-atomics
C^+	H_2	H_2O	NH_3	HCOOH
Ca^{2+}	OH	H_2S	H_2CO	NH_2CN
H^+	CO	HCN	HNCO	HC_3N
	CN	HNC	HNCS	C_4H
	CS	SO_2	C_3N	CH_2NH
	NS	OCS		CH_4
	SO	N_2H^+		
	SiO	HCS^+		
	SiS	HCO^+		
	C_2	NaOH		
	CH^+			
	NO			

Hexa-atomics	Hepta-atomics	Octa-atomics	Nona-atomics	Others
CH_3OH	CH_3CHO	$HCOOCH_3$	CH_3CH_2OH	HC_9N
NH_2CHO	CH_3NH_2		CH_3OCH_3	$HC_{11}N$
CH_3CN	H_2CCHCN		CH_3CH_2CN	
CH_3SH	CH_3C_2H		HC_7N	
CH_2CCH				

Table 5 Some chemical species in the dense gas clouds.

EL5 *The molecules of life*

Chemicals between the stars

You have heard a lot about atoms so far, but humans are made up of molecules and some ions, rather than single atoms. So what are the molecules of life, and how did they come into existence?

Molecules are formed in the colder parts of the universe when individual atoms happen to meet and bond to one another. (Molecules do not exist in stars because the bonds connecting the atoms cannot survive at the high temperatures there.) Molecules and fragments of molecules have been detected in dense gas clouds, both by radio and infrared telescopes on Earth and by spectroscopic instruments carried by rockets.

Table 5 shows some of the chemical species found in dense gas clouds; some will look familiar, but many will look strange.

Chemical Ideas 3.1 tells you about the ways in which elements can combine with each other.

Many of the substances in Table 5 can be described as **organic** species. This means that they contain carbon atoms bonded to elements other than just oxygen.

There is something familiar about the elements in these species – they are the elements which are the major constituents of the human body.

Molecules (and ions) that contain covalent bonds take up definite shapes. You can read about the shapes of molecules in Chemical Ideas 3.3.

Balloons can be used to give good illustrations of molecular shapes. You can try using them like this in Activity EL5.

Where did the molecules of life come from?

Some scientists have suggested that the molecules in the dense gas clouds were the building blocks which reacted together to make the molecules that form the basis of life on Earth. They believe that the energy needed to make these reactions take place came from ultraviolet radiation, X-rays and cosmic rays, and closer to Earth from lightning flashes.

In 1950 an American scientist, Stanley Miller, put methane (CH_4), ammonia (NH_3), carbon dioxide (CO_2) and water – simple molecules like those present in the dense gas clouds – into a flask and heated them. He also subjected the mixture to an electrical discharge to simulate the effect of lightning. On analysing the products, Miller found that some of the reaction mixture had been converted into amino acids. Amino acids are the building blocks of proteins, which are formed from long chains of amino acids linked together.

Figure 22 Stanley Miller used this apparatus to make amino acids from simple molecules.

In a separate experiment Leslie Orgel, another scientist in the US, made a very dilute solution of ammonia and hydrogen cyanide (HCN) and froze it for several days. When he analysed the 'ice' he identified amino acids and the compound adenine. Adenine is one of four compounds which, together with phosphate units and a sugar called deoxyribose, make up DNA (deoxyribonucleic acid) – the substance which contains the genetic code for reproduction.

Both Miller and Orgel showed that molecules like those in the dense gas clouds could react together, under conditions similar to the ones which existed during the early history of the Earth, to form some of the molecules of life. Their experiments give added weight to the suggestion that life on Earth has its origin in molecules from outer space.

There are other theories on how molecules of life may be formed in outer space, theories that may be supported by the space mission *Stardust*, launched by NASA in 1999. One of its aims is to collect a sample from a dense gas cloud.

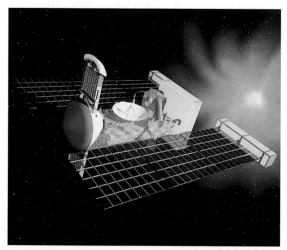

Figure 23 An artist's impression of the spacecraft Stardust *approaching the comet* Wild 2. *The spacecraft should come within 100 km of the comet's nucleus in early 2004.* Stardust *will sample interstellar dust particles and material from the gas and dust envelope surrounding the comet's nucleus.*

One theory suggests that the key is held by the interstellar dust thrown out by stars, which is protected from the most intense ultraviolet radiation in space by the dense gas clouds. The dust contains minute dust particles. At the centre of each particle is a hard core made up of graphite (carbon), silica (silicon dioxide), iron and other substances. Around this core are solid compounds with simple structures, such as water, ammonia, methane and carbon dioxide.

The temperatures in gas clouds are too low for 'normal' chemical reactions to occur. However, the ultraviolet light which does penetrate the clouds can break the covalent bonds in these simple molecules and so reactions can take place, leading to the formation of larger molecules. These molecules, in turn, can react at slightly higher temperatures to form biological compounds, such as amino acids and the bases that link together to form DNA.

Figure 24 Dense molecular clouds and dust in the Eagle nebula. The 'fingers' emerging from the pillar of molecular hydrogen and dust contain small, very dense regions which are embryonic stars. This image was taken by the Hubble Space Telescope.

These theories are being tested in a series of experiments using the *Stardust* spacecraft. Scott Sandford, one of its leaders, points out that "Even if we only find three grains, they will be all the only three we have – all that science has to study."

You can take a more detailed look at proteins, DNA and other 'molecules of life' later in the course, in the **Engineering Proteins** storyline.

EL6 *Summary*

You began this unit by looking at the elements present in human beings. This led to a more detailed study of two of the elements found in the body, iron and calcium. To work out the proportions of the different kinds of atoms in a human body, you needed to know about moles of atoms. You learned how to convert the masses of elements that combine into moles of atoms, and so work out chemical formulae. You also learned how to use chemical formulae to write balanced chemical equations.

Having come across a number of different elements, you then found out about the ways in which chemists developed a system for classifying the elements and you looked in detail at the modern Periodic Table.

You then went on to consider the origin of these elements in stars. This allowed you to revise and develop your ideas about atomic structure, and you learned to write nuclear equations for the reactions taking place. You learned about atomic spectra and how spectroscopy is used to tell us about the composition of the universe.

Finally, you studied the ways chemical elements combine together to form compounds and saw how the chemical elements present in outer space could have combined to form the 'molecules of life'.

Activity EL6 will help you to check the notes you have made on Sections **EL4** and **EL5**.

DEVELOPING FUELS

Why a unit on DEVELOPING FUELS?

This unit tells the story of petrol: what it is and how it is made. It also describes the work of chemists in improving fuels for motor vehicles, and in searching for and developing alternative fuels for the future.

To achieve this, some fundamental chemistry is introduced. There are two main areas. First it is important to understand where the energy comes from when a fuel burns. This leads to a study of enthalpy changes in chemical reactions, the use of energy cycles, and the relationship between energy changes and the making and breaking of chemical bonds. Secondly, the unit provides an introduction to organic chemistry. Alkanes are studied in detail, and other homologous series such as alcohols and ethers are mentioned.

Isomerism is looked at in connection with octane ratings of petrol, and simple ideas about catalysis arise out of the use of catalytic converters to control exhaust emissions. There is also a brief qualitative introduction to entropy, which follows from a consideration of why the liquid components of a petrol blend mix together. All these topics will be developed and used in later units.

Overview of chemical principles

In this unit you will learn more about …

ideas you will probably have come across in your earlier studies

- balancing equations
- simple organic chemistry and homologous series
- useful products from crude oil
- combustion of alkanes
- exothermic reactions
- catalysis

ideas introduced in the **The Elements of Life** storyline

- moles
- empirical and molecular formulae
- covalent bonding
- polar bonds
- molecular shape

… as well as learning new ideas about

- calculating reacting quantities using a balanced chemical equation
- enthalpy changes and enthalpy cycles
- bond enthalpies
- nomenclature of organic compounds
- properties of alkanes
- isomerism
- alcohols and ethers
- entropy
- molar volumes of gases.

DF1 *Petrol is popular...*

Why don't we all drive electric vehicles? After all, they are clean, quiet and cause no exhaust pollution. Despite these advantages, out of the UK's 30 000 electric vehicles, 27 000 are milk-floats!

One problem with electric vehicles is their poor performance. This is mainly because of the heavy batteries they have to carry around. When were you last overtaken by a milk-float?

Another problem is that it takes so long to recharge their batteries. An electric vehicle has to be recharged overnight for several hours. Electrical energy trickles into the batteries at a rate of about 55 joules per second (55 W).

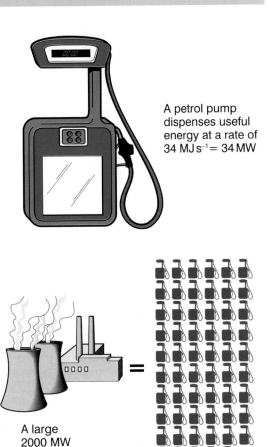

A petrol pump dispenses useful energy at a rate of 34 MJ s^{-1} = 34 MW

A large 2000 MW power station

= 59 Petrol pumps

Figure 2 Just 59 petrol pumps can match the power output of a large power station.

Figure 1 This is what car parks for electric cars may look like in the future – while the driver is out shopping or at work, the car can be 'filled up'.

Now compare this with the rate at which you can fill up a petrol-engined car with its source of energy. A petrol pump delivers petrol at about 1 litre per second, and a litre of petrol transfers 34 000 000 J of energy when you burn it. So the petrol pump replaces the car's energy supply over 600 000 times faster than the battery charger. These numbers are illustrated in a different way in Figure 2.

Petrol is a highly concentrated energy source. Couple that with the ease and cheapness of building petrol engines, and it's not surprising that petrol is popular.

. . . . . *but there are problems*

One problem with petrol is that it's a finite resource, because crude oil supplies are limited. They probably won't last more than about another 100 years.

We need crude oil for more than petrol alone. Crude oil provides the starting materials or **feedstocks** for the petrochemicals industry. For example, it is needed for making synthetic fibres, detergents and pharmaceuticals. As we use up supplies, oil products may become too valuable to burn in car engines.

In **Activity DF1.1** you can look at some of the possibilities for replacing petrol.

Another problem is pollution. Petrol produces carbon dioxide when it burns, and that's a major contributor to the greenhouse effect (see **The Atmosphere** storyline, Section **A6**). Petrol also produces other kinds of polluting emissions, as you'll see later in this unit.

So there's a real need to improve the performance of petrol in car engines so that it burns as cleanly and efficiently as possible – and to find suitable fuels to replace petrol for road transport in the future. How chemists and chemistry can help is one of the major themes of this unit.

First, though, we need to look at the chemistry behind combustion, and ask where the energy comes from when a fuel burns.

You will need to find out how to deal with energy changes in chemistry. **Chemical Ideas 4.1** will help you with this.

You will need to use balanced equations and chemical calculations with confidence throughout your course. If you feel in need of revision, **Chemical Ideas 1.3** will help.

In **Activities DF1.2** and **DF1.3** you can measure the energy changes involved when fuels burn.

DF2 *Getting energy from fuels*
The role of oxygen

Even the crude experimental method in **Activity DF1.2** shows that different fuels have different enthalpy changes of combustion. Let's compare five important fuels – look at Figure 3.

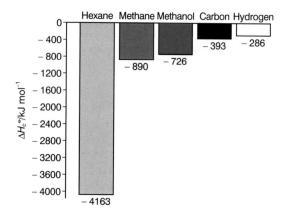

Figure 3 Standard enthalpy changes of combustion for some important fuels.

Note that the enthalpy changes are *negative* – this is because during combustion the reactants *lose* energy to their surroundings.

You can see that the enthalpy changes of combustion vary widely. Why should this be? What decides how much energy you get when you burn a mole of a fuel?

We think of fuels as energy sources, but they can't release any energy until they have combined with oxygen. So really we should think of *fuel–oxygen systems* as the energy sources. As you can see in **Chemical Ideas 4.2**, when substances burn the energy released comes as bonds with oxygen are formed.

Chemical Ideas 4.2 will tell you about making and breaking bonds.

ASSIGNMENT I

a Write down balanced equations for the complete combustion of methane (CH_4), hexane (C_6H_{14}) and methanol (CH_3OH).

b Use the ideas of bond making and bond breaking to explain the following. (Think in terms of the bonds that have to be broken and the new bonds that have to be made when the fuel burns in oxygen.)

   i Why is $\Delta H_c^{\ominus}$ for methane so much less negative than $\Delta H_c^{\ominus}$ for hexane?

   ii Why is $\Delta H_c^{\ominus}$ for methanol less negative than $\Delta H_c^{\ominus}$ for methane? After all, they both have the same number of C and H atoms.

The enthalpy change of combustion of a fuel depends on two things. First, there is the *number* of bonds to be broken and made – and that depends on the size of the molecule involved. That's why larger molecules such as hexane have a more negative $\Delta H_c^{\ominus}$ than smaller ones such as methane.

But $\Delta H_c^{\ominus}$ also depends on the *type* of bonds involved. Let's take your answer to **b (ii)** in **Assignment 1** a bit further. The equations on the next page show the bonds involved in the burning of methane and methanol. The products are the same, but the key difference is that *methanol already has an O–H bond*. In other words, one of the bonds to oxygen is already made – unlike with methane, where all the new bonds have to be made, starting from scratch. Another way of looking at this is to say that methanol is methane which is already partly oxidised.

The energy released during combustion comes from the making of bonds to oxygen. If methanol already has one bond made, it will give out less energy when it burns.

(a) Combustion of methane

$$CH_4 + 2O_2 \longrightarrow CO_2 + 2H_2O$$

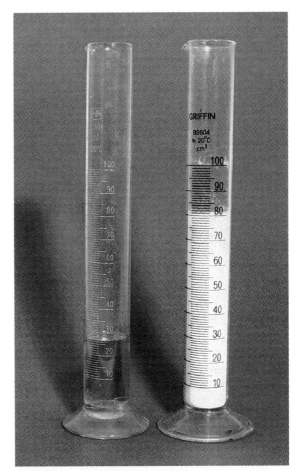

(b) Combustion of methanol

$$CH_3OH + 1\tfrac{1}{2}O_2 \longrightarrow CO_2 + 2H_2O$$

As a general rule, the more oxygen a fuel has in its molecule, the less energy it will give out when one mole of it burns. Oxygenated fuels such as alcohols and ethers are less energy-rich than hydrocarbon fuels. However, that's not to say that they are poor fuels. In some ways oxygenated fuels are better, because they are less polluting than hydrocarbons, and they often have a high octane number. More about all this later.

You can use the idea of bond breaking and making by designing a spreadsheet to calculate $\Delta H_c^{\ominus}$ in **Activity DF2.1**.

Important news for slimmers

When you eat too much of an energy food, the excess energy gets stored in your body as fat. The more energy-rich the food, the more fattening it is.

Compare a carbohydrate such as glucose with a fat such as olive oil. Here are the formulae:

glucose $C_6H_{12}O_6$
glyceryl trioleate $C_{57}H_{104}O_6$
(the main constituent of olive oil)

Per carbon atom, glucose is much more oxygenated than olive oil, so it is much *less* energy-rich. From 1 gram of a carbohydrate such as glucose you can get about 17 kJ. From 1 gram of a fat such as olive oil you can get about 39 kJ. Gram for gram, fats are more than twice as fattening as carbohydrates.

Alcohol is neither a fat nor a carbohydrate – but it too is bad news for slimmers. In fact there is a whole series of related compounds called **alcohols**, but the particular alcohol present in drinks is ethanol, C_2H_5OH. The same substance is used as an alternative to petrol for cars in some countries. It burns in the car engine releasing energy – and it also releases energy when metabolised in the body (Figure 5).

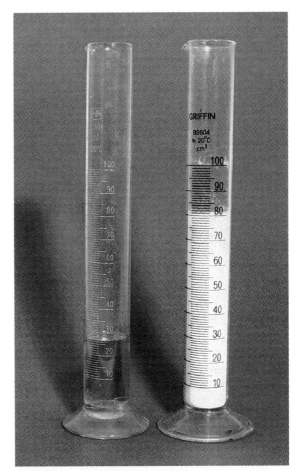

Figure 4 Fats and oils are more energy-rich than carbohydrates – the oil on the left will provide the same quantity of energy as the solid glucose on the right.

1 single measure of spirits   *1 glass of wine* $\tfrac{1}{2}$ *pint of beer or lager*

Figure 5 Alcohol can be fattening: each of these drinks provides about 300 kJ of energy (about 70 Calories), equivalent to $1\tfrac{1}{2}$ slices of bread.

But be thankful that you are not a potato. Humans store energy in fat, but potatoes store it in starch. To store a given amount of energy, you need to have more than twice the mass of starch compared with fat. Think how oversize *that* would make you!

Figure 6 Why are these foods considered to be less fattening?

Carrying fuels around

For a practical fuel, the enthalpy change of combustion may not be the most important thing to consider. What really matters is the **energy density** – how much energy you get per kilogram of fuel. After all, you have to carry the stuff around with you. We can work this out from the enthalpy change of combustion, using the relative molecular mass. We've done this for five fuels and the results are in Table 1.

Figure 7 The energy density of coal is relatively low – but it is a lot higher than for wood.

ASSIGNMENT 2

Look at the values for energy density of the different fuels in Table 1.

a On the basis of energy density, which is the best fuel in the table? What are the practical difficulties involved in using this particular fuel?

b Compare hydrogen and hexane. Explain why hydrogen has the higher energy density, even though it has the lower (least negative) enthalpy change of combustion.

c Here are some data for octane, C_8H_{18}, and decane, $C_{10}H_{22}$, both of which are components of petrol:

	$\Delta H_c^{\ominus}$ /kJ mol^{-1}	Relative molecular mass
octane, C_8H_{18}	−5470	114
decane, $C_{10}H_{22}$	−6778	142

Use the data to calculate the energy density for each of these compounds.

Compare your two answers. How do they compare with the values of energy density given for the fuels in Table 1?

It is important that you make notes as you work through each section. **Activity DF2.2** will help you to check that your notes cover the main points in Sections **DF1** and **DF2**.

This might be a good time to review progress in your preparations for your presentation in **Activity DF1.1**. The work you have covered so far may include some useful information about the fuel you have been allocated.

Later in this unit we will come back to some of the fuels that may replace petrol in the future. But for now, let's look more closely at petrol itself.

DF3 *Focus on petrol*

"Sorry I'm late, the car wouldn't start …"

What do you blame if the car won't start on a cold morning? Almost certainly the car itself – probably not the driver and certainly not the petrol. Yet the right petrol is actually very important.

Fuel	Formula	Standard enthalpy change of combustion, $\Delta H_c^{\ominus}$/kJ mol^{-1}	Relative molecular mass	Energy density (energy transferred on burning 1 kg of fuel)/kJ kg^{-1}
hexane	C_6H_{14}(l)	−4163	86	−48 400
methane	CH_4(g)	−890	16	−55 600
methanol	CH_3OH(l)	−726	32	−22 700
carbon	C(s)	−393	12	−32 800
hydrogen	H_2(g)	−286	2	−143 000

Table 1 Energy densities of some important fuels.

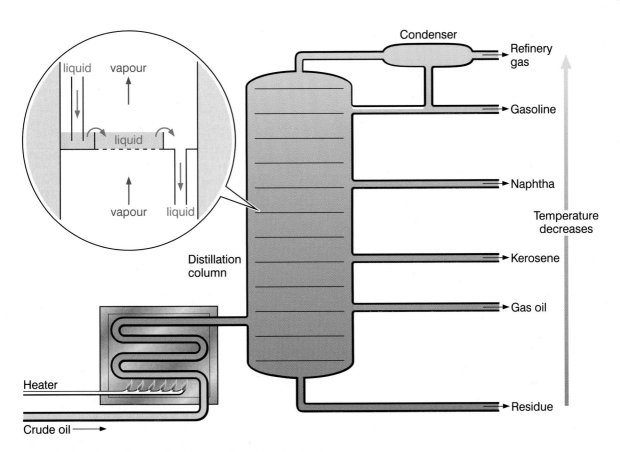

Figure 8 The primary fractional distillation of crude oil is a continuous process. Vapour rises up through the column and liquids condense and are run off at different levels, depending on their volatility.

What is petrol?

Petrol is a complex mixture of many different compounds carefully blended to give the right properties. The compounds are obtained from **crude oil** in several ways.

Crude oil is a mixture of many hundreds of **hydrocarbons**. It is a thick black liquid (*black gold* as it used to be called) but dissolved in it are gases and solids. Oil from the North Sea is pumped along pipes on the seabed to the UK refineries and special tankers bring crude oil from distant oilfields, such as those in the Middle East and Alaska. The refineries are either close to the shore (such as Fawley, near Southampton) or the oil is off-loaded into a pipeline leading to the refinery (such as from Finnart on the west coast of Scotland to the refinery at Grangemouth near Edinburgh).

At the refinery, the crude oil is heated and the vapour passes into a distillation column. In the column there is a temperature gradient, coolest at the top, hottest at the bottom. There are trays across the column with holes through which the rising vapour passes. The less volatile hydrocarbons condense on the trays and the more volatile pass through. This process is known as **fractional distillation** (Figure 8).

The oil is separated into fractions, each having a specific boiling range. The fractions do not have an exact boiling point because they are *mixtures* of many different hydrocarbons. For example, the *gasoline* fraction is a mixture of liquids, mostly **alkanes** with between five and seven carbon atoms, boiling in the range 25 °C–75 °C.

The gasoline and *gas oil* fractions are sources of petrol components. Another important fraction, *naphtha*, is also converted into high-grade petrol as well as being used in the manufacture of many organic chemicals. You can find information about the other fractions produced in the distillation of crude oil in Figure 8 and Table 2.

You can find out more about hydrocarbons and alkanes in **Chemical Ideas 12.1**.

In **Activity DF3.1** you can investigate how physical properties change along the alkane series.

Name of fraction	Boiling range/°C	Composition	% Crude oil	Use
refinery gas	<25	$C_1–C_4$	1–2	liquid petroleum gas (propane, butane) blending in petrol feedstock for organic chemicals
gasoline	25–75	$C_5–C_7$	20–40	petrol for cars
naphtha	75–190	$C_6–C_{10}$		production of organic chemicals converted to petrol
kerosene	190–250	$C_{10}–C_{16}$	10–15	jet fuel heating fuel (paraffin)
gas oil	250–350	$C_{14}–C_{20}$	15–20	diesel fuel central heating fuel converted to petrol
residue	>350	$>C_{20}$	40–50	fuel oil (eg power stations, ships) lubricating oils and waxes bitumen or asphalt for roads and roofing

Table 2 Fractions obtained from the fractional distillation of crude oil.

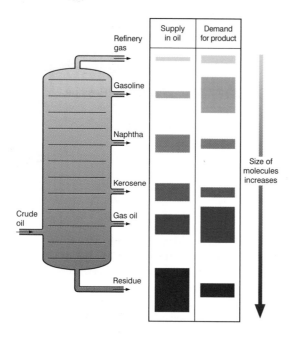

Figure 9 Supply and demand for different fractions of crude oil.

There are two problems. The first is that the *'straight-run'* gasoline from the primary distillation makes poor petrol. Some is used directly in petrol but most is treated further. The second is a problem of supply and demand. Crude oil contains a surplus of the high boiling fractions such as the gas oil and the residue and not enough of the lower boiling fractions such as gasoline. Figure 9 compares the supply and demand for different fractions of crude oil.

The job of the refinery is to convert crude oil into useful components. In order to do this, the structure of the alkane molecules present must be altered to produce different alkanes. The alkanes are also converted into other types of hydrocarbon that are used in petrol. These include **cycloalkanes**, which contain rings of carbon atoms, and **arenes** (**aromatic hydrocarbons**), which contain a benzene ring. The processes involved (isomerisation, reforming and cracking) are discussed later in the unit. The products are blended to produce high-grade petrol.

Other products formed in refinery processes (for example, in cracking) include the **alkenes**, which contain a C=C double bond. Alkenes are needed to make some very useful compounds, for example, plastics, medicines and dyes.

You can look up the structures of alkenes and arenes in **Chemical Ideas 12.2** and **12.3**. You will study these sections in detail later. For now, you just need to be able to recognise these compounds from their structures.

The residue can also be used to make useful products. First it is distilled again, this time under reduced pressure, by **vacuum distillation**. This avoids the high temperatures that would be needed for distillation at atmospheric pressure, since such temperatures would tend to crack the hydrocarbons. The more volatile oil distils over, leaving behind a tarry residue. The oils are used as fuel oils in power stations or ships; some fractions are used as the 'base' oil in lubricating oils.

Vacuum distillation

This is a technique for distilling a liquid which decomposes when heated to its boiling point. Distillation cannot be carried out at atmospheric pressure, but if the external pressure is reduced the boiling point can be brought below the decomposition temperature.

ASSIGNMENT 3

a Which fraction from the primary distillation of crude oil would be most likely to contain the following hydrocarbons?

   i $C_{35}H_{72}$

   ii C_4H_{10}

   iii $C_{20}H_{42}$

   iv C_8H_{18}

b What type of hydrocarbon is each of the above? Explain your answer.

c The following hydrocarbons are all found in petrol. In each case, draw out one possible structural formula and state the type of hydrocarbon.

   i C_7H_8

   ii C_4H_8

   iii C_5H_{12}

Winter and summer petrol

About 30%–40% of each barrel of crude oil goes to make petrol. But you can see that it's not as simple as just distilling off the right bit at the refinery and sending it to the petrol stations. Petrol has to be blended to get the right properties. One important property is the **volatility**.

In a car engine, a mixture of petrol vapour and air is ignited in a cylinder. The vapour–air mixture is made in the carburettor (see Figure 10). When the weather is very cold, the petrol is difficult to vaporise, so the car is difficult to start.

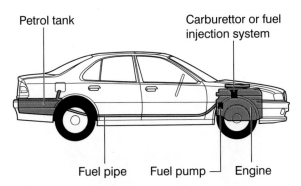

Petrol tank

Carburettor or fuel injection system

Fuel pipe   Fuel pump   Engine

Figure 10 The fuel supply system. When you start a car engine, the fuel pump sucks petrol out of the petrol tank and pushes it to the carburettor. The carburettor partially vaporises the petrol, mixes it with air and sends it to the cylinders in the car engine (nowadays the carburettor is often replaced by an electronic fuel injection system).

To get over this problem, petrol companies make different blends for different times of a year. In winter,

they put more volatile components in the petrol so it vaporises more readily. This means putting in more of the hydrocarbons with small molecules, such as butane and pentane.

Figure 11 Winter and summer petrol is not a new idea. These advertisements for seasonal blends of petrol were produced about 70 years ago.

On the other hand, in hot weather you don't want too many of these more volatile components, or the petrol would vaporise too easily. For one thing, you'd lose petrol from your tank by evaporation – a process which is costly and polluting. Also, if the fuel vaporises too readily, pockets of vapour form in the fuel supply system. The fuel pump then delivers a mixture of liquid and vapour to the carburettor instead of mainly liquid. This means that not enough fuel gets through to keep the engine running. It's called *vapour lock*.

Any petrol is a blend of hydrocarbons of high, medium and low volatility. As well as altering the petrol blend for the different seasons in one country, the blend will be different in different countries, depending on the climate. The colder the climate, the more volatile are the components added to the blend. Petrol companies change their blend four times a year – and you don't even notice. But you'd notice if they didn't!

You can compare the volatility of winter and summer blends of petrol in **Activity DF3.2**.

ASSIGNMENT 4

Petrol blenders talk about the 'front-end' (high volatility), 'mid-range' and 'tail-end' (low volatility) components of a blend.

a What differences will there be between
  i a winter and a spring blend for the UK?
  ii a summer blend for Spain and a summer blend for The Netherlands?

b Use the Data Sheets to look up the densities of some alkanes.
  i How does the density of alkanes change as their molecular mass increases?
  ii Which will have the greater mass at a given temperature: 1 litre of petrol bought in The Netherlands or 1 litre bought in Spain? Explain your answer.

c 'Spanish people get a better bargain than Dutch people when they buy petrol.' Discuss.

The problem of knocking

Another important property which blenders must take into account is the **octane** rating of the petrol (see box). This is a measure of the tendency of the petrol to cause a problem known as '**knock**'.

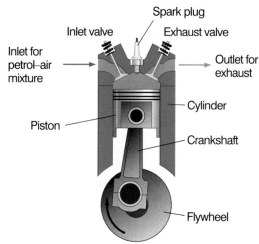

Figure 12 How a four-stroke petrol engine works. The compression stroke is shown here. The piston compresses the petrol–air mixture, then a spark makes the mixture explode, pushing the piston down and turning the crankshaft.

In a petrol engine, the petrol–air mixture has to ignite at the right time, usually just before the piston reaches the top of the cylinder.

Look at Figure 12. As the fuel–air mixture is compressed it heats up and the more it is compressed the hotter it gets. Modern cars achieve greater efficiency than in the past by using *higher compression ratios*, often compressing the gases in the cylinder by about a factor of 10.

Many hydrocarbons **auto-ignite** under these conditions. The fuel–air mixture catches fire as it is compressed. When this happens, *two* explosions occur: one due to the compression and another when the spark occurs. This produces a 'knocking' or 'pinking' sound in the engine. The thrust from the expanding gases is no longer occurring at the proper time so engine performance is lowered, and the inside of the combustion cylinder can be damaged.

The auto-ignition of hydrocarbons is explored in **Activity DF3.3**.

Octane numbers

The tendency of a fuel to auto-ignite is measured by its **octane number**. 2,2,4-Trimethylpentane (which used to be called 'iso-octane' – hence the name of the scale) is a branched alkane with a low tendency to auto-ignite. It is given an octane number of 100. Heptane, a straight-chain alkane, auto-ignites easily and is given an octane number of 0.

The octane number of any fuel is the percentage of 2,2,4-trimethylpentane in a mixture of 2,2,4-trimethylpentane and heptane which knocks at the same compression ratio as the given fuel. For example, four-star petrol has an octane number of 97 and knocks at the same compression ratio as a mixture of 97% 2,2,4-trimethylpentane and 3% heptane.

$$CH_3-\underset{\underset{CH_3}{|}}{\overset{\overset{CH_3}{|}}{C}}-CH_2-\underset{\underset{CH_3}{|}}{CH}-CH_3$$

2,2,4-trimethylpentane low tendency to auto-ignition scores 100

$$CH_3-CH_2-CH_2-CH_2-CH_2-CH_2-CH_3$$

heptane high tendency to auto-ignition scores 0

Figure 13 In diesel engines, the fuel ignites spontaneously when compressed, without the need for spark plugs.

DF4 *Making petrol – getting the right octane rating*

Different cars have different compression ratios. High performance engines usually have a high compression ratio, and they need a high octane fuel – otherwise there would be knocking and the engine would suffer. There are two ways of dealing with the problem of knock. One is to put special additives in the petrol which discourage auto-ignition. The other is to blend high-octane compounds in with the ordinary petrol.

Look – no lead

Anti-knock additives are substances which reduce the tendency of alkanes to auto-ignite. They increase the octane rating of the petrol.

From the 1920s, until recently, small amounts of lead compounds were used as economical and effective anti-knock additives. Exactly how they worked is still not fully understood, but they helped to prevent the reactions which cause knocking.

However, concern over environmental effects led to a gradual phasing-out of leaded petrol. For one thing, the lead compounds present in the exhaust fumes are toxic. They also poison the metal catalysts in the catalytic converters installed to reduce the levels of other pollutants in exhaust fumes.

Rather than developing alternative lead-free additives, the petrol companies turned their attention to using refining and blending to get high octane ratings without using lead.

Refining and blending

The hydrocarbons which give the best performance in a petrol engine are not the ones which are most plentiful in crude oil. So it is the job of the refinery to 'doctor' the hydrocarbons to suit our needs.

Mostly, it is a case of getting the right kind of alkane – although other types of hydrocarbons and some oxygenated compounds are also important, as you'll see later.

Which alkanes?

The structure of an alkane has an important influence on its tendency to auto-ignite – in other words, on its octane number. In general, the shorter the alkane chain, the higher the octane number. Short-chain alkanes are also more volatile, of course, so they can be used both to increase the octane number and to improve cold-starting. Even gaseous alkanes such as butane can be used – they just dissolve in the petrol.

The idea of isomerism is important in this section. You can find out about it in **Chemical Ideas 3.4**.

But the petrol blenders are limited in the proportion of short-chain alkanes they can include in the blend. Too much and the petrol blend would be *too* volatile.

The other factor that affects octane number is the *degree of branching* in the alkane chain. Quite simply, the more branched the chain, the higher the octane number.

Crude oil contains both straight-chain and branched alkanes. Unfortunately, it does not contain enough of the branched isomers to give it a naturally high octane number. The octane number of 'straight-run' gasoline is about 70.

To get around this, the petrol companies have a number of clever ways to increase the octane number of petrol. These include **isomerisation**, **reforming** and **cracking**.

You can make models of alkane isomers and practise naming them in **Activity DF4.1**.

In **Activity DF4.2** you can look in more detail at the effect of structure on the octane numbers of alkanes.

Isomerisation

Isomerisation involves taking straight-chain alkanes, heating them in the presence of a suitable catalyst so the chains break, and letting the fragments join together again. When the fragments join again, they are more likely to do so as *branched* rather than as *straight* chains.

Oil refineries do this on a large scale with pentane (C_5H_{12}) and hexane (C_6H_{14}), both products of the distillation of crude oil. It is one important way of increasing the octane quality of petrol.

One isomerisation reaction which can happen with pentane is shown below:

$$CH_3-CH_2-CH_2-CH_2-CH_3$$

pentane
octane number 62

$$CH_3-\underset{\underset{CH_3}{|}}{CH}-CH_2-CH_3$$

2-methylbutane
octane number 93

Isomerisation reactions like this do not go to completion and a mixture is formed containing all the possible isomers. We say the reaction has come to a **state of equilibrium** when no further change is possible under the reaction conditions. The arrows ($\rightleftharpoons$) indicate that the reaction is reversible and forms an equilibrium mixture. (You will study equilibrium reactions in detail later in the course.)

ASSIGNMENT 5

a Explain why 2-methylbutane is an isomer of pentane.

b A second isomer of pentane is also formed in the isomerisation process. Draw out its structural formula.

c What products might you expect to obtain from the isomerisation of hexane (C_6H_{14})? Draw out the structural formulae and give the name of each of your products.

In a modern plant, the isomerisation takes place in the presence of a catalyst of aluminium oxide on which platinum is finely dispersed. The products then pass over a form of **zeolite** which acts as a **molecular sieve** and separates the straight-chain from the branched isomers (see Figure 15). The straight-chain alkanes are then recycled over the platinum catalyst. Figure 14 gives a flow diagram for the process.

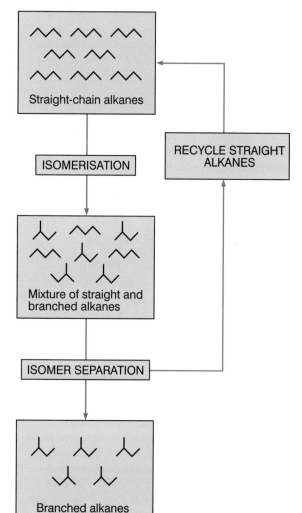

Figure 14 The isomerisation process.

Zeolites

Zeolites belong to a large family of naturally occuring minerals containing mainly aluminium, silicon and oxygen. They can also be made synthetically.

There are many different zeolites because of the different ways the atoms can be arranged in the crystal structure. They all contain an extensive network of interlocking pores and channels, which provide a large surface area. These pores and channels are of different sizes in different zeolites.

Zeolites are used widely in industry as **catalysts** and also as **molecular sieves** to sort out molecules by size and shape (Figure 15).

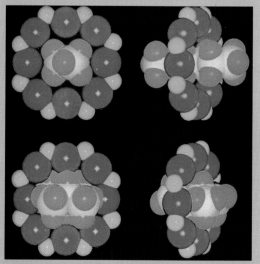

Figure 15 This computer-generated picture shows a molecular sieve in action. In the top pictures, a straight-chain molecule of hexane passes through a molecular sieve. In the bottom pictures, a branched-chain molecule of dimethylbutane cannot pass through. Zeolite molecular sieves are used to separate straight-chain and branched-chain isomers in the isomerisation process.

You can find out more about zeolites and molecular sieves by making models of zeolites in **Activity DF4.3**.

Reforming

Reforming is another trick used by oil refiners to increase the octane quality of petrol components. Naphtha, which is mainly made up of straight-chain alkanes with 6–10 carbon atoms, is heated to about 500 °C and passed over a catalyst. The straight-chain alkanes are converted to ring compounds – first to cycloalkanes and then to aromatic hydrocarbons. Some typical reforming reactions are shown in Figure 16.

The catalyst is platinum which is finely dispersed on the surface of aluminium oxide. The process is called

platforming. During the process, some of the hydrocarbons decompose to carbon, which decreases the efficiency of the catalyst. There can be £5 million worth of platinum inside one platformer, so it is very important to regenerate the catalyst and keep it in prime working order. To suppress the formation of carbon, excess hydrogen is mixed with the naphtha going into the process.

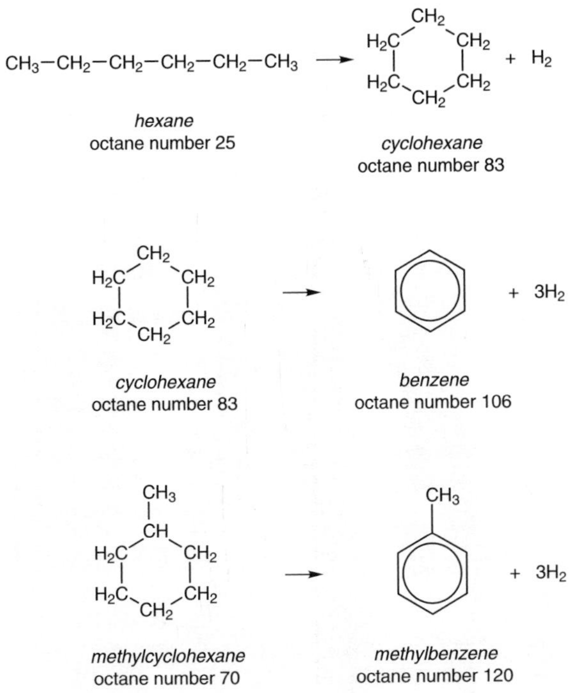

$$CH_3-CH_2-CH_2-CH_2-CH_2-CH_3 \longrightarrow$$

hexane
octane number 25

cyclohexane
octane number 83

$+ H_2$

cyclohexane
octane number 83

benzene
octane number 106

$+ 3H_2$

methylcyclohexane
octane number 70

methylbenzene
octane number 120

$+ 3H_2$

Figure 16 Some typical reforming reactions, showing octane numbers of reactants and products.

Cracking: using the whole barrel

Cracking is one of the most important reactions in the petroleum industry. It starts with alkanes that have large molecules and are too big to use in petrol – for example alkanes from the gas oil fraction. These large molecules are broken down to give alkanes with shorter chains that can be used in petrol. What's more, these shorter-chain alkanes tend to be highly branched, so petrol made by cracking has a higher octane number.

There's another benefit from cracking: it helps solve the supply and demand problem (see Figure 9).

Cracking: how is it done?

Much of the cracking carried out to produce petrol is done by heating heavy oils, such as the gas oil, in the presence of a catalyst. It is called catalytic cracking or cat cracking for short. The molecules in the feedstock can have 25–100 carbon atoms, although most will usually have 30–40 carbon atoms.

Cracking reactions are quite varied. Some of the types of reactions are

- alkanes → branched alkanes + branched alkenes

An example of this is:

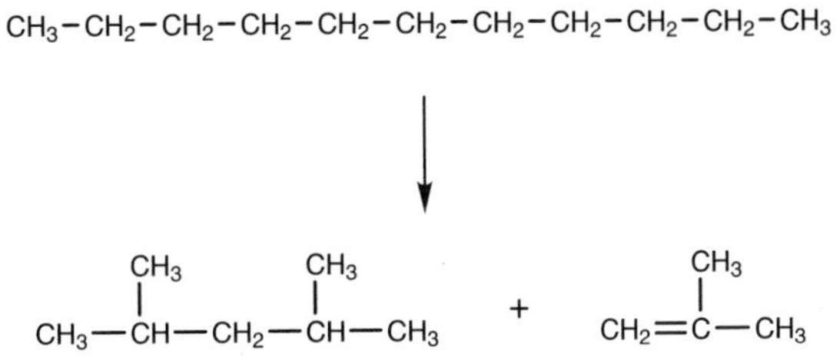

$$CH_3-CH_2-CH_2-CH_2-CH_2-CH_2-CH_2-CH_2-CH_2-CH_2-CH_3$$

$$\downarrow$$

- alkanes → smaller alkanes + cycloalkanes
- cycloalkanes → alkenes + branched alkenes
- alkenes → smaller alkenes.

The alkenes which are produced are important feedstocks for other parts of the petrochemicals industry.

Cracking always produces many different products, which need to be separated in a fractionating column.

Once again, zeolites play an important role but this time as catalysts. Type Y zeolite is particularly effective in producing good yields of high octane number products.

In a modern cat cracker the cracking takes place in a 60-m-high vertical tube about 2 m in diameter (Figures 17 and 18). It is called a **riser reactor** because the hot vaporised hydrocarbons and zeolite catalyst are fed into the bottom of the tube and forced upwards by steam. The mixture is a seething **fluidised bed** in which the solid particles flow like a liquid.

It takes the mixture about 2 seconds to flow from the bottom to the top of the tube – so the hydrocarbons are in contact with the catalyst for a very short period of time.

One of the problems with cat cracking is that, in addition to all the reactions you have already met, coke (carbon from the decomposition of hydrocarbon molecules) forms on the catalyst surface so that the catalyst eventually becomes inactive. To overcome this problem, the powdery catalyst needs to be **regenerated**.

After the riser reactor, the mixture passes into a separator where steam carries away the cracked products leaving behind the solid catalyst. This then goes into the regenerator, where it takes about 10 minutes for the hot coke to burn off in the stream of air which is blown through the regenerator. The catalyst is then reintroduced into the base of the reactor ready to repeat the cycle.

The energy released from the burning coke heats up the catalyst. The catalyst transfers the energy to the feedstock so that cracking can occur without additional heating.

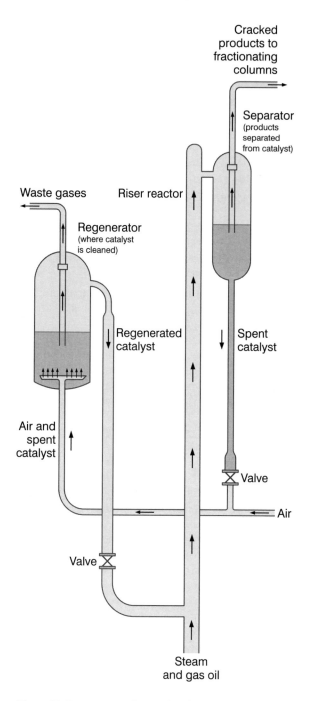

Figure 17 How a cat cracker works. The feedstock is gas oil and the cracking reaction takes place in the riser reactor.

You can try cracking alkanes for yourself in **Activity DF4.4**.

Cat crackers have been in operation since the late 1940s – and have become very flexible and adaptable. They can handle a wide range of different feedstocks. The conditions and catalyst can be varied to give the maximum amount of the desired product – in this case branched alkanes for blending in petrol.

Figure 18 A catalytic cracker which converts gas oil into branched alkanes for blending to produce high octane petrol. The process is shown in detail in Figure 17.

ASSIGNMENT 6

The riser reactor uses a fluidised bed of solid catalyst and reactants. Fluidised beds have the following useful properties:

- the solid can flow along pipes

- as the solid can flow, it is easier to make the reaction into a continuous process

- there is very good contact between the gas and the large surface area of the solid catalyst

- energy transfer between the solid and the gas is very efficient.

Suggest why each of these properties is useful or important in the cat cracker.

ASSIGNMENT 7

a Using molecular formulae, write out a balanced equation for one of the conversions shown in Figure 16.

b Explain why these reactions are not isomerisations.

c Below is a list of alkanes. For each one, say whether you think

   A It could be used unchanged in petrol

   B It should go through a reforming process before blending to form petrol

   C It should be cracked before blending to form petrol.

   i $CH_3CH_2CH_2CH_2CH_2CH_3$

   ii $CH_3(CH_2)_{15}CH_3$

   iii $(CH_3)_3CCH(CH_3)CH_3$

   iv $CH_3CH_2CH_2CH_3$

Adding oxygenates

'Oxygenates' is the name the petrol blenders use for fuels containing oxygen in their molecules. Two types of compounds are commonly used: **alcohols** and **ethers** (see box).

Alcohols and ethers

Alcohols all have an OH group in their molecule. The best known alcohol is ethanol, commonly called just 'alcohol':

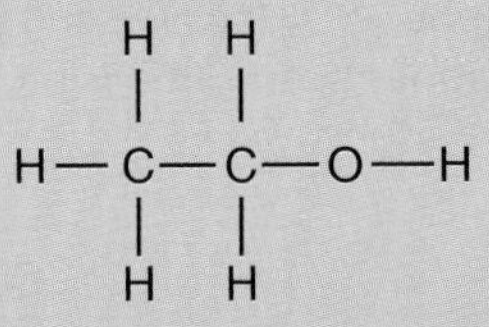

Ethers all have an oxygen atom bonded to two carbons. An example is ethoxyethane (sometimes just called 'ether'):

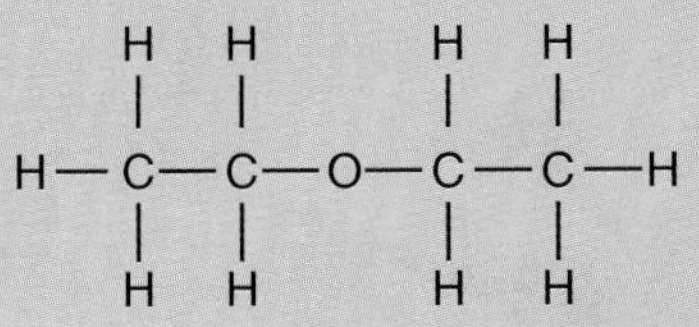

You can find out more about alcohols and ethers in **Chemical Ideas 13.2**.

You can practise naming alcohols and investigate some of their physical properties in **Activity DF4.5**.

Table 3 shows the oxygenates most commonly used for blending with petrol, together with 'straight-run' gasoline for comparison.

One oxygenate that is commonly used is MTBE. (MTBE stands for 'methyl tertiary butyl ether', the name that was once used for this compound; its modern systematic name is 2-methoxy-2-methylpropane.)

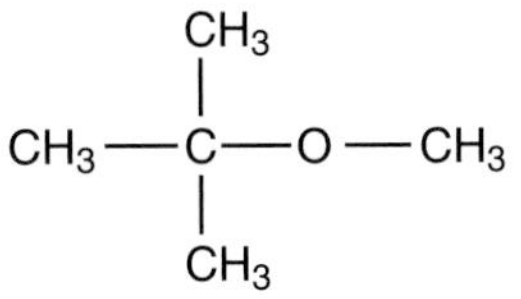

Figure 19 shows the percentages of MTBE added to different grades of petrol.

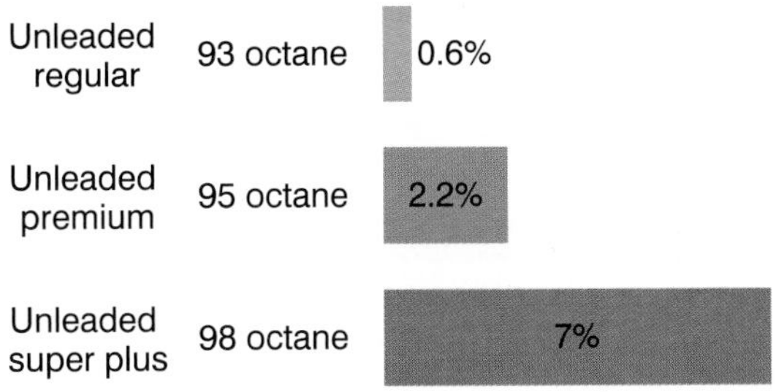

Unleaded regular	93 octane	0.6%
Unleaded premium	95 octane	2.2%
Unleaded super plus	98 octane	7%

Figure 19 Percentages by volume of MTBE typically used in some countries in different blends of petrol.

Oxygenates are added to petrol for two reasons. Firstly, they increase the octane number, as you can see from Table 3. Secondly, some believe that they tend to cause less pollution when they burn. In particular, levels of carbon monoxide in the exhaust are reduced. However, these claims are not universally accepted. Moreover, some states in the US have banned the use of MTBE in petrol. It is soluble in water and spillages can pollute water supplies.

Table 3 Oxygenates commonly used for blending with petrol ('straight-run' gasoline is given for comparison).

Name	Formula	Homologous series	Octane number	Boiling point/°C	Relative cost per litre (approx)
methanol	CH_3OH	alcohols	114	65	1.0
ethanol	CH_3CH_2OH	alcohols	111	79	5.8
MTBE	$CH_3OC(CH_3)_3$	ethers	113	55	1.6
'straight-run' gasoline	—	—	70	—	1.1

The perfect blend

Figure 20 shows the octane numbers of some of the key ingredients at the petrol blender's disposal.

The blender's job is to produce petrol from these ingredients at minimum cost. Of course, the petrol must meet specifications about volatility, octane number, density, etc. Prices and availability fluctuate and refiners are helped in their decisions by computer models of their refinery. They must take into account not only the prices of crude oil and the final petrol, but prices of their other products from the refinery and energy costs too.

Blending is done in batches of around 20 000 000 litres at a time, and can take as long as 20 hours to complete. Thorough mixing is important to give a homogenous liquid.

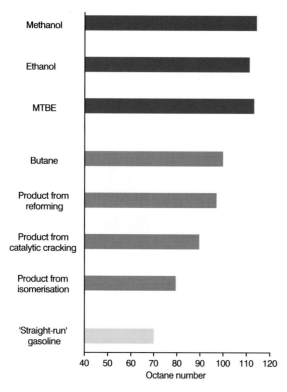

Figure 20 Octane numbers of some of the key ingredients used in petrol blending. Note the high octane numbers of the three oxygenates.

The idea of entropy is discussed in more detail in **Chemical Ideas 4.3**.

You can try your hand at blending petrol in **Activity DF4.6**.

You can find out more about the probability of two substances mixing in **Activity DF4.7**.

Activity DF4.8 will help you pull together all the work you have done on petrol so far.

Why do hydrocarbons mix?

Hydrocarbons are completely miscible with each other and can be blended in any proportions. Have you ever wondered why some liquids mix easily and others, like oil and vinegar, don't mix at all? Mixing of two substances is a natural change – it happens by chance with no external help. The result is an increase in disorder. The idea of disorder or randomness in a system is very important in determining the direction in which changes occur.

We give the amount of disorder in a system a name – **entropy**. When the system gets more disordered, like a pack of cards being shuffled, or two liquids mixing, its entropy increases. Natural changes are always accompanied by an overall increase in entropy.

All substances tend to mix with one another, unless there is something stopping them. In the case of oil and vinegar, there are attractive forces between the particles in vinegar which prevent them mixing with the oil particles.

DF5 *Trouble with emissions*

A serious problem

We'll look next at a problem of motor fuels which is causing worldwide concern: exhaust emissions. Figure 21 shows what goes into – and what comes out of – a car engine. The nitrogen oxides, NO and NO_2, are grouped together as NO_x, although the NO_x emitted by vehicle exhausts contains mainly NO with only small quantities of NO_2. Similarly, SO_x represents the oxides of sulphur, SO_2 and SO_3, and C_xH_y represents the various hydrocarbons present in the exhaust fumes.

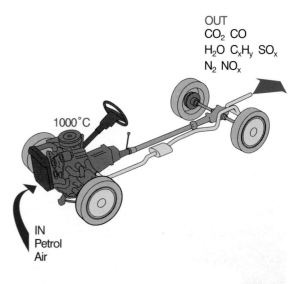

Figure 21 What goes into – and comes out of – a car engine.

Another problem arises because petrol is so volatile. A parked car on a warm day gives off hydrocarbon fumes, mostly butane, from the petrol tank and the carburettor. This is called *evaporative emission* and accounts for about 10% of emissions of volatile organic compounds from vehicles.

The oxides of sulphur in vehicle exhausts come from sulphur compounds in the fuel. In **Activity DF5** you can see how hydrogen is used to remove most of the sulphur from the fuel.

ASSIGNMENT 8

a For each of the substances emitted in the exhaust, as shown in Figure 21, use your knowledge of chemistry to explain where the substance comes from. Write chemical equations where appropriate.

   Some will be easier to explain than others! There are two things to bear in mind: the temperature inside a car engine's combustion chamber is about 1000 °C, and air is a *mixture* of gases.

b For each of the emissions, say which of the following environmental problems you think it may contribute to:

   A Acid rain

   B The greenhouse effect

   C Damage to the ozone layer

   D Toxic effects on humans

   E Photochemical smog.

You will need to use your knowledge of air pollution from your earlier studies, or find the information in a textbook.

Figure 22 Views of Denver, Colorado, US. (a) A computer-produced view showing what the visibility would be like without photochemical smog. (b) A photograph showing the current smoggy conditions.

Photochemical smog

Unfortunately, the substances shown in Figure 21 are not the only pollutants caused by motor vehicles. **Ozone** is a *secondary pollutant*, because it is not released directly into the atmosphere. It is formed as a result of chemical reactions that take place when sunlight shines on a mixture of two of the *primary pollutants*, nitrogen oxides (NO_x) and hydrocarbons (C_xH_y), together with oxygen and water vapour. Also produced are irritating and eye-watering compounds formed by the breakdown and further reaction of the hydrocarbons. These reactions all occur in **photochemical smogs**, which are a great cause for concern. (A photochemical reaction occurs when a molecule absorbs light energy and then undergoes a chemical reaction.)

Ozone – a molecule of many parts

Ozone is a highly reactive substance whose molecules contain three oxygen atoms. Its formula is O_3. Ozone occurs in the atmosphere, both high up in the *stratosphere* and close to the ground in the *troposphere*. Its presence in these two regions affects us in very different ways.

In the stratosphere, ozone acts as a sunscreen, filtering out dangerous ultraviolet light from the incoming solar radiation. Its presence there is vital to our survival, though there is now great concern because chlorofluorocarbons (CFCs) are reaching the stratosphere and are causing depletion of the protective ozone layer. You will find out more about this in **The Atmosphere**.

Meanwhile, in the troposphere, ozone has an important role in the production of hydroxyl radicals (HO). These are short-lived species, formed from water, which bring about the breakdown of many substances that would otherwise build up to hazardous levels in the atmosphere. However, ozone is an irritant toxic gas and high concentrations near ground level are detrimental to human health. It weakens the body's immune system and attacks lung tissue. Also, ozone in the troposphere acts as a greenhouse gas and so contributes to global warming.

Photochemical smog contains a mixture of primary and secondary pollutants (Figure 23). Its exact composition varies enormously and depends, for example, on the nature of the primary pollutants, the local geography, weather conditions, the time of day and the length of the smog episode.

Photochemical smogs normally occur in the summer during high pressure (anticyclonic) conditions. The still air means that there is much less mixing with high altitude air and the pollutants are trapped near ground level. Figure 24 summarises the formation of photochemical smog. Note that, even in clean relatively 'unpolluted' air, there is a small background concentration of ozone, and this ozone is involved in the series of reactions that produces the photochemical smog; its concentration is *increased* during smog formation.

If there is a light prevailing wind, the polluted air mass will be transported from the built-up urban area where it was generated and will move across rural areas. In fact the highest ozone readings are often recorded over rural areas because the chemical reactions producing the secondary pollutants take place relatively slowly.

What are the effects of photochemical smog?

Photochemical smogs cause haziness and reduced visibility in the air close to the ground. For many people they can cause eye and nose irritation and some difficulty in breathing, but for vulnerable groups – people such as asthmatics, who already have respiratory problems, very young children and many old people – the effects can be more serious.

Ozone is not the only substance in smog to cause health problems. The polluted air contains a whole cocktail of harmful chemicals but the exact links between photochemical smog and health problems such as asthma are very difficult to establish and are still the subject of much research. There is no simple relationship between the two because so many other factors are involved.

People are not the only ones to suffer. High ozone concentrations affect animals and plants too. Ozone is a highly reactive molecule which attacks most organic matter. Compounds with carbon–carbon double bonds are particularly vulnerable, so many materials, such as plastics, rubbers, textiles and paints, can be damaged.

How can chemists help?

Research work is taking place on several fronts to unravel and explain the complex chemistry involved.

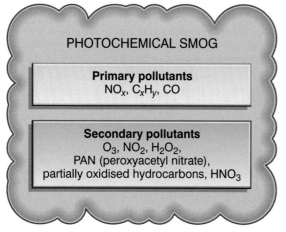

Figure 23 *The main constituents of photochemical smog.*

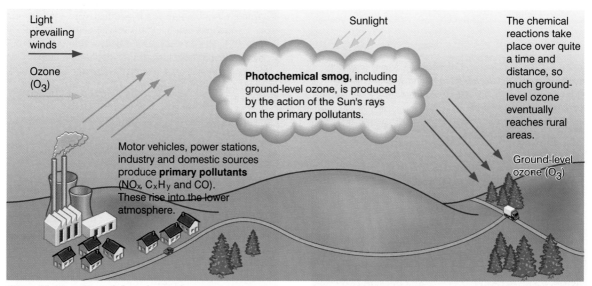

Figure 24 *Formation of photochemical smog.*

Monitoring pollutants

An important first step is to find out exactly which pollutants are present in the troposphere and how their concentrations vary. For example, ozone concentrations in urban areas show regular fluctuations which reflect changes in sunlight and vehicle emissions during the day (Figure 25). There are now several networks of monitoring stations across the UK recording pollutant concentrations. Figure 26 shows the location of the ozone monitoring sites in the UK.

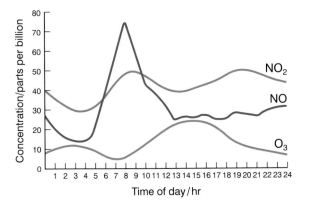

Figure 25 *Variation of the concentrations of ozone and oxides of nitrogen on a summer weekday in London. The high NO peak coincides with the early morning traffic rush hour, and is quickly followed by a build-up of NO_2, as reactions in the air convert NO into NO_2. Photochemical reaction of NO_2 leads to formation of O_3, which reaches its peak in the early afternoon. The evening rush hour generates more NO and NO_2, but decreasing sunlight slows down photochemical reactions and there is no corresponding build-up of ozone.*

Figure 26 *A map showing ozone automatic monitoring sites in the UK.*

Studying individual reactions in the laboratory

The air around us may look very still and unchanging but this is deceptive. The troposphere acts as a huge reaction mixture with a vast array of interrelated chemical reactions taking place all the time. Many of these reactions involve 'broken down' fragments of molecules, called **radicals**, and the reactions can take place very quickly indeed. Other reactions take place more slowly. To make predictions about pollution, chemists need to know what reactions can take place and how quickly they occur. This means investigating each reaction in the laboratory and measuring the rate at which it takes place under a variety of carefully controlled conditions.

Modelling studies

Much of the information on rates of reactions is used in computer simulation studies which aim to reproduce and predict the behaviour of pollutants during a smog episode.

Smog chamber simulations

These are laboratory experiments on a grand scale. Primary pollutants are mixed in a huge plastic bag called a smog chamber and exposed to sunlight under carefully controlled conditions (Figure 27). Analytical probes monitor the concentrations of various species as the photochemical smog builds up.

Figure 27 *The EU smog chamber in Valencia, Spain. The reaction chamber has a volume of approximately 200 m^3 (200 000 litres). It has to be this big to reduce to a minimum any 'surface effects' where reactions take place on the walls of the container instead of in the gas phase.*

DF6 *Tackling the emissions problem*

Concern about air pollution from motor vehicles is mounting worldwide and many countries are bringing in legislation to limit emissions. The US has led the way, and cars are put through a rigorous emissions test cycle before they are allowed on the road. Figure 28 shows how emission limits have become increasingly severe.

There are two ways of tackling the emissions problem directly. One involves changing the design of cars. The other involves changing the fuel used by the car.

Of course, there are important indirect methods for tackling the emissions problem, such as limiting the traffic entering towns and encouraging car-sharing schemes for people travelling to and from work.

However, in this section we shall concentrate on methods of reducing the emissions from a single car by changing the technology of the car engine and by modifying the exhaust system.

Changing the engine technology

How much air does a petrol engine need? You can work this out by doing **Assignment 9**.

ASSIGNMENT 9

For the purposes of this assignment, assume that petrol is pure heptane, C_7H_{16}.

a Write an equation for the complete combustion of heptane vapour.

b How many moles of oxygen are needed for the combustion of 1 mole of heptane?

c How many moles of heptane are there in 1 g?

d What mass of oxygen is needed to burn 1 g of heptane (A_r: O, 16)?

e What mass of air is needed to burn 1 g of heptane? (Assume that air is 22% oxygen and 78% nitrogen by *mass*.)

f What *volume* of air (measured at 25 °C and 1 atmosphere pressure) is needed to burn 1 g of heptane? (Assume that 1 mole of oxygen has a volume of 24 dm³ under these conditions and that air is 21% oxygen and 79% nitrogen by *volume*.)

Chemical Ideas 1.4 will help you to carry out calculations involving gases.

The ratio of air to fuel which you worked out in Assignment 9 is the ratio needed to make the petrol burn completely. It is called the **stoichiometric ratio**. (*Stoichiometric* means *involving the exact amounts shown in the chemical equation*.)

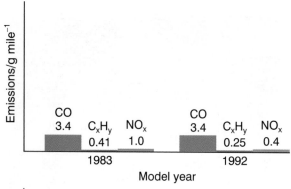

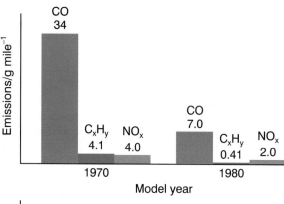

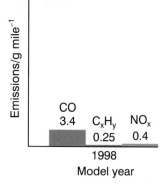

Figure 28 US Federal emission limits for new vehicles have become increasingly severe.

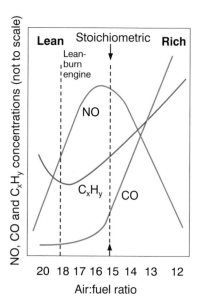

Figure 29 How gases in exhaust emissions vary with the composition of the air–fuel mixture.

For an average car engine, the stoichiometric ratio is usually taken to be about 15:1 (by mass).

If you have *less* air than the stoichiometric ratio, it is called a 'rich' mixture, because it's rich in petrol. You burn a rich mixture when you use the choke to start a car on a cold morning. A 'lean' mixture has *more* air than the stoichiometric ratio: it produces less NO_x and CO but the levels of C_xH_y can actually increase (Figure 29). The trouble is, if the mixture is *too* lean, the engine misfires and emissions increase.

'Lean-burn' engines use an air:fuel ratio of around 18:1 and have specially designed combustion chambers to get over the problem of misfiring.

Another advantage of a lean-burn engine is that it gives better fuel economy, because you are using less fuel for each firing of the cylinder.

Using catalysts

Catalysts speed up chemical reactions without getting used up themselves. We can use them to speed up reactions which involve pollutants in car exhausts. Look at these reactions:

$2CO(g) + O_2(g) \rightarrow 2CO_2(g)$ (reaction 1)
$C_7H_{16}(g) + 11O_2(g) \rightarrow 7CO_2(g) + 8H_2O(g)$ (reaction 2)
$2NO(g) + 2CO(g) \rightarrow N_2(g) + 2CO_2(g)$ (reaction 3)

(Here we've used C_7H_{16} as an example of an unburnt hydrocarbon and NO as the main component of NO_x in car emissions.)

In these reactions, pollutants are being converted to CO_2, H_2O and N_2, which are all naturally present in the air. These reactions go of their own accord, but under the conditions inside an exhaust system they go too slowly to get rid of the pollutants.

ASSIGNMENT 10

Look at reactions 1–3.

a For each of the pollutants CO, C_7H_{16} and NO, say whether it is being oxidised or reduced.

b Which of these reactions would be important in controlling pollutants from
  i an ordinary engine?
  ii a lean-burn engine? (Look at Figure 29.)

A catalyst made of a precious metal such as platinum or rhodium speeds up these reactions in the exhaust system. Such catalysts are used in **catalytic converters**.

You can find out more about catalysts in **Chemical Ideas 10.4**.

A lean-burn engine uses an *oxidation catalyst system* which removes CO and C_xH_y. The exhaust gases are rich in oxygen, so CO and C_xH_y are oxidised to CO_2 and H_2O on the surface of the catalyst. This kind of catalyst system does little to reduce NO, because NO needs reducing, not oxidising, to turn it to harmless N_2. But that doesn't matter too much in a lean-burn engine, because this type of engine produces less NO anyway (see Figure 29).

The three-way catalyst system

Catalytic converters can be fitted to ordinary engines too. But for an ordinary engine, a simple oxidation catalyst won't do – it wouldn't remove the NO, which is in much higher concentration in the exhaust than from a lean-burn engine. Here, a *three-way catalyst system* is needed, which both oxidises CO and C_xH_y, *and* reduces NO. This kind of catalyst system *only* works if the air–petrol mixture is carefully controlled so that it's exactly the stoichiometric mixture for the fuel. If the mixture is too rich, there is not enough oxygen in the exhaust fumes to remove CO and C_xH_y.

This means that cars fitted with three-way catalyst systems need to have oxygen sensors in the exhaust gases, linked back to electronically controlled fuel injection systems. You can see the arrangement in Figure 30.

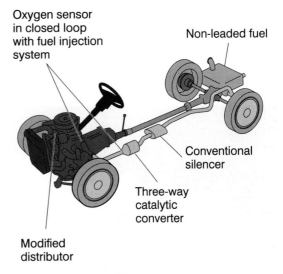

Figure 30 The three-way catalyst system.

Figure 31 shows the effect of catalysts on exhaust emissions from ordinary and lean-burn engines. A three-way catalyst system is the most efficient way of simultaneously lowering all three emissions.

All catalytic converters work only when they are hot. A platinum catalyst starts working around 240 °C, but by alloying the platinum with rhodium you can get the catalyst to start working at about 150 °C. These catalysts are poisoned by lead, so the converters can only be used with lead-free fuel.

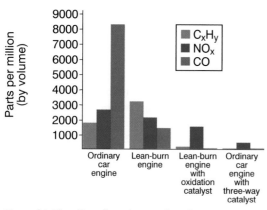

Figure 31 The effect of catalysts on the exhaust emissions from ordinary and lean-burn engines.

The catalyst is used in the form of a fine powder spread over a ceramic support whose surface has a network of tiny holes (Figure 32). The surface area of the catalyst exposed to the exhaust gases is about the same as two or three football fields.

ASSIGNMENT II

a Why is it important that catalytic converters start working at as low a temperature as possible?

b What is meant by a *catalyst poison*?

c Why is the catalyst used in the form of a fine powder?

d Suggest a reason why the catalytic converter has eventually to be replaced.

e Catalytic converters convert the pollutants CO, C_xH_y and NO_x into harmless gases. This is still only a partial solution to the emissions problem. Why?

f Why is it possible to fit oxidation catalytic converters to existing cars, but not three-way converters?

DF7 *Changing the fuel*

The other approach to tackling the emissions problem is to change the fuel used by cars.

Aromatic hydrocarbons make up as much as 40% of lead-free petrol. Aromatic hydrocarbons may cause higher CO, C_xH_y and NO_x emissions, and some of them may cause cancer. Benzene is the worst, and is strictly controlled, but others may also be controlled in the future.

Butane content too will probably be lowered in the future. Butane is volatile and is responsible for evaporative emissions leading to ozone formation and photochemical smogs.

However, both butane and aromatic hydrocarbons are high octane components of petrol. If they are removed, their octane quality must be replaced by something else. This is why the petrol companies looked to the oxygenates as a possible solution.

In the US, there is a very active policy to reduce the emissions caused by petrol itself and by its combustion products.

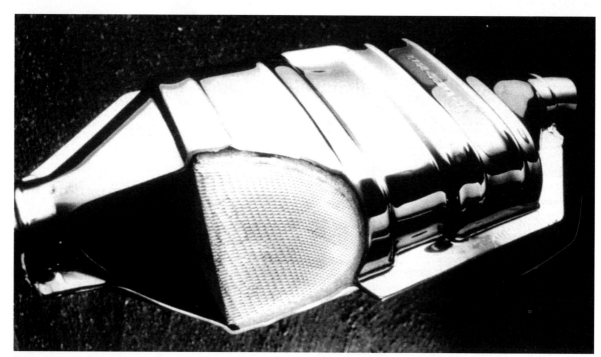

Figure 32 A catalytic converter for a car exhaust. Note the honeycomb structure, giving a high surface area.

There are attempts to produce 'reformulated gasolines' that have

- low volatility, thus reducing the concentration of hydrocarbons in the atmosphere
- reduced concentrations of benzene (a carcinogen)
- added oxygenates (to improve burning and so reduce the concentration of pollutants such as CO).

All this has to be achieved without reducing the efficiency of the fuel.

Methanol – one key to future fuels?

The pros

Methanol burns cleanly in a car engine. Its combustion is more complete than that of petrol so emissions, particularly of carbon monoxide, are reduced. Nor does it release carcinogenic benzene vapour or other aromatic hydrocarbons into the air. It has a high octane number (114), and its use requires only small changes to engine and petrol pump design. What's more, it's cheap.

So, why don't we use methanol as a fuel? After all, Indy racing drivers in the US have been using it for years. Methanol is less volatile than petrol and less likely to explode in a collision.

In fact, some countries, such as Germany and parts of the US, have used methanol, but as part of a mixture with petrol.

Figure 33 Cars in the CART Championship World Series also run on methanol.

The cons

The main problem when using a mixture of methanol and petrol is getting the two liquids to mix. They tend to separate into layers, like oil and vinegar, unless a co-solvent is used. In some countries, a higher alcohol than methanol, called TBA (from the older name for it, tertiary butyl alcohol), is mixed with the methanol for this purpose. TBA dissolves in both methanol and petrol.

$$CH_3-\overset{\displaystyle CH_3}{\underset{\displaystyle CH_3}{C}}-O-H$$

TBA

Another problem is that the petrol–alcohol mixture is **hygroscopic** and absorbs water vapour from the air. The wet methanol corrodes parts of the engine (Figure 34).

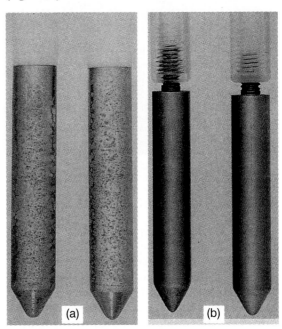

Figure 34 Methanol absorbs water, making it corrosive. The steel rods in (a) were immersed in petrol containing methanol. Compare them with the rods in (b), which were immersed in normal unleaded petrol.

When the water content reaches a critical level (about 0.5%) two layers separate, giving a layer of petrol sitting on top of a water–alcohol mixture.

Although polluting emissions are reduced, methanol itself is toxic. Long-term exposure can cause blindness and brain damage – that's why 'meths' drinkers go blind. So there is concern over its possible effect on mechanics and petrol-pump attendants at filling stations.

One litre of methanol produces about 40% less energy than 1 litre of petrol. This means larger and heavier fuel tanks and more refuelling trips.

Making methanol

Methanol is manufactured from synthesis gas, a mixture of carbon monoxide and hydrogen. In turn, synthesis gas is manufactured from natural gas, oil and coal.

Figure 35 A petrol station in Santos, Brazil. The fuel is a mixture of petrol and ethanol.

Adding ethanol

Ethanol is often added to petrol. It has a high octane number and is thought to be less polluting to the atmosphere. It is manufactured from ethene, which is obtained from the cracking of naphtha. However, ethanol can also be obtained from a renewable source – the fermentation of cane sugar juice. In the 1970s Brazil had to import oil, but the climate is good for growing cane sugar. A programme was set up to produce ethanol. By the 1990s Brazil made enough ethanol to power about one-third of the country's 12 million cars – either using gasohol (a mixture of gasoline and ethanol) (Figure 35) or ethanol alone.

Burning ethanol produces less carbon monoxide, sulphur dioxide and nitrogen oxides than petrol, and it is believed that unburned ethanol does not contribute so much as hydrocarbons to photochemical smogs. The carbon dioxide produced is balanced by absorption of carbon dioxide in new cane, and so using ethanol as a fuel adds less to the greenhouse effect.

Problems arise when oil prices drop because ethanol from fermentation is then less competitive. Large amounts of energy are needed for intensive cultivation of the crop, and there are queries about the overall energy efficiency of the process.

DF8 *Hydrogen – a fuel for the future?*

Petrol has been developed to work well as a transport fuel. But problems remain. You have already studied the problem of emissions. Changing the fuel used by cars could help to solve the emissions problem. Figure 36 shows past and projected rates of world consumption of oil and coal. The use of petrol will decline in the future as oil supplies run out, so a search for alternative fuels could be vital to our transport system.

People who favour using hydrogen as a fuel see water as a plentiful source of hydrogen. If hydrogen could be extracted from water without consuming fossil fuels it would reduce our dependence on these fuels, and help to reduce the amount of carbon dioxide released into the atmosphere.

Hydrogen could be distributed as we now distribute natural gas, and burned as a heating fuel, or used in internal combustion engines, or converted into electricity in a **fuel cell**.

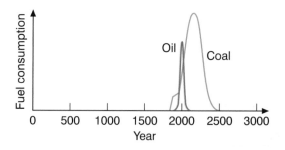

Figure 36 Past and projected rates of world consumption of oil and coal: the use of fossil fuels will decline in the future as supplies run out.

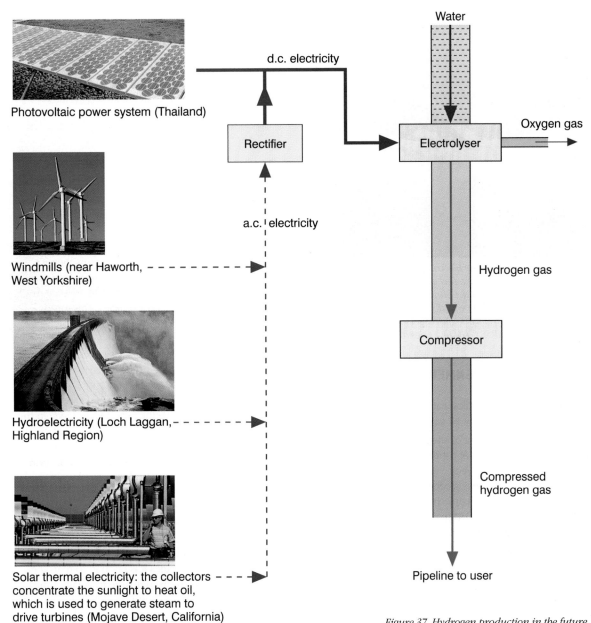

Photovoltaic power system (Thailand)

Windmills (near Haworth, West Yorkshire)

Hydroelectricity (Loch Laggan, Highland Region)

Solar thermal electricity: the collectors concentrate the sunlight to heat oil, which is used to generate steam to drive turbines (Mojave Desert, California)

Figure 37 Hydrogen production in the future.

But how can we make the hydrogen? The most likely large-scale method of producing hydrogen seems to be by electrolysis of water, obtaining the energy needed for this from some renewable source, such as solar cells. But why produce electricity and then use it to make hydrogen? There are two major advantages:

- hydrogen can be stored
- it can be used in the internal combustion engine.

A possible scheme is shown in Figure 37. All the technologies needed already exist. We have large-scale electrolysis plants for other purposes, large storage tanks, and a network of gas pipelines.

The **hydrogen economy** would use hydrogen as a way of storing and distributing energy. If systems are costed over whole lifetime use in terms of money and energy, then distributing hydrogen by pipeline may be cheaper than transmitting electricity.

Fuel cells are being used to generate electricity on a small scale in cars (Figure 38).

The main problem in the design of such a car is the storage of hydrogen. A large volume of gaseous hydrogen is required to get equivalent mileage to a fuel tank of petrol. We need to find some way to store it more compactly. One solution is to store it as a liquid in a high-pressure fuel tank.

Schemes for developing alternative fuel use on a large scale depend on long-term and large-scale investments in new infrastructures, and so success will depend on political as well as economic factors. The reward could be cleaner and renewable fuels.

Figure 38 The Necar4 (New Electric Car) is being developed by Mercedes-Benz and uses a fuel cell. The hydrogen is stored in a cylinder cell as a liquid. Cylinders will be returned and exchanged for full ones at centrally located hydrogen filling stations. The fuel economy is equivalent to almost 20 km per litre (90 miles per gallon). The first sales are expected in 2004.

ASSIGNMENT 12

The petrol tank of a typical car holds about 45 litres of petrol (approximately 10 gallons).

a Calculate the amount of energy that is released by burning 45 litres of petrol. Use the following information to help you.
- Assume that petrol is octane (C_8H_{18}).
- The standard enthalpy change of combustion of C_8H_{18} is $-5500\,kJ\,mol^{-1}$.
- The density of octane is $0.70\,g\,cm^{-3}$.
- 1 litre is $1000\,cm^3$.

b Calculate the mass and volume (at 20 °C and 1 atmosphere pressure) of hydrogen needed to provide the same amount of energy as 45 litres of octane.
- The standard enthalpy change of combustion of H_2 is $-286\,kJ\,mol^{-1}$.
- 1 mole of a gas at 20 °C and atmospheric pressure has a volume of about 24 litres.

(You can remind yourself about calculations involving gases by reading **Chemical Ideas 1.4**.)

Hydrogen engines are efficient. In motorway driving conditions a hydrogen engine can be over 20% more efficient than a petrol engine. In city 'stop-go' driving conditions the hydrogen engine is about 50% more efficient than a petrol engine.

c Taking efficiencies into account, what mass and volume of hydrogen are needed to give the same mileage as 45 litres of petrol in
  i motorway driving conditions?
  ii city driving conditions?

Now is the time for you to present your reports on the prospects for some alternatives to petrol in **Activity DF1.1**. Try to present the most convincing case for your fuel so that it has a fair hearing. You can then discuss the relative merits of each fuel in class and decide on what you think is the most promising way (or ways) forward.

DF9 *Summary*

This unit began by considering some of the desirable properties of a good fuel. This led you to compare the energy released when different fuels are burned in oxygen and to a general study of energy changes in chemical reactions. You then found out how petrol is made and how chemists and chemical engineers are helping to develop better fuels for motor vehicles. This means producing fuels which give improved performance and greater fuel economy, so that we use our precious reserves of crude oil as economically as possible.

Petrol is a mixture of hydrocarbons, many of which are alkanes. This led you to find out more about the properties and reactions of alkanes. But the hydrocarbons which give the best performance in the petrol engine are not the ones which are most plentiful in crude oil. So, after the primary distillation of crude oil into fractions, there is a whole range of chemical processes carried out in the refinery to 'doctor' the hydrocarbons to suit our needs.

Petrol must have the correct octane number to avoid auto-ignition. Branched-chain alkanes have higher octane numbers than straight-chain alkanes, which led you to study structural isomerism in more detail. For some molecules it is enough to rearrange the position of the atoms, but large molecules must be broken down into more useful smaller ones, and some smaller ones are joined together. The final petrol is a blend of many hydrocarbons and possibly some oxygenates too. This led you to find out about the structures of compounds in other homologous series: alkenes, arenes, alcohols and ethers. The mixing of petrol components to form the final 'blend' introduced you to simple ideas about entropy.

You went on to consider the problems of emissions from motor vehicles, the difference between primary and secondary pollutants and the formation of photochemical smog. You examined ways in which car manufacturers and oil companies are tackling the problem of emissions – by changing engine technology to control the way the fuel burns, by changing the composition of the fuel, and by using catalytic converters to speed up reactions in the exhaust that convert the pollutant to less harmful gases. You then looked in more detail at the role of catalysts in chemical reactions.

In the immediate future, it seems that the use of oxygenates will play a large part in fuel development. However, the story did not end with petrol. Our supplies of crude oil are finite and are needed for more than petrol alone, so you were invited to think about alternative fuels, such as hydrogen, for motor vehicles for the more distant future.

Much of the chemistry you have covered in this unit is fundamental to other areas and you will need to use these ideas in other parts of the course.

Activity DF9 will help you to check your notes on this unit.

FROM MINERALS TO ELEMENTS

Why a unit on MINERALS AND ELEMENTS?

The first unit in the course – **The Elements of Life** – told the story of how the elements were formed. The theme is taken further in this unit, which tells how we have learned to win back some elements from the minerals which contain them and turn them into useful substances. Two non-metals (bromine and chlorine) and one metal (copper) are chosen. The elements are not evenly distributed around the world – we are dependent on other countries for our supply. Thus events in other parts of the world can affect our own mineral industry.

The unit introduces three major types of inorganic chemical reaction: acid/base, redox and ionic precipitation. Some important halogen chemistry is covered, and transition metal chemistry is introduced.

Three areas of chemistry that you met in earlier units are revisited and taken further. The concept of amount of substance is extended to include concentrations of solutions. Ideas about the electronic structure of atoms are developed to include the distribution of electrons in atomic orbitals, and the properties of covalent compounds are linked to their bonding and structure.

Overview of chemical principles

In this unit you will learn more about …

ideas you will probably have come across in your earlier studies
- the halogens
- ions in solids and solutions
- dissolving
- acids
- precipitation
- useful products from metal ores and rocks

ideas introduced in earlier units in this course
- amounts in moles (**The Elements of Life**)
- electronic structure of atoms (**The Elements of Life**)
- covalent bonding (**The Elements of Life** and **Developing Fuels**)
- the Periodic Table (**The Elements of Life**)

… as well as learning new ideas about
- redox reactions
- acids and bases.

M
FROM MINERALS TO ELEMENTS

M1 Chemicals from the sea

The lowest point on Earth

The Dead Sea is the lowest point on Earth, almost 400 m below sea level in the rift valley which runs from East Africa to Syria. It is like a vast evaporating basin: water flows in at the north end from the River Jordan but there is no outflow. The countryside around it is desert and in the scorching heat so much water evaporates that the air is thick with haze, making it hard to see across to the mountains a few kilometres away on the other side. Steady evaporation of the water for thousands of years has resulted in huge accumulations of salts so that the water in the Dead Sea is much denser than usual.

The Dead Sea is such a natural curiosity that surveys of the salt concentration were conducted as early as the 17th century, even though many of the elements in the salts were then unknown.

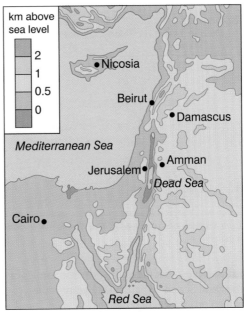

Figure 1 The region around the Dead Sea in relief.

Figure 2 The Dead Sea – a natural curiosity.

The water contains about $350\,g\,dm^{-3}$ of salts compared with $40\,g\,dm^{-3}$ in water from the oceans. A comparison of the most abundant ions present in Dead Sea water and typical ocean water is shown in Table 1.

	Ocean water		Dead Sea water	
Ion	Mass in 1 dm³ of water/g		Ion	Mass in 1 dm³ of water/g
Na^+	11		Na^+	39
K^+	0.4		K^+	6.9
Mg^{2+}	1.3		Mg^{2+}	39
Ca^{2+}	0.4		Ca^{2+}	17
Cl^-	19		Cl^-	208
Br^-	0.07		Br^-	5.2
HCO_3^-	0.1		HCO_3^-	trace
SO_4^{2-}	2.5		SO_4^{2-}	0.6

Table 1 Compositions of samples of typical ocean water and Dead Sea water.

Estimates suggest that there are 43 billion tonnes of salts in the Dead Sea and a particular feature is the relatively high proportion of bromides.

The sea is the major source of minerals in the region. A chemical industry has grown up around the Dead Sea in Israel and it has become one of the largest exporters of bromine compounds in the world. The annual production of bromine compounds in Israel exceeds 230 000 tonnes.

You can learn more about ionic compounds as solids and in solutions in **Chemical Ideas 5.1**.

Activity M1.1 investigates some of the ideas about dissolving and precipitation you will be reading about in this unit.

Figure 3 Despite the low concentration of bromide ion in the ocean, it is economic to extract the element from sea water.

ASSIGNMENT I

Dead Sea water is certainly more salty than ocean water, but Table 1 shows that there are differences in ionic composition between the two.

a What do you notice about the abundances of the ions of Group 1 and Group 2 in the two samples?

b Compared with ocean water, Dead Sea water has a particularly high proportion of one ion. Which ion is this? How many times more abundant is this ion in Dead Sea water than in ocean water?

Extracting bromine from sea water

There is no similar source of bromine in the UK. The only source available is the ocean (see Table 1).

Bromine is extracted from the Irish Sea at Amlwch in Anglesey. Currently, just under 40 000 tonnes of bromine are produced annually and this figure is rising as uses for new bromine compounds are found.

ASSIGNMENT 2

a Use the data in Table 1 to calculate the mass in tonnes of ocean water that contain 1 tonne of bromide ion. (Assume that 1 dm³ of ocean water has a mass of 1 kg.)

b At Amlwch approximately 22 000 tonnes of ocean water are used to produce 1 tonne of bromine. Using your answer in part a, calculate the percentage yield of the extraction process at Amlwch.

Turning bromide ions into bromine is simple chemistry: you can do it in the laboratory by adding chlorine water to a solution containing bromide ions.

This is an example of a **redox reaction**. You can find out more about such reactions in **Chemical Ideas 9.1**.

Industrially, it is more complicated than that and involves some clever chemical engineering, as you will see later. Because the concentration of bromide ions in the ocean is so low, it is necessary to use a four-stage process to extract bromine:

1 oxidation of bromide ions (Br^-) to bromine (Br_2)

2 removal of bromine vapour

3 reduction of bromine to hydrobromic acid (HBr)

4 oxidation of hydrobromic acid to bromine.

Figure 4 The main pump house at Amlwch – a third of a million gallons of sea water are pumped every minute.

1 Oxidation of bromide ions to bromine

At Amlwch, water is pumped from the ocean (Figure 4). After marine debris is removed, the water passes through a large pipeline (Figure 5) where it is acidified with sulphuric acid. The pH of the sea water is changed from 8 (weakly alkaline) to 3.5 (acidic). Acidification is necessary because, at higher pH, bromine and chlorine react with water.

$$Br_2(aq) + H_2O(l) \rightarrow HBr(aq) + HBrO(aq)$$
$$Cl_2(aq) + H_2O(l) \rightarrow HCl(aq) + HClO(aq)$$

The first reaction would reduce the overall yield of the process considerably and chlorine would be wasted by the second reaction, adding to the overall costs.

A slight excess of chlorine is injected into the acidified water in the sea water main to displace the bromine by a redox reaction:

$$Cl_2(aq) + 2Br^-(aq) \rightarrow 2Cl^-(aq) + Br_2(aq)$$

The pH scale

The pH scale measures the acidity or alkalinity of a solution.

- For pure water, pH = 7.
- Solutions with pH<7 are acidic.
  The lower the number, the more acidic the solution.
- Solutions with pH>7 are alkaline.
  The higher the number, the more alkaline the solution.

You will find out more about the pH scale in **The Oceans**.

Representing this reaction by two half-equations helps to make clear what has been oxidised and what has been reduced.

$$2Br^-(aq) \rightarrow Br_2(aq) + 2e^-$$
$$Cl_2(aq) + 2e^- \rightarrow 2Cl^-(aq)$$

Electrons appear on the product side of the first half-equation so it is clear that bromide ions have been oxidised to bromine because they have lost electrons. Electrons appear on the reactant side of the second half-equation. Chlorine has been reduced because it has gained electrons to form chloride ions.

$Cl_2(aq)$ is the oxidising agent because it has removed electrons from the bromide ions. $Br^-(aq)$ is the reducing agent because it has given electrons to the chlorine.

Figure 5 The sea water main at Amlwch.

2 Removal of bromine vapour

The treated sea water then passes into a 'blowing-out tower' (Figures 6 and 7). A strong current of air passes up the tower as the water is sprinkled over the packing. Bromine is volatile so the air removes most of the bromine and some chlorine from the water. At this point the bromine concentration in the air is too low for effective separation of liquid bromine.

Figure 6 The blowing-out towers – here bromine is blown from the treated water by air and then reduced to hydrobromic acid.

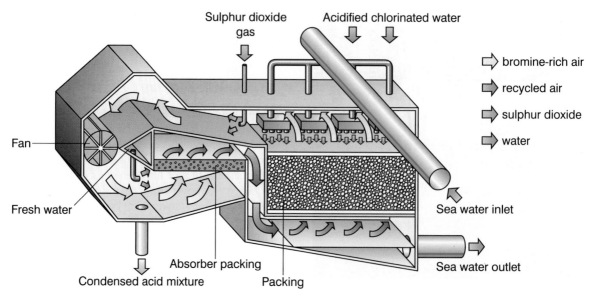

Sulphur dioxide gas

Acidified chlorinated water

⇨ bromine-rich air

⇨ recycled air

⇨ sulphur dioxide

⇨ water

Fan

Fresh water

Absorber packing

Condensed acid mixture

Packing

Sea water inlet

Sea water outlet

Figure 7 Inside a blowing-out tower.

3 Reduction of bromine to hydrobromic acid

To raise the concentration of bromine further the element is first reduced to hydrobromic acid by injecting sulphur dioxide gas into the blowing-out tower (Figure 7). The reaction also produces sulphuric acid:

$$Br_2(aq) + SO_2(g) + 2H_2O(l) \rightarrow$$
$$2HBr(aq) + H_2SO_4(aq)$$

Injection of fresh water into the blowing-out tower produces a fine mist of the acids which is condensed in the absorber section of the tower. The acid mixture at the end of this stage contains about 13% by mass of bromine as hydrobromic acid.

4 Oxidation of hydrobromic acid to bromine

The collected acid is passed into the 'steaming-out tower'. Here, chlorine regenerates the bromine, which is then distilled out of the tower with steam.

$$2HBr(aq) + Cl_2(g) \rightarrow Br_2(g) + 2HCl(aq)$$

The hot vapour mixture is condensed to form an aqueous layer and a lower layer of bromine (Figure 8). The resulting damp bromine is dried with concentrated sulphuric acid.

The spent sea water is returned to the Irish Sea. Any chlorine or bromine present is destroyed by the presence of a small excess of sulphur dioxide before the water leaves the blowing-out tower. On discharge it has a pH of 3.5 (acidic) but this rapidly falls to pH 8 (alkaline) within 100 m of the outfall.

Sulphur dioxide, sulphuric acid and hydrochloric acid are all produced at Amlwch. Chlorine is produced in Cheshire by the electrolysis of brine and transported to Amlwch by road or rail.

A flow diagram for the manufacture of bromine is given in **Activity M1.2**. Doing this activity will help you to understand the processes underlying the production.

Figure 8 After leaving the steaming-out tower, bromine condenses to a deep red liquid.

ASSIGNMENT 3

The bromide ion concentration in the Dead Sea is much higher than that in the ocean. Water from the Dead Sea is evaporated and sodium chloride, potassium and magnesium chlorides and then calcium chloride crystallise out in turn, leaving a solution rich in bromide ions. This solution is then treated with chlorine. Compare this with the techniques used in the first stages of the manufacture of bromine in Anglesey, giving one advantage and one disadvantage in making bromine at the two sites.

Dense, dark and dangerous – but useful

Bromine was discovered in 1826 by a French chemist, Jérôme Balard, who later became a professor at the Sorbonne in Paris. His discovery helped Johann Döbereiner spot the idea of 'triads', which was a step on the way to the development of the Periodic Table.

Bromine has a dense and choking vapour – hence its name which is based on a Greek word, *bromos*, meaning *stench*. The liquid produces painful sores if spilled on the skin. You will use bromine several times in this course and you must take great care when you handle it.

The company at Amlwch as well as independent bodies such as the Health and Safety Executive, the Environment Agency and the Pollution Inspectorate make regular checks on the quality of the air and the fresh and ocean water around the site.

The extraction process is a continuous process and takes place in sealed equipment. When maintenance or repair is required, a section is isolated by valves. A continuous process enables close monitoring of the whole plant. This is done by a system of sensors connected to a computer in a control room, with a system of automatic alarms and shutdowns providing further protection.

Great care has to be taken when transporting bromine (Figure 9). Most of it is carried in lead-lined steel tanks supported in strong metal frames. Each tank holds several tonnes of the element. International regulations control the design and construction of road and rail tankers. The industry's safety record is good but there have been some accidents. One such accident is discussed in Figure 13 (p.50) and **Assignment 4** (p.51).

Bromine may once have been just a laboratory curiosity but is now an important industrial commodity which is manufactured on a large scale. One of its most important uses is in the production of flame retardants (Figure 10). For example, tetrabromobisphenol A (TBBA) can be incorporated into some polymers (*via* the –OH group), making the polymer much less prone to combustion.

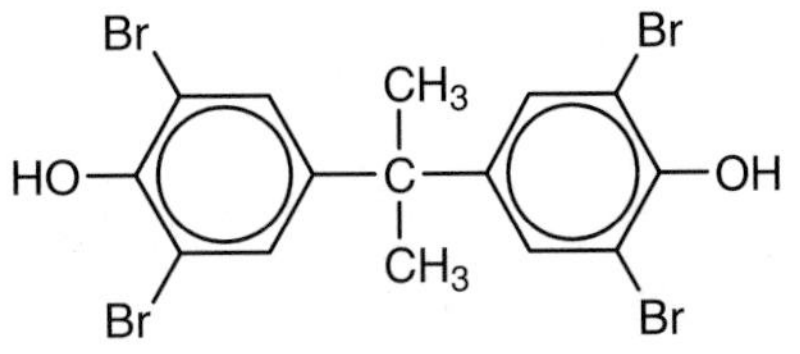

TBBA

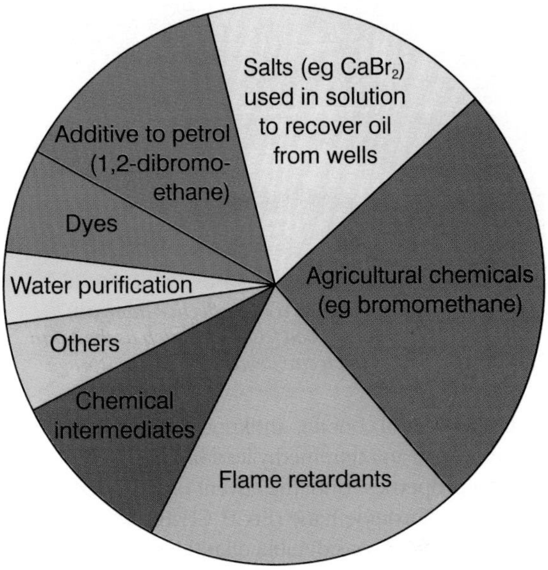

Figure 10 A pie chart showing the uses of bromine.

Figure 9 A road tanker for the transport of bromine – note the protective clothing worn by the operator.

Bromine compounds are used in agriculture. Bromomethane (CH_3Br) is used as a fumigant against many pests (Figure 11). Unfortunately, if bromomethane gets into the upper atmosphere it forms bromine atoms which destroy ozone. (You will find out more about ozone depletion in **The Atmosphere**.) For this reason, attempts are being made to phase out bromomethane and other fumigants are being developed.

Figure 11 Bromomethane is a very effective fumigant to protect food, grain and wood from pests such as the Asian long-horned beetle which can devastate them in storage.

Leaded petrol contains antiknock agents such as tetraethyllead and tetramethyllead to improve its ignition properties. Until the advent of unleaded petrol, 1,2-dibromoethane ($BrCH_2CH_2Br$) was manufactured in considerable quantities as it is added to leaded petrol as a scavenger for the lead. It prevents lead oxides being deposited in the engine by converting them to volatile lead bromides which are removed in the exhaust gases. Now that leaded petrol has been phased out, the amount of bromine used to make 1,2-dibromoethane is decreasing rapidly.

Figure 12 The multi-product bromination reactor.

At the Amlwch site, there is a batch reactor (Figure 12) which can be used for the manufacture of different organic bromine compounds. Many of the organic bromine produces are specialised chemicals in the sense that they are required by other companies in relatively small amounts and at irregular intervals. The company is able to manufacture one product and then switch the reactor to a different product. It may be a few years before there is a need to produce more of any one product.

CHEMICAL SCARE SHAKES NEGEV

The driver of the truck that overturned near here early yesterday morning, sending an orange cloud of poisonous bromine gas into the sky over the Arava, said before he died that the accident occurred because he fell asleep at the wheel. The cloud of gas formed by the evaporating bromine, which had a diameter of some 10 kilometres later in the morning, forced scores of residents to leave their homes for several hours. Negev police closed off the road for the day.

According to a senior member of the Dead Sea Works rescue crew, the driver, Yisrael Taib, 32, of Dimona, was trapped when his flatbed truck flipped over on its side at about 5.15 am after rounding a bend some 20 kilometres south of the Arava junction.

Eight people were injured as a result of the accident, most of them rescue workers who tried frantically but in vain to extricate Taib, who was crushed by the steering wheel and the engine block in the mangled cabin.

A senior police officer on the scene praised the rescue crew for the speed with which it reached the spot.

The rescuers arrived shortly after 6 am and worked for several hours, even after a wind change sent the poisonous gas straight towards them.

A safety expert from a major industrial concern in the Negev told *The Jerusalem Post* that the bromine had apparently been packed according to international safety standards, but that yesterday's accident highlighted the need for more careful monitoring of the transportation of dangerous chemicals.

The expert, who sits on a national safety committee, told *The Post* that repeated efforts over the years to have 'black boxes' installed in the cabins of trucks carrying dangerous materials had been thwarted by trucking officials. The expert, who asked not to be identified, conceded that the black boxes wouldn't have prevented yesterday's accident, but would have told safety officials how fast the driver was travelling and how many hours he had been working.

The truck, on its way from the Dead Sea bromine plant south of Beersheva to the port in Eilat, was carrying two containers with over 20 tons of liquid bromine.

They said it was the first accident ever in Israel involving bromine, a deep rust-coloured liquid used in industry, agriculture and medicine. The bromine evaporated quickly in the desert heat. Bromine's boiling point is 58 °C.

Figure 13 Report from The Jerusalem Post.

ASSIGNMENT 4

According to *The Jerusalem Post* report (Figure 13), a safety expert recommended more careful monitoring of the transportation of dangerous chemicals. In the light of the report, what do you think the chemical industry does to reduce the risk of injury or death when chemicals such as bromine are transported?

Photography: the almost impossible process

People were aware that silver salts darken on exposure to light as long ago as the 16th century. The darkening is caused by the production of specks of silver metal, but it takes a long time – millions of times longer than you would like to sit for when your photograph is being taken. Fortunately, chemistry can cut down the time needed to take a photograph to just a fraction of a second.

Photographic film is a transparent plastic strip coated with emulsion: a layer of gelatin in which are scattered millions of tiny crystals of silver halides, particularly silver bromide. The emulsion is similar for black and white or colour film. Colour film just has three layers of emulsion, each layer containing a different dye.

When light hits a silver bromide crystal, a bromide ion is converted into a bromine atom and an electron. The electron migrates to the surface of the crystal. The bromine atom is 'trapped' by the gelatin while the electron combines with a silver cation to produce a silver atom.

This process has to happen several times, so that a cluster of a few silver atoms is produced on the surface of the crystal, before the next stage – development – can occur. In development, the film is soaked in a solution of a chemical 'developer'. The silver atom cluster seems to act as a kind of 'conducting bridge' between the developer and the unreacted part of the crystal of silver bromide.

If we represent the developer by the formula DH_2, the reaction between the developer and the silver bromide is

$$DH_2 + 2AgBr \longrightarrow D + 2HBr + 2Ag$$

The end result of developing is that the whole crystal is converted to silver – much more quickly and far more conveniently than if you had relied on light alone.

Next, unreacted AgBr crystals are removed in the fixing process. This leaves behind the dark silver. The more light, the more crystals turn to silver, and the darker the colour. The film is then washed and dried to produce the final negatives.

Several special conditions have to be fulfilled before the little clusters of silver atoms are able to form on silver bromide crystals. It has been estimated that the chances are about 1 : 100 000 000 of finding a substance which satisfies these conditions and which can be used to record photographic images. There are only about 100 000

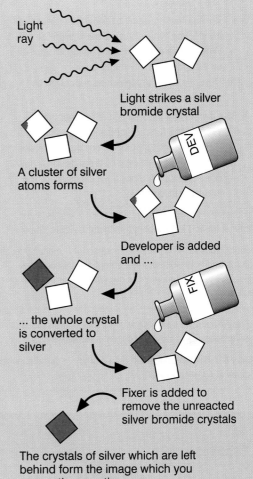

Light strikes a silver bromide crystal

A cluster of silver atoms forms

Developer is added and ...

... the whole crystal is converted to silver

Fixer is added to remove the unreacted silver bromide crystals

The crystals of silver which are left behind form the image which you see on the negative

Figure 14 The chemical changes which are involved in the production of a photographic print.

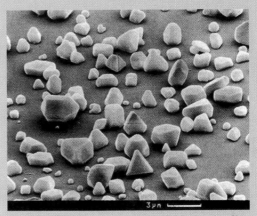

Figure 15 Silver bromide crystals in photographic film.

suitable materials to choose from, so the odds against photography being possible are extremely high. We are very lucky that silver bromide exists and it remains the basis of photography despite its high cost.

ASSIGNMENT 5

a i Write a half-equation to show the release of an electron when light strikes a bromide ion.

ii Write a half-equation to show the capture of an electron by a silver ion.

iii Describe these processes as reduction or oxidation.

b The developing process is a redox reaction.

i What is the function of the developer?

ii What happens to the silver ions during developing?

iii From the point of view of bromide ions, how does developing differ from exposure of the emulsion to light?

Chemical Ideas 11.4 is about the chemistry of the halogens.

Chemical Ideas 2.4 will tell you more about the energy levels in atoms and how electrons are arranged in these energy levels. This will explain why the halogens are in the p-block of the Periodic Table.

You can investigate some of the properties of bromine and other halogens and their compounds in **Activity M1.3**.

Activity M1.4 looks in detail at how bromine is handled when it arrives at a chemical plant.

Making chlorine

Although the sea contains high concentrations of sodium chloride, almost pure sodium chloride is found as rock salt. It can be recovered either by mining the solid underground (Figure 16) or pumping water into the salt and collecting the salt solution at the surface.

On electrolysing sodium chloride solution, chlorine and hydrogen are generated at the electrodes:

positive electrode: $2Cl^-(aq) \rightarrow Cl_2(g) + 2e^-$

negative electrode: $2H^+(aq) + 2e^- \rightarrow H_2(g)$

Remaining in the solution are sodium ions, hydroxide ions and some chloride ions. Purification leads to the production of sodium hydroxide, another important compound.

An essential requirement is to find an economic and effective means of separating the hydrogen and chlorine as they are produced.

One modern way in which the electrolysis of sodium chloride solution can be carried out is described and discussed in **Activity M1.5**.

Figure 16 The Winsford mine in Cheshire, where solid rock salt is cut from the walls of an underground cavern 200 m below the surface. The cavern is over 7 m high and about 3500 tonnes are mined each day.

Chlorine – friend or foe?

Chlorine has a poor image. In the public mind, it is associated with pollution – pollution of the land through pesticides which contain chlorine and pollution of the upper atmosphere through CFCs (chlorofluorocarbons). You may also have heard about the use of chlorine as a poison gas. Both chlorine and a compound derived from it, known as phosgene, $COCl_2$, were used with deadly effect in the trenches in the First World War. It is believed they have been used against civilian populations in recent years.

However, chlorine is used in many ways to make our lives safe and comfortable. About 50 million tonnes of chlorine are produced annually worldwide. The best known use is in water treatment but it is also used to make poly(chloroethene), PVC. It is needed for the manufacture of polyurethanes and is present in a wide variety of solvents (such as trichloroethene) used both in dry cleaning of clothes and in industry to clean grease off metals.

M2 Copper from deep in the ground

A gift from the Sun?

Copper was probably discovered by chance over 11 000 years ago. Potters, using glazes containing highly coloured copper minerals, accidentally reduced the copper compounds with the hot carbon in their fires.

Copper could be shaped easily by beating or moulding. Its beautiful appearance – like flames or a burning Sun – made it valuable for ornaments and

Figure 17 Tools, utensils and coins have been made using copper for thousands of years. This bison from Turkey dates from about 2300 BC and is made of solid copper.

jewellery. Its durability made it superior to wood and clay for pots and other utensils. The discovery that copper could be alloyed with tin to make bronze – a harder and even more useful and versatile material – gave us the name for a whole era in our history.

Where did it come from?

Copper minerals were formed along with minerals containing silver, lead, iron, zinc, molybdenum, tungsten, tin and other metals, in **hydrothermal deposits**. In other words, they formed from hot water. Deep underground, water is under pressures which are very much greater than atmospheric pressure. It's like a giant pressure cooker, and the high pressures prevent the water boiling. The high temperatures encourage dissolving, and solutions form even though the temperature is well over 100 °C. Many compounds which are insoluble under familiar laboratory conditions are soluble in superheated underground water and some of these compounds can reach high concentrations.

Make sure you understand about concentrations and how to calculate them by working through **Chemical Ideas 1.5**.

In the mineral deposits of southwest England and those you will be finding out about from the US, hot solutions like these were left over when magma cooled under the Earth's surface. Magma is a molten mixture of rocks, water and other components of the Earth's crust: a kind of 'crustal soup'.

Around 300 million years ago there were great upheavals in the Earth's crust. Magma was forced up under the sandstones, shales and limestones which had been laid down about 50 million years earlier.

Most of the magma solidified to form granite intrusions, leaving hot mineral-rich solutions near the top. The rocks in the crust cracked as they cooled, and the solutions streamed out along the cracks. Some were only a fraction of a millimetre wide; others were nearly 1 m across.

The solutions cooled as they came into contact with the colder rocks near the surface and as they moved further away from the hot granite. Minerals began to crystallise. Cassiterite (SnO_2) is one of the first minerals to crystallise in hydrothermal deposits when hot solutions cool. Then a variety of metal sulphides form, such as pyrite (FeS_2), chalcopyrite ($CuFeS_2$), sphalerite (ZnS) and galena (PbS). A vein of minerals which fills a crack in rock is called a **lode**.

In **Activity M2.1** you can look for evidence of these minerals in a sample taken from a lode in The Pennines.

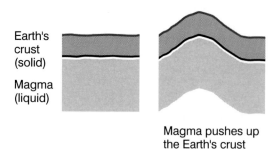

Earth's crust (solid)

Magma (liquid)

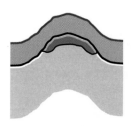

Magma pushes up the Earth's crust

Granite and hot solutions form where magma cools down

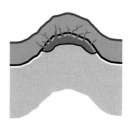

The crust cracks as it cools and the cracks become filled with minerals from the hot solution

Figure 18 How hydrothermal mineral deposits form.

Ancient miners

Chalcopyrite ($CuFeS_2$) is the principal copper-containing mineral in hydrothermal lodes. When the lodes are exposed to air and water, at the Earth's surface, chalcopyrite becomes altered to oxides, hydroxides and carbonates of copper and iron. Four examples are

- cuprite, Cu_2O
- tenorite, CuO
- malachite, $Cu_2(OH)_2CO_3$
- azurite, $Cu_3(OH)_2(CO_3)_2$.

Early civilisations, like the Ancient Egyptians and Phoenicians, could supply their small needs for copper by mining the richest deposits of these altered minerals and reducing them with hot carbon. It has been estimated, for example, that the Egyptians used only 10 000 tonnes of copper in 1500 years.

Sometimes these rich deposits contained as much as 15% by mass of copper. Miners in those days did not know how to process the sulphide materials which lay unaltered 30 m or so below the surface.

But a way had to be found when the surface minerals were used up, and miners were forced to extract the minerals which lay underground. The answer was to roast the sulphide minerals in air. Roasting involves heating in oxygen. A simple example is

$$CuS + 1.5O_2 \rightarrow CuO + SO_2$$

The metal oxide can then be reduced to the metal. More advanced roasting techniques which lead directly to copper metal are used for copper minerals today.

In **Activities M2.2** and **M2.3** you can process a sample of roasted copper ore and produce some copper from it. The method was once used but has now been replaced by more economical processes. When you have done the activities you will have the opportunity to compare your technique with one which is used in a modern process.

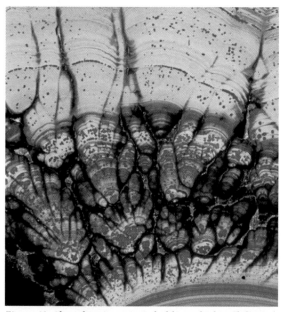

Figure 19 Altered copper minerals, like malachite (left) and azurite (right), often have attractive green or blue colours which are typical of copper compounds.

ASSIGNMENT 6

Azurite is a deep blue mineral and was probably used as a glaze by potters thousands of years ago. When it is heated with carbon (such as the charcoal which would have been present in wood fires), copper, carbon dioxide and steam are produced.

a Write a balanced equation for the reaction of azurite with hot carbon.

b Use oxidation states to explain why this reaction is an example of a redox reaction.

c Calculate the maximum mass of copper that could be obtained by heating 1 kg of azurite with carbon.

The biggest hole in the world

Towards the end of the 19th century, the demand for copper rose rapidly in response to the need of 'the age of electricity' for wires and other articles. There were not enough rich copper deposits to meet this demand. Deposits with only 0.5%–1% copper in the rock had to be used. One place where this was profitable was at Bingham Canyon in the US.

Bingham Canyon is south of Salt Lake City in Utah. Copper has been mined there for about a hundred years. Although the percentage of copper in the rock is small, the total quantity of the mineral is enormous. So it was worth building up a process which could handle the low-grade material and pick out the small proportion of chalcopyrite present. Over 5 000 000 000 tonnes of material have been removed so far.

Figure 20 Bingham Canyon mine, south of Salt Lake City, Utah, US.

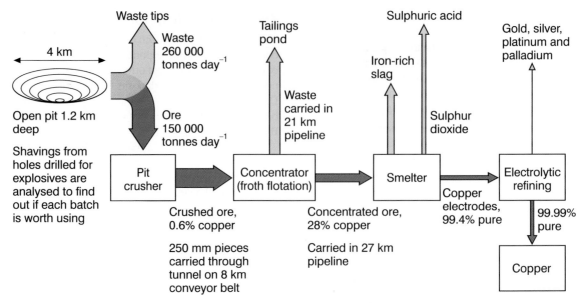

Figure 21 Outline of the Bingham Canyon process.

Getting the copper

Figure 21 describes the Bingham Canyon process in outline. There are four main stages: **mining**, **concentration**, **smelting** and **electrolytic refining**. The scale is huge.

Concentration

The first step is to turn the pieces of rock into a fine powder in a series of *grinding mills*. By doing this, the grains of copper mineral are *liberated* – a large lump of rock made up of lots of crystals of different substances is turned into a mixture of individual tiny crystal grains (Figure 22). **Froth flotation** then separates the grains of copper mineral from the rest of the mixture.

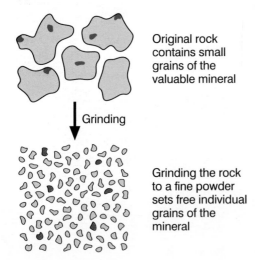

Figure 22 Liberation of grains: valuable mineral is set free from worthless rock by reducing the size of the lumps.

The job that froth flotation has to do is to separate the very small percentage of ground-up rock which is valuable chalcopyrite from the other material of no value. The minerals are of two different types: chalcopyrite is a sulphide; the others are silicates containing negative ions made from silicon and oxygen atoms.

In froth flotation, a chemical called a *collector* is added to a water/mineral mixture and binds to the surface of the chalcopyrite grains, giving them a water-repellent, hydrocarbon coating. Reagents, including detergents, are added, and air is then blown into the mixture, causing it to froth. Because they have a water-repellent coating, the chalcopyrite grains become concentrated in the froth and can be removed with it. In this case, the percentage of copper increases from 0.6% to about 30%, producing material in which the copper is nearly 50 times more concentrated than in the ore which is mined. Figure 23 illustrates the froth flotation process.

Froth flotation can be very selective: first one mineral and then others can be removed from quite complicated mixtures.

The *slurry* of copper mineral and water produced from froth flotation is pumped to the *smelter*; the waste slurry is disposed of in a *tailings pond*. Tailings ponds require careful management: in cold countries, the sight of large areas of dirty water is unattractive; in arid Utah, the ponds can dry out and produce dust storms. To prevent this, the entire surface area of the tailings pond is kept wet. The sides of the pond are eventually revegetated.

Surface of water

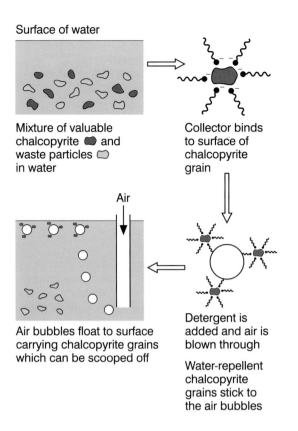

Mixture of valuable
chalcopyrite ● and
waste particles ◯
in water

Collector binds
to surface of
chalcopyrite
grain

Air

Air bubbles float to surface
carrying chalcopyrite grains
which can be scooped off

Detergent is
added and air is
blown through

Water-repellent
chalcopyrite
grains stick to
the air bubbles

Figure 23 Froth flotation.

*Figure 24 Froth flotation: the lightweight particles of copper
sulphide are floating on the froth.*

Smelting

Production of copper from the chalcopyrite
concentrate is carried out in a series of three reactions.
In the first, the concentrate, together with some silica,
is passed down the heated reaction shaft of the smelter
(Figure 25). Air enriched with oxygen is passed down
with this solid feed.

The reaction that occurs is very exothermic and
violent. Indeed the process is known as *flash smelting*.
The iron oxides and sulphides are converted by the
silica to a molten slag. The bottom layer is copper(I)
sulphide, Cu_2S, and is known as *copper matte*. This is
tapped off and run into another furnace where it is
'blown' with air to produce copper.

$$Cu_2S(l) + O_2(g) \rightarrow 2Cu(l) + SO_2(g)$$

The purity of this copper, known as *blister copper*, is
about 99%. It is further refined to 99.4% before the
final purification. The copper is cast into electrodes
(see Figure 27) ready for **electrolytic refining**.

Smelting produces large quantities of sulphur
dioxide which would lead to serious pollution if its
release into the atmosphere was not strictly controlled.
Nowadays, the sulphur dioxide is collected and
converted into sulphuric acid, but in the past its release
into the atmosphere did cause problems.

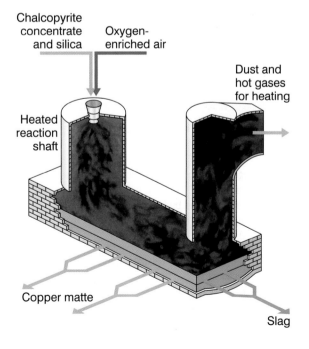

Chalcopyrite
concentrate
and silica

Oxygen-
enriched air

Dust and
hot gases
for heating

Heated
reaction
shaft

Copper matte

Slag

Figure 25 A copper smelter.

Breathing air which contains sulphur dioxide causes respiratory damage, and sulphur dioxide is also one of the gases responsible for acid deposition (or 'acid rain' as it is sometimes called). Coal-burning power stations are now one of the main causes of acid rain. Acid deposition is known to have damaged the ecology of many lakes in parts of Europe, Scandinavia and North America, and to be partly to blame for disease and death among trees in large areas of forest in the same regions.

ASSIGNMENT 7

a Below is a list of some of the environmental hazards which may be associated with mining and mineral processing:
- noise
- rock and solid waste
- dust and smoke
- contaminated water.
- dangerous gases

Set up a table which contains a 5×3 grid, and use it to make a list of the hazards which could be associated with each of the following operations:
- mining
- concentrating
- smelting.

b What steps could be taken by a responsible mining company to reduce or eliminate the effects of the hazards listed in part **a**?

ASSIGNMENT 8

This assignment makes you think about the waste created at a larger copper mine each day. Assume that the company mines 200 000 tonnes of rock of average density 2.5 tonnes m^{-3} each day. The rock contains 1% copper by mass on average.

a What volume of rock is removed each day?

b What is the approximate volume of waste, including tailings, which has to be disposed of each day?

c Which of the stages – mining, concentration and smelting – contribute significantly to the production of waste?

d A large office block built recently has a ground floor area of 1200 m^2 and is 50 m high. Compare the volume of the building with the volume of waste produced each day at the copper mine.

Reading **Chemical Ideas 8.1** will tell you more about acids and their reactions.

The analytical technique of titration is often used in chemistry. In **Activity M2.4** you can work with a solution which is similar to a sample of acid rain, and you can find out the concentration of acid in it.

ASSIGNMENT 9

This assignment is about froth flotation which produces a concentrate containing 28% of copper by mass in the form of chalcopyrite ($CuFeS_2$).

a What is the percentage of copper in pure chalcopyrite?

b What is the percentage of chalcopyrite in the concentrate produced by froth flotation?

c Comment on the effectiveness of the technique of froth flotation.

Electrolytic refining

The copper electrodes which are produced from the smelter are 99.4% pure, but this still isn't pure enough for the copper which will be used to conduct electricity. This has to be at least 99.98% pure for the copper to be sold for electrical wire.

The final purification is done using electrolysis. Figure 26 shows you an illustration of how this is done.

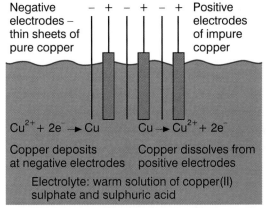

Figure 26 The processes involved in the electrolytic refining of copper.

The half-equation for the reaction occurring at the negative electrode is

$$Cu^{2+}(aq) + 2e^- \rightarrow Cu(s) \quad \text{reduction}$$

and at the positive electrode is

$$Cu(s) \rightarrow Cu^{2+}(aq) + 2e^- \quad \text{oxidation}$$

Electrolysis takes about 2 weeks and during this time about 120 kg of copper are deposited on the negative electrodes. Some modern cells use thin sheets of stainless steel rather than pure copper for the negative electrodes. The negative electrodes are removed and washed. The copper is stripped from the stainless steel sheets which can be re-used many times. Some of the insoluble impurities are valuable (eg silver and gold) and these are extracted from the sludge which forms. Other impurities are soluble and contaminate the solution, which has to be replaced as electrolysis goes on.

Figure 27 Electrolytic refining of copper.

Rock-eating bacteria

Ten per cent of all copper produced in the US comes from bacteria which 'feed off' chalcopyrite. The bacteria actually make use of the Fe^{2+} and S^{2-} ions rather than the copper in the ore. Bacteria like *Thiobacillus ferro-oxidans* and *Thiobacillus thio-oxidans* obtain the energy they need for life by a series of reactions which involve the oxidation of Fe^{2+} and S^{2-} ions:

$$Fe^{2+} \rightarrow Fe^{3+} + e^-$$
$$S^{2-} + 4H_2O \rightarrow SO_4^{2-} + 8H^+ + 8e^-$$

The microorganisms do not really 'feed' off the minerals because the ions do not enter the bacteria. Electrons are transported into their cells and are used in *biochemical processes* to reduce oxygen molecules to water. The overall result is that the bacteria convert insoluble chalcopyrite into a solution of iron(III) sulphate and copper(II) sulphate:

$$4CuFeS_2 + 17O_2 + 4H^+ \rightarrow$$
$$4Cu^{2+} + 4Fe^{3+} + 8SO_4^{2-} + 2H_2O$$

The optimum conditions for bacteria used in this process are pH 2–3 and 20 °C–55 °C.

Extracting copper by this process of *bacterial leaching* is cheaper, quieter and less polluting than the smelting process. Unfortunately it is also much slower. It is used at Bingham Canyon on the waste dump. Acidified water is sprayed onto the top surface of the dump. The aerated water then slowly percolates through the pieces of broken rock with their bacterial colonies. The colonies have established themselves naturally – the bacteria are all around us, and are responsible for the brown, iron(III) colour of some streams.

The solutions of copper sulphate produced by bacterial leaching are dilute and impure. The copper needs to be concentrated and separated from other metal ions which are also present. At Bingham Canyon this is done by displacing the copper from solution by using scrap iron to reduce the Cu^{2+} ions.

$$Cu^{2+}(aq) + Fe(s) \rightarrow Cu(s) + Fe^{2+}(aq)$$

The impure copper is purified further before being turned into anodes for the final electrolysis step.

At present, about 3% of the world's copper is produced using this technique.

ASSIGNMENT 10

a The reactions which occur in smelting and bacterial leaching of chalcopyrite are both redox reactions. Explain this by naming the elements which are oxidised and reduced, and describing their changes in oxidation state.

b Minerals such as chalcopyrite are often described as reduced ores by the mineral industry. Minerals such as malachite are said to be oxidised. Use a chemical explanation to show why, with regard to the negative ions present, the use of these terms is valid.

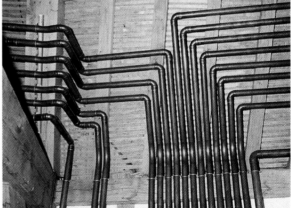

Figure 28 The Globe Theatre in London has a thatched roof and needs strict fire precautions. The copper tubing used to supply the water to the sprinklers is resistant to corrosion and has a long life.

Copper: a vital element

You read about iron in **Elements of Life**. In many ways, copper and iron have similar chemistries. They both belong to a family of elements called **transition metals**. Three important features of transition metals include

- the elements can exist in a variety of oxidation states in their compounds
- the elements and their compounds are often effective catalysts
- their compounds are almost always coloured.

Variable oxidation state

Cuprite (Cu_2O) is copper(I) oxide and tenorite (CuO) is copper(II) oxide. CuO is the black powder used as the starting point for making blue copper(II) sulphate crystals in a school laboratory. You may also be familiar with Cu_2O – it is the red precipitate you see as a positive response to tests for reducing sugars with Benedict's solution or Fehling's solution. Have you noticed what happens when you heat a piece of copper? You get a mixture of red and black colours – Cu_2O and CuO.

The same oxidation states of copper help to keep some molluscs and Crustacea alive. Oxygen is not transported round their bodies by haemoglobin (as in humans) but by haemocyanin – a blue substance which contains copper. The action of haemocyanin depends on the change

$$Cu^{2+} + e^- \rightleftharpoons Cu^+$$

Catalytic properties

Oxygen began to build up in the Earth's atmosphere about 1.5 billion years ago. It was a deadly posion to organisms used to the previously anoxic (without oxygen) environment, and these organisms had to develop substances to protect themselves from the effects of oxygen. One problem was the accumulation of superoxide ions, O_2^-, in cells. Superoxide ions are powerful oxidising agents and cells had to be equipped with an enzyme which could bring about their rapid destruction. The enzyme *superoxide dismutase* contains copper and still protects our cells today.

Copper and its compounds are also important industrial catalysts. For example, ethene is oxidised to ethanal using a catalyst containing copper(II) chloride.

Coloured compounds

Copper compounds are often coloured blue or green, as shown in Figure 19.

Using copper

Copper chemistry also plays an essential role in the processes of life, but the everyday uses of copper which are most familiar to us rely on copper's *lack* of chemical reactivity!

Because it does not react with water, even when heated, copper is used for domestic water pipes and hot water cylinders. Copper reacts very slowly with air and retains its untarnished, attractive, metallic appearance when used for objects of art or decoration. Often it is combined with other metals to form **alloys**, for example *bronze* (95% Cu, 5% Sn) and *brass* (70% Cu, 30% Zn).

Our coinage is based on copper: 'copper' coins contain 97% copper mixed with tin and zinc, although since 1992 1p and 2p coins have been made of copper-plated steel; 'silver' coins are really made from *cupronickel* (75% Cu, 25% Ni).

Pure copper has a high electrical conductivity and is tough – it can be bent and twisted without breaking. That is why it has to be electrolytically refined: tiny quantities of impurity cause it to snap easily or to get dangerously hot when current is passed through it. Almost 60% of all the copper produced is used in electrical applications.

Although bacterial leaching is only used as a secondary way of producing copper, the method can be used as a primary way of extracting gold. You can find out about this in **Activity M2.5**.

In this section, you have met a range of substances that occur in the Earth's crust. You have seen why ionic compounds (such as sodium chloride and copper minerals) are solids.

But what is it that makes some covalent compounds (such as silica, SiO_2) solids, while other covalent compounds (such as carbon dioxide, CO_2, and sulphur dioxide, SO_2) are gases at normal temperatures?

Activity M2.6 will help you to find out.

You can read about the relationship between the properties of covalent substances and their bonding and structure in **Chemical Ideas 5.2**.

M3 *Summary*

In this unit you have learned about the production of three elements (bromine, chlorine and copper) from their minerals. The processes involved all rely on redox chemistry. Ideas about redox are also important in explaining many of the important reactions of the halogens. To understand the chemistry of the halogens you needed to know more about the energy levels in atoms and how electrons are arranged in these energy levels.

The processes by which bromine, chlorine and copper are extracted all involve hazards. In the case of bromine and chlorine, these arise from the dangerous nature of the elements themselves. In the case of copper, the production of large quantities of sulphur dioxide at the smelting stage can pose a serious problem if not dealt with properly. Sulphur dioxide is an acidic gas and provides a link with acid/base chemistry.

Most of the bromine, chlorine and copper compounds in this unit have ionic structures, and much of the chemistry takes place in solution. You have had to learn more about ionic compounds and the processes of dissolving and precipitation, as well as how to calculate concentrations of substances in solution. Large quantities of silica and silicates occur with copper-containing minerals in the Earth's crust. This led you to look at the structure of silica and other substances with giant covalent (network) structures.

Finally, the unit gave you an insight into some aspects of the chemical industry: for example, the quantities of materials which are involved, the emphasis on safety and the importance of economic factors.

Activity M3 will help you to check your notes on this unit.

THE ATMOSPHERE

Why a unit on THE ATMOSPHERE?

The chemical and physical processes going on in the atmosphere have a profound influence on life on Earth. They involve a highly complex system of interrelated reactions, yet much of the underlying chemistry is essentially simple. The focus of this unit is *change*: change in the atmosphere brought about by human activities, and the potential effects on life. Two major problems are explored: the depletion of the ozone layer and the influence of human activities on global warming through the greenhouse effect.

In considering these phenomena, some important chemical principles are introduced and developed. In particular, the effect of radiation on matter, the factors that affect the rate of a chemical reaction, the formation and reactions of radicals and the idea of dynamic equilibrium are introduced, as well as the specific chemistry of species such as oxygen, carbon dioxide, methane and organic halogen compounds that are met with in the context of atmospheric chemistry.

Overview of chemical principles

In this unit you will learn more about …

ideas you will probably have come across in your earlier studies

- the effect of chlorofluorocarbons on the ozone layer
- rates of reactions
- the greenhouse effect
- reversible reactions

ideas introduced in earlier units in this course

- the electromagnetic spectrum (**The Elements of Life**)
- covalent bonding (**The Elements of Life** and **Developing Fuels**)
- electronegativity and bond polarity (**The Elements of Life**)
- the chemistry of simple organic molecules (**Developing Fuels**)
- the use of moles and quantitative chemistry (**The Elements of Life**, **Developing Fuels** and **From Minerals to Elements**)
- enthalpy changes, enthalpy cycles and bond enthalpies (**Developing Fuels**)
- catalysis (**Developing Fuels**)

… as well as learning new ideas about

- the interaction of matter with electromagnetic radiation
- the formation and reactions of radicals
- the properties of organic halogen compounds
- factors which affect the rate of a reaction
- the nature of chemical equilibrium.

A

THE ATMOSPHERE

A1 *What's in the air?*

The atmosphere is a relatively thin layer of gas extending about 100 km above the Earth's surface. If the world were a blown-up balloon, the rubber would be thick enough to contain nearly all of the atmosphere. Thin though it is, this layer of gas has an enormous influence on the Earth.

A simplified picture of the lower and middle parts of the atmosphere is shown in Figure 2. The two most chemically important regions are the troposphere and the stratosphere. Note the way that temperature changes with altitude.

The atmosphere becomes less dense the higher you go. In fact, 90% of all the molecules in the atmosphere are in the troposphere. Mixing is easy in the troposphere because hot gases can rise and cold gases can fall. The reverse temperature gradient in the stratosphere means that mixing is much more difficult in a vertical direction. Horizontal circulation, however, is rapid in the stratosphere, particularly around circles of latitude.

Figure 1 The Earth seen from space.

Figure 2 The structure of the atmosphere and the change in temperature with altitude.

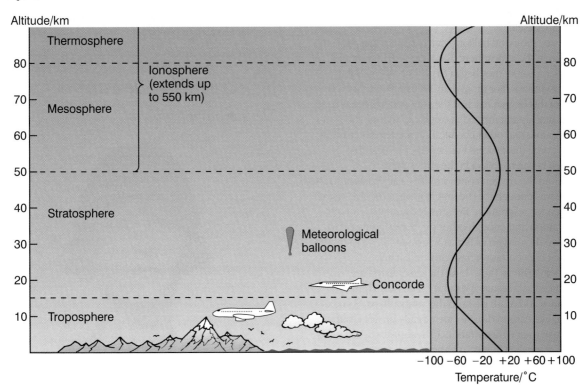

Gas	Concentration (by volume)
	Per cent
nitrogen	78
oxygen	21
argon	1
	Parts per million
carbon dioxide	367*
neon	18.0
helium	5.0
methane	1.8*
krypton	1.1
hydrogen	0.5
dinitrogen oxide, N_2O	0.3*
carbon monoxide	0.1*
xenon	0.09
nitrogen monoxide, NO $\}$ NO_x nitrogen dioxide, NO_2	0.003*

Table 1 Composition by volume of dry tropospheric air from an unpolluted environment (*variable).

Table 1 shows the average composition by volume of dry air from an unpolluted environment and is typical of the troposphere.

The concentrations of some of these substances are measured in **parts per million (ppm)** by volume. This measure is often used where small concentrations are involved; 367 ppm corresponds to a percentage concentration of 0.0367 %.

The atmosphere hasn't always had this composition. The first atmosphere was lost altogether during the upheavals in the early life of the solar system. The next atmosphere consisted of compounds such as carbon dioxide, methane and ammonia which bubbled out of the Earth itself.

Three thousand million years ago there was very little oxygen in the atmosphere. But when the first simple plants appeared, they began to produce oxygen through photosynthesis. For more than 1000 million years very little of this oxygen reached the atmosphere. It was used up as quickly as it formed in oxidising sulphur and iron compounds and other chemicals in the Earth's crust. It wasn't until this process was largely complete that oxygen began to collect in the atmosphere.

When the oxygen concentration reached about 10%, there was enough for the first animals to evolve, using oxygen for respiration. Eventually there was enough respiration and enough other processes going on to remove the oxygen as fast as it formed. Since then, the oxygen concentration has remained at about 21%.

Look again at Table 1. All the gases listed are produced as a result of natural processes. Human activities add more gases to the atmosphere. Some of them, like carbon dioxide, are already present, but we increase their concentration. (These gases are marked by an asterisk in Table 1.) Other gases in the atmosphere, like the chlorofluorocarbons, are *only* produced as a result of human activity.

The first plants

The first plants appeared about 3000 million years ago and were probably single-celled organisms called *cyanobacteria*, or *blue-green bacteria*.

They lived in the surface waters of the oceans under anaerobic conditions, and used sunlight as their energy source to drive the chemical reactions needed to maintain their growth.

Similar organisms are still found on Earth today.

Gases always mix together completely and this natural diffusion process is greatly speeded up by air currents. So in time, pollutant gases spread throughout the atmosphere. Atmospheric pollution is a global problem: it affects us all. In this unit, we shall be looking at two global problems in particular:

- the destruction of the ozone layer
- global warming arising from the greenhouse effect.

ASSIGNMENT I

Use Table 1 to answer the following questions.

a How many parts per million (by volume) of argon are there in a typical sample of tropospheric air?

b In 1 dm^3 of tropospheric air, what is
  i the volume of methane present?
  ii the percentage of methane *molecules* in the sample?

c For each of the gases marked with an asterisk, suggest *one* way in which human activities increase its concentration.

A2 *Screening the Sun*

The sunburn problem

Figure 3 The trend-setting clothes designer Coco Chanel, who set a new fashion for suntanned skin among white-skinned Europeans in the 1920s.

Until the 1920s a suntan was something a white-skinned person could not avoid if they had to work outdoors in the Sun. Those who didn't preferred to distinguish themselves by remaining pale. It was the clothes designer Coco Chanel who made sunbathing fashionable. She appeared with a golden tan after a cruise on the yacht belonging to the Duke of Westminster, who was one of the world's wealthiest men.

As we discover more about the effects of the Sun's radiation on the chemical bonds in living material, sunbathing seems less of a good idea.

ASSIGNMENT 2

a Look at the scatterplot in Figure 4. Suggest a reason for the variation of death rate from skin cancer in the US.

b Skin cancer is more common among office workers than among farmers. Suggest a reason why.

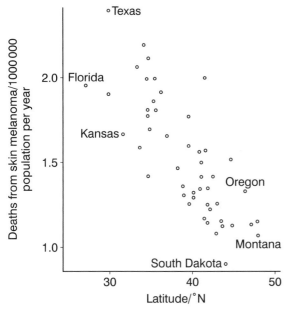

Figure 4 Deaths from skin melanoma in the US.

The Sun radiates a wide spectrum of energy. Part of this spectrum corresponds to the energy required to break chemical bonds. Sunlight can therefore break bonds, including those in molecules such as DNA in living material. This can cause damage to genes and lead to skin cancer. For example, the number of cases of skin cancer caused by melanoma has doubled in the US over the last 20 years, with about 7000 deaths a year.

On a less serious level, it can damage the proteins of the connective tissue beneath the skin, so that years of exposure to the Sun can make people look wrinkly and leathery. Even brief exposure to the Sun may cause irritation of the blood vessels in the skin, making it look red and sunburnt.

Many of the ideas in this story are linked to the interaction of radiation with matter. **Chemical Ideas 6.2** covers the different ways that radiation and matter can interact.

Chemical sunscreens

The diagram in Figure 5 shows the effect of different parts of the Sun's spectrum on the skin.

You can see that the most damaging region of this spectrum is in the ultraviolet. Fortunately, there are chemicals which absorb much of this radiation.

Have you ever wondered why people don't get sunburnt indoors? You can sit by a window for hours on a sunny day without burning. The glass in the window absorbs the damaging ultraviolet radiation, so it never reaches your skin. (Perspex does let through some ultraviolet, so it is possible to burn through Perspex.)

Chemists have found materials that do a similar job to glass, but can be spread onto the skin. They are called *sunscreens*, and tonnes of them are sold every summer.

But the best sunscreen of all is not made by chemists. It has always been with us. It is the atmosphere.

In **Activity A2.1** you can examine the effect of some substances on radiation from the Sun.

Activity A2.2 investigates the effectiveness of sunscreens.

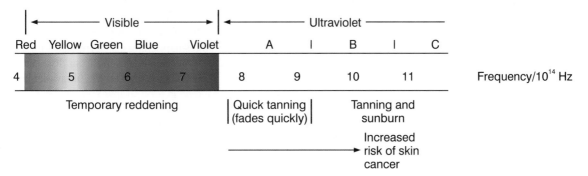

Figure 5 The effects of sunlight on the skin.

Figure 6 A sunscreen's ability to protect skin is expressed as a 'Sun Protection Factor' or SPF number — SPFs indicate the time it will take for the Sun to produce a certain effect on your skin.

Why is the atmosphere such a good sunscreen?

Certain atmospheric gases absorb ultraviolet radiation strongly. They act as a global sunscreen, preventing much of the Sun's harmful radiation from reaching the Earth.

Most of this absorption goes on in the upper part of the atmosphere, the stratosphere. Particularly important is the gas **ozone**, which absorbs ultraviolet radiation in the region 10.1×10^{14} Hz – 14.0×10^{14} Hz. This is the region which is most damaging to the skin. (Look back at Figure 5 which shows the effect on skin of radiation of different frequencies.) Much of the damaging ultraviolet radiation is absorbed by ozone in the stratosphere.

There is no life in the stratosphere, because the powerful ultraviolet radiation would break down the delicate molecules of living things. Indeed, even simple molecular substances get broken down in the stratosphere. Some of the covalent bonds break to give fragments of molecules (atoms or groups of atoms) called **radicals**.

Higher up in the atmosphere, above the stratosphere, the radiation is powerful enough to knock electrons out of the atoms, molecules and radicals. Ions are produced which lead to the name of that part of the atmosphere – the *ionosphere*.

ASSIGNMENT 3

Look back at Figure 2 on page 63 which shows the different regions of the atmosphere. Suggest the *types* of chemical particles – atoms, molecules, radicals and ions – which can be found in

a the troposphere

b the stratosphere

c the ionosphere.

Activity A2.3 investigates the screening effect of different gases on solar radiation.

A3 Ozone: A vital sunscreen

Ozone is present in the atmosphere in only tiny amounts, dispersed among other atmospheric gases. If all the ozone in the atmosphere were collected and brought to the Earth's surface at atmospheric pressure, it would form a layer only 3 mm thick.

High up, in the stratosphere, ozone *protects* us by absorbing harmful ultraviolet radiation; however, lower down, in the troposphere, it can be a real nuisance if concentrations near ground level become too high (see **Developing Fuels** storyline, Section **DF5**).

It isn't really surprising that there is so little ozone in the atmosphere. It reacts so quickly with other substances and gets destroyed. In fact, we might ask why the ozone in the atmosphere hasn't run out. Some reactions must be producing it too.

You can find out more about ozone in **Activity A3.1**.

A lot of the ideas in this part of the story are concerned with radicals. You can find out more about these by studying **Chemical Ideas 6.3**.

How is ozone formed in the atmosphere?

Ozone is formed when an oxygen atom (which is an example of a radical) reacts with a dioxygen molecule:

$$O + O_2 \rightarrow \underset{\text{ozone}}{O_3}$$

One way to make oxygen atoms is by splitting up (**dissociating**) dioxygen molecules. This requires quite a lot of energy – remember that the bond energy of the oxygen–oxygen bond in dioxygen is $+498$ kJ mol^{-1}. In this case the energy can be provided by ultraviolet radiation, or by an electric discharge.

As soon as oxygen atoms have been produced they react with the dioxygen molecules which are always present in the air. You can often smell the sharp odour of ozone near electric motors or photocopiers. The electric discharges happening inside the machine make some of the dioxygen molecules in the air dissociate into atoms. Have you ever noticed the ultraviolet lamps used to kill bacteria in food shops? You can often smell ozone near them too.

Some of the ozone in the troposphere is formed in the complex series of reactions taking place in photochemical smogs. These develop in bright sunlight over large cities which are heavily polluted by motor vehicle exhaust fumes (see the **Developing Fuels**

Figure 7 (a) Helium-filled balloons are sent up into the stratosphere to measure ozone concentrations. (b) A cord almost 10 miles in length attaches the balloon to measuring instruments on the launch vehicle; the cord can be reeled in and out to obtain measurements at different altitudes.

storyline). In this case, oxygen atoms are produced by the action of sunlight on the pollutant gas, nitrogen dioxide.

In the stratosphere, oxygen atoms are formed by the **photodissociation** of dioxygen molecules. This happens when dioxygen absorbs ultraviolet radiation of the right frequency.

ASSIGNMENT 4

a What is the bond energy, in *joules per molecule*, of the oxygen–oxygen bond in dioxygen?

b Use the expression $E = h\nu$ to calculate the frequency of radiation that would break this bond.

c What you have calculated is the *minimum* frequency to cause the bond to break. Explain why it is a minimum value.

The reaction can be summarised as:

$$O_2 + h\nu \rightarrow O + O \quad \text{(reaction 1)}$$

In this reaction, $h\nu$ indicates the photon of ultraviolet radiation that is absorbed.

The oxygen atoms produced can do a number of things when they meet another particle and collide with it. The least interesting of these is that the particles just bounce apart again. But even when the oxygen atom collides with a particle it can react with, not every collision results in a reaction.

More interesting is when the oxygen atom sticks onto the particle it collides with. This could be one of three things: O_2, another O or O_3. The three possible outcomes can be represented by reactions 2–4 below:

$$O + O_2 \rightarrow O_3 \qquad \Delta H^\ominus = -106\,\text{kJ mol}^{-1} \text{ (reaction 2)}$$
$$O + O \rightarrow O_2 \qquad \Delta H^\ominus = -498\,\text{kJ mol}^{-1} \text{ (reaction 3)}$$
$$O + O_3 \rightarrow O_2 + O_2 \quad \Delta H^\ominus = -392\,\text{kJ mol}^{-1} \text{ (reaction 4)}$$

Reaction 2 is of course the one which produces ozone.

When the ozone absorbs radiation in the $10.1 \times 10^{14}\,\text{Hz} - 14.0 \times 10^{14}\,\text{Hz}$ region, some molecules undergo photodissociation and split up again:

$$O_3 + h\nu \rightarrow O_2 + O \quad \text{(reaction 5)}$$

It is this reaction which is responsible for the vital screening effect of ozone, since it absorbs the radiation which is responsible for sunburn.

ASSIGNMENT 5

Look at reactions 1–5 on page 67.

a Which reaction or reactions *remove* ozone from the atmosphere?

b Which reaction or reactions absorb ultraviolet radiation?

c Which reaction or reactions are exothermic?

d What relationship is there between
   i reactions 1 and 3?
   ii reactions 2 and 5?

e Explain why this series of reactions has the net effect of using the Sun's ultraviolet radiation to heat the stratosphere, and why the stratosphere is hottest at the top and coolest at the bottom.

Photodissociation and the subsequent reactions of the radicals produced are investigated in **Activities A3.2** and **A3.3**. You cannot use the damaging radiation needed to break down O_2 or O_3, so in the activities you will be working with Br_2 which absorbs light in the visible/near ultraviolet region.

Ozone – here today and gone tomorrow

You can see from reactions 1–5 on page 67 that ozone is being made and destroyed all the time. Left to themselves, these reactions would reach a point where ozone was being made as fast as it was being used up:

rate of producing ozone = rate of destroying ozone

At this point, the concentration of ozone would remain constant. This is called a **steady state**.

It's like the situation in Figure 8 when you are running water into a basin with the plug out of the waste pipe. Before long you get to the point where water is running out as fast as it's running in, and the level of water in the basin stays constant. If you turned the tap on more, the level of the water would rise – but that would make the water run out faster, because of the higher pressure. Before long you would get to a steady state again, but this time with more water in the basin. What would happen if you made the waste pipe larger?

Figure 8 One example of a steady state situation.

The water coming out of the tap is like the reactions producing ozone, and the water going down the waste pipe is like the reactions that destroy it.

To estimate the concentration of ozone in the stratosphere, you need to know the **rates** of the reactions that produce and destroy ozone.

Chemists have studied these reactions in the laboratory and are able to write mathematical equations giving the rates of all the reactions involved in producing and destroying ozone. So, for example, taking reactions 1–5, the rate of producing ozone will be the rate of reaction 2. If there is a steady state, this will be balanced by the rate of destruction of ozone: the rate of reaction 4 *plus* that of reaction 5. From these relations chemists can work out what the concentration of ozone *should* be at different altitudes, at different times of day and at different times during the year.

However, when chemists compared their *calculated* concentrations of ozone with *measured* values, they found that the actual concentration of ozone was in fact a good deal *less* than expected.

ASSIGNMENT 6

A number of factors can affect the rate of ozone production and destruction. Remembering that the reactions are taking place in the gas phase and in strong sunlight, suggest *three* factors which could affect the rates of these reactions.

Explain how altering each factor would affect the rate of reaction.

This suggests that the ozone is being removed faster than expected. Going back to the analogy of the basin and the running tap, it's as if the waste pipe had been made larger. But by what?

You can read about the factors which affect the rate of a chemical reaction in **Chemical Ideas 10.1**.

The effect of temperature on the rate of a chemical reaction is discussed in **Chemical Ideas 10.2**.

What is removing the ozone?

We have seen that ozone is very reactive and reacts with oxygen atoms. But oxygen atoms aren't the only radicals to be found in the stratosphere. There are other radicals which can remove ozone by reacting with it.

Two important examples of radicals which react in this way are the **chlorine atom** (Cl) and the **bromine atom** (Br). In the case of chlorine, small amounts of chloromethane, CH_3Cl, reach the stratosphere as a result of natural processes (see Figure 9).

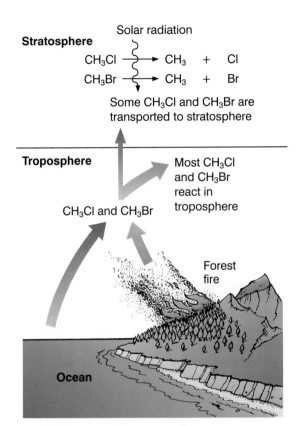

Stratosphere

Solar radiation

$CH_3Cl \longrightarrow CH_3 + Cl$

$CH_3Br \longrightarrow CH_3 + Br$

Some CH_3Cl and CH_3Br are transported to stratosphere

Troposphere

Most CH_3Cl and CH_3Br react in troposphere

CH_3Cl and CH_3Br

Forest fire

Ocean

Figure 9 Chloromethane (CH₃Cl) and bromomethane (CH₃Br) are given off from the oceans and by burning coal and vegetation. They are responsible for the small amounts of naturally produced chlorine and bromine in the stratosphere.

Once in the stratosphere, chloromethane is split up by solar radiation giving chlorine atoms.

Other chlorine-containing compounds reach the stratosphere, in greater concentrations, as a result of human activities. These also absorb high energy solar radiation and break down to give chlorine atoms.

Chlorine atoms react with ozone like this:

$$Cl + O_3 \rightarrow ClO + O_2 \quad \text{(reaction 6)}$$

The ClO formed is another reactive radical and can react with oxygen atoms:

$$ClO + O \rightarrow Cl + O_2 \quad \text{(reaction 7)}$$

So now we have two reactions competing with each other to remove ozone from the stratosphere:

$$O + O_3 \rightarrow O_2 + O_2 \quad \text{(reaction 4)}$$

and

$$Cl + O_3 \rightarrow ClO + O_2 \quad \text{(reaction 6)}$$

The concentration of Cl atoms in the stratosphere is much less than the concentration of O atoms. So how significant is reaction 6?

ASSIGNMENT 7

Chlorine atoms are particularly effective at removing ozone. A single Cl atom can remove about 1 million ozone molecules.

Add equations 6 and 7 together to produce the equation for the *overall* reaction caused by chlorine atoms.

Comment on the result. What role are Cl atoms playing in the overall reaction?

It is in situations like this that it is very important for chemists to know something about the *rates* at which reactions occur.

The reaction of O_3 with Cl atoms would not matter much if it took place a lot more slowly than the reaction of O_3 with O atoms.

Chemists measured the rates of these two reactions in the laboratory under different conditions. They showed that, at temperatures and pressures similar to those in the stratosphere, the reaction of O_3 with Cl atoms takes place more than 1500 times *faster* than the reaction of O_3 with O atoms.

Even when they took into account the fact that Cl atoms have a much lower concentration than O atoms in the stratosphere, the chemists still found that the reaction with Cl atoms takes place sufficiently quickly to make a very large contribution to removal of ozone.

What's more, the Cl atoms are regenerated in a **catalytic cycle** (reactions 6 and 7 on this page) and can go on to react with more O_3. So you can see why their effect could be devastating.

You can write a similar catalytic cycle involving bromine atoms. In fact, although the concentration of bromine atoms is much smaller than that of chlorine atoms, bromine is about 100 times more effective in destroying ozone than chlorine.

You can read more about the role of catalysts in **Chemical Ideas 10.5.**

Other ways ozone is removed

Chlorine and bromine atoms aren't the only radicals present in the stratosphere which can destroy ozone in a catalytic cycle in this way.

If we represent the radical by the general symbol X, we can rewrite reactions 6 and 7 as:

$$X + O_3 \rightarrow XO + O_2$$
and
$$XO + O \rightarrow X + O_2$$
Overall reaction: $\overline{O + O_3 \rightarrow O_2 + O_2}$

Competing reactions

Activation enthalpies are very important when *comparing* the rates of two competing reactions taking place under similar conditions. If the activation enthalpy of a reaction is large, only a small proportion of colliding particles will have enough energy to react, so the reaction proceeds slowly. If, however, the activation enthalpy is very small, most of the colliding particles will have sufficient energy to react and the reaction occurs very quickly.

Figure 10 shows the activation enthalpies for the reactions of O atoms and Cl atoms with ozone (O_3). The reaction with the lower activation enthalpy (ie O_3 + Cl) will proceed more quickly.

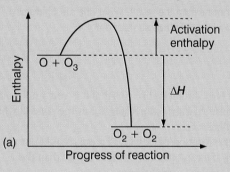
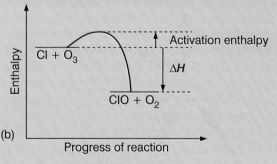

Figure 10 *Plots of enthalpy changes as the reactants come closer together and form products.*

Two other important radicals (HO, the hydroxyl radical, and NO, nitrogen monoxide) which can destroy ozone in this way are described below.

Hydroxyl radicals (HO)

These are formed by the reaction of oxygen atoms with water in the stratosphere. They react with ozone like this:

$$HO + O_3 \rightarrow HO_2 + O_2$$

The HO_2 radicals then go on to react with oxygen atoms to re-form the HO radicals:

$$HO_2 + O \rightarrow HO + O_2$$

So this is another example of a catalytic cycle, and the HO radicals released can go on to react with more O_3 molecules.

The spectra below were taken in the early days of the technique. They were taken at different times following the flash photolysis of a mixture of chlorine and oxygen. The dark lines indicate the absorption of radiation by the radical. Initially there is no ClO present, but its concentration rises very rapidly after the flash and then decays as it reacts.

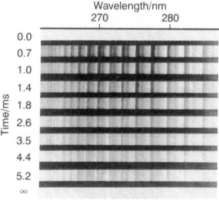

Figure 11 *George Porter shared the 1967 Nobel Prize for Chemistry for his work on very fast reactions using a technique called* **flash photolysis***. Nowadays, a brief intense flash from a laser starts the reaction. The composition of the mixture is measured spectroscopically with a carefully timed second flash. Reactions which take place in nanoseconds (1 ns = 1 × 10⁻⁹ s)* *or even picoseconds (1 ps = 1 × 10⁻¹² s) can be studied in this way.*

Figure 12 An increasing number of people are supplementing their diet with vitamins. One, vitamin E, reacts with the hydroxyl (OH) radical, which might otherwise damage DNA. Vitamin E is acting as an anti-oxidant.

Nitrogen monoxide (NO)

Nitrogen monoxide reacts with ozone to form nitrogen dioxide (NO_2) and dioxygen. Nitrogen dioxide can then react with oxygen atoms to release nitrogen monoxide and dioxygen to complete the catalytic cycle.

NO and NO_2 are both radicals. They are unusual radicals because they are relatively stable and they can be prepared and collected like ordinary molecular substances.

ASSIGNMENT 8

a Write an equation to show the formation of HO radicals from O atoms and water.

b Write equations to show how nitrogen monoxide can destroy ozone in a catalytic cycle.

Nitrogen monoxide in the stratosphere – where does it come from?

Most of the nitrogen monoxide (NO) in the stratosphere is formed naturally. But an increasing proportion is produced as a result of the reaction of dinitrogen oxide (N_2O) with oxygen atoms

$$N_2O + O \rightarrow 2NO$$

Dinitrogen oxide is the most abundant oxide of nitrogen in the atmosphere (see Table 1 on page 64). It is produced by biological reactions on Earth and some is carried up into the stratosphere. Dinitrogen oxide is released by bacteria which break down nitrogen compounds in the soil and also by bacteria in the oceans. But not all comes from natural processes. The increased use of fertilisers contributes to increasing dinitrogen oxide concentrations.

The radicals mentioned in this section (Cl, Br, HO and NO) are important, but they are only part of the whole picture. In all, hundreds of reactions have been suggested which affect the gases in the stratosphere.

Many of these reactions have been going on since long before there were humans on Earth. But human activities can have a serious effect on certain key reactions, and so lead to dramatic changes in the concentration of ozone in the stratosphere.

A4 *The CFC story*

For many years, chemists have been concerned that substances put into the atmosphere by human activities may be destroying the ozone layer. In the early 1970s there was concern about high-flying jet aircraft. Jet engines release nitrogen oxides (mostly NO) in their exhaust gases: could this make a significant difference to the amount of NO_x (NO + NO_2) in the stratosphere and so damage the ozone layer? In the end, it turned out that this wasn't a significant problem – the number of aircraft concerned was then too small to make much difference. The problem, however, may reappear because large aircraft are being designed in the US which will fly at supersonic speeds in the stratosphere.

In 1974 another concern was raised.

To understand some of the ideas and activities in this section you will need to know something about organic halogen compounds. **Chemical Ideas 13.1** gives you an introduction to these.

Sherry Rowland's predictions

Figure 13 Professor Sherwood ('Sherry') F Rowland.

Professor Sherry Rowland is the American scientist who, with Professor Mario Molina, predicted back in the early 1970s that chlorofluorocarbons (CFCs) would damage the ozone layer. He described to us how he made his discovery.

"I originally started looking at the CFC compounds as an interesting *chemical* problem in an environmental setting: whether we could predict from our laboratory knowledge what the fate of the CFCs would be in the Earth's atmosphere. We were not at the beginning thinking of the destruction of the CFCs as an environmental *problem*, but rather as just something that would be happening there.

"I've always been attracted to chemical problems – my first research experiment after graduate school involved putting a powdered mixture of lithium carbonate and ordinary glucose in the neutron flux of a nuclear reactor. A nuclear reaction in lithium produces tritium, a radioactive isotope of hydrogen, and I wanted to see if the tritium atoms could replace hydrogen atoms in glucose. (They did – and this led to many further interesting experiments in a field called *hot atom* chemistry.)

"With CFCs, we wanted to find out how quickly they would break down in the atmosphere, and by what chemical process. I was working with a young postdoctoral research associate called Mario Molina – this was in 1973 at the University of California at Irvine – and we knew that CFCs are very stable compounds. But how stable? How long could they hang around in the atmosphere? One year or 100 years?

"Mario and I looked at all of the processes that could conceivably affect CFCs in the troposphere, and calculated how rapidly such reactions could occur. The answer was very slowly indeed. CFCs remain unreacted for *many decades* – even centuries.

CFC-11   CFC-12

"So now we knew that CFCs survive unchanged for a very long time. But we also knew that when they eventually reach the stratosphere they must be broken down by the fierce ultraviolet radiation there – everything is! The CFCs contain atoms of chlorine, fluorine and carbon, and their ultraviolet breakdown releases free chlorine atoms.

For example

$$CCl_3F \rightarrow CCl_2F + Cl$$

"So we did some calculations to find out *how many* chlorine atoms would be formed now and in the future, and then asked what would happen to the chlorine atoms. In the stratosphere, we found that chlorine atoms are about a thousand times more likely to react with ozone than with anything else, leaving still another chlorine-containing chemical, chlorine oxide (ClO).

"So, once more we asked the same kind of question: what was going to happen to chlorine oxide in the stratosphere? And we found that it would react with oxygen atoms, releasing atomic chlorine again. The two reactions seem to go around in circles – chlorine atoms form chlorine oxide; chlorine oxide forms chlorine atoms – a seemingly endless chain. But with ozone being destroyed at every step! We then calculated how much ozone could be destroyed – each chlorine atom on average destroys about 100 000 molecules of ozone, and mankind has been putting about 1 million tonnes of CFCs into the atmosphere every year since the 1970s.

"We couldn't believe the answer! The calculated ozone loss was so high that we thought we must have moved a decimal point by mistake! But we checked very carefully and couldn't find any errors – there really was that much chlorine up there, with much more expected in the future, and the ozone losses would eventually be enormous!

"We had started on this problem at the beginning of October 1973, and by mid-December we realised we were onto something very important. When a scientist makes a discovery, the first instinct is to publish it so that other scientists can learn about it, and test the ideas with their own experiments. But the calculated ozone loss was so large that we wanted to be extra certain that no mistake had been made.

"So we visited Hal Johnson at the University of California at Berkeley, because he had played a major role in showing how nitrogen oxides from high-flying aircraft – such as Concorde – could affect stratospheric ozone. From Hal we learned that the chlorine chain reaction with ozone had just been discovered, but without a source for chlorine. And we had discovered that the CFCs would be an enormous source of chlorine released directly into the stratosphere!

"Now, we were sure that we had discovered something really significant – now it was a major environmental problem. We published our results and conclusions in the scientific journal *Nature*, and waited to see if others could pick holes in them. Many tried – because that's the way science works – but none succeeded. The issue of CFCs and stratospheric ozone had really arrived.

"The state of Oregon banned CFCs as propellant gases in aerosols in 1975, and the whole of the US, as well as Canada, Norway and Sweden, followed in the next 2 or 3 years. Unfortunately, the rest of Europe and Japan did not ban aerosol propellant use, and the other major applications – refrigeration, insulation, cleaning electronics, etc – continued to grow until the appearance of the hole in the Antarctic ozone layer resulted in international action in 1988–1990."

For their work in atmospheric chemistry., Professors Sherry Rowland and Mario Molina, together with Professor Paul Crutzen, were awarded the Nobel Prize for Chemistry in 1995.

As the unit develops you will see how their worries have been heeded.

_____ **ASSIGNMENT 9** _____

a Rowland and Molina's first question was *how long will chlorofluorocarbons stay in the atmosphere?* Why was it important to know this?

b Why did they not publish their results as soon as they knew how much ozone would be destroyed by chlorine in the stratosphere?

c There is a debate about the value of *pure* scientific research – ie research that has no obvious applications or uses. What does Rowland and Molina's story say about this?

The predictions come true

Science is all about making predictions, then testing them experimentally. The problem with Sherry Rowland's predictions is that they involved a long time scale. What is more, they needed a large laboratory to test them: in fact they needed the largest laboratory in the world – the Earth's atmosphere.

But in 1985, scientists examining the atmosphere above the Antarctic made a momentous discovery.

Large losses of total ozone in Antarctica reveal seasonal ClO_x/NO_x interaction

J. C. Farman, B. G. Gardiner & J. D. Shanklin

British Antarctic Survey, Natural Environment Research Council, High Cross, Madingley Road, Cambridge CB3 0ET, UK

Recent attempts[1,2] to consolidate assessments of the effect of human activities on stratospheric ozone (O_3) using one-dimensional models for 30°N have suggested that perturbations of total O_3 will remain small for at least the next decade. Results

Figure 14 The headline-making paper which appeared in the scientific journal Nature *in May 1985, reporting the discovery of a 'hole' in the ozone layer over Antarctica.*

Joe Farman's story

Figure 15 Dr Joe Farman.

Dr Joe Farman is the British scientist whose group first discovered the 'hole' in the ozone layer. We went to Cambridge to talk to him, and this is how he described their discovery.

"To stand up and make a fuss you need to have your background secure.

"We were measuring ozone concentrations over the Antarctic as part of our work with the British Antarctic Survey. We use ultraviolet spectroscopy: ozone absorbs ultraviolet radiation of a particular frequency. If you measure how strongly the atmosphere is absorbing ultraviolet of that frequency, you can work out the concentration of ozone in that part of the atmosphere. Our measurements started in 1981, and over 2 or 3 years I became convinced that there was something seriously wrong. The concentrations of ozone were

much lower than expected, particularly in October, which is the Antarctic spring. In 1984 we put in a new instrument to check our readings, and that gave us confirmation. There really was a 'hole' in the ozone layer.

"We were making our measurements from *below*, by looking up into the atmosphere from the Earth's surface at our base in Antarctica. The Americans were making similar measurements, but from *above*: NASA satellites were using ultraviolet spectroscopy to measure ozone concentrations by looking down from above.

"So why didn't NASA spot the 'hole' first? The trouble is, NASA satellites make *too many* measurements. They collect enormous quantities of data on all sorts of things, not just ozone concentrations.

"They can't possibly process all the data, so their computers are programmed to ignore any data that seem impossibly inaccurate. In the case of the ozone measurements, this meant that they ignored most of them, because they were so far out from what was expected. About 80% of the data for October were discarded by the computer because no-one believed ozone concentrations could get that low.

"Later, when our own measurements showed the concentration of ozone really was that low, NASA went back and re-examined the discarded data, which confirmed our own measurements.

Figure 16 The British Antarctic Survey's Halley Base.

"In 1985 we published our findings in the scientific journal *Nature*. Then it was a matter of convincing the world how serious the problem is. Once a scientist has made a momentous discovery of this kind, there is a duty to tell everyone about it.

"To me the real horror is the sheer speed with which it has happened. In 1985 the United Nations published a report saying there was plenty of time to study the ozone problem. At about the same time we published our own results which show that the ozone gets turned over in a period of about *5 weeks*! We still don't know exactly what the effects of ozone depletion will be, but when you are affecting a system that fast you have to be aware that almost anything could happen.

"We may think we understand a lot about the way that nature works, but to my mind we're so ignorant that anything could happen. However good your models are, you have to *keep making measurements* to make sure that nature really does work the way you think.

"It's easy to see now, with the benefit of hindsight, where we went wrong with CFCs. The original idea was to look for something *very stable* which would not be flammable, poisonous or corrosive. But the trouble with very stable substances is that they stay around for a long time – and when they do eventually break down they form something *very unstable* and reactive, which is likely to cause trouble."

ASSIGNMENT 10

a Explain how concentrations of ozone in the atmosphere are measured.

b Why did NASA discard 80% of their October ozone concentration measurements?

c Joe Farman's group were first aware of the low concentrations of ozone around 1983, but did not publish their findings until 1985. Why did they delay publication?

d Joe Farman gives two reasons why the great stability of CFCs makes them an environmental problem. What are the two reasons?

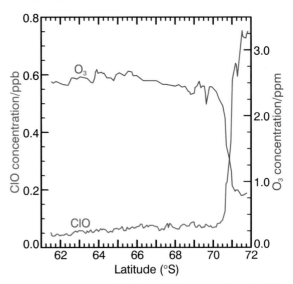

Figure 17 ER–2 aircraft fly through the stratosphere loaded with scientific instruments. These graphs show measurements of ClO radicals (in parts per billion) and ozone (in parts per million) recorded at 18 km altitude. (Note that the concentrations of the two species are about 10^3 different.) The measurements are convincing evidence that Cl radicals are involved in ozone depletion.

Since the mid-1980s, monitoring of ozone concentrations has continued with even greater urgency. Satellite readings and measurements from balloons and high-altitude planes have supported the measurements taken from the ground, and have confirmed the presence of the 'hole' over the Antarctic.

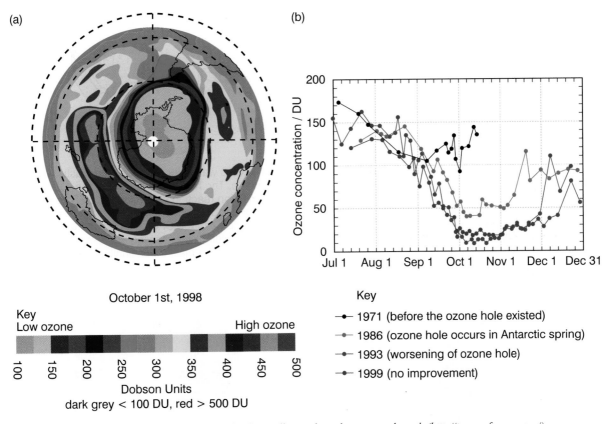

(a) October 1st, 1998

Key
Low ozone — High ozone

Dobson Units
dark grey < 100 DU, red > 500 DU

(b)

Key
—•— 1971 (before the ozone hole existed)
—•— 1986 (ozone hole occurs in Antarctic spring)
—•— 1993 (worsening of ozone hole)
—•— 1999 (no improvement)

Figure 18 (a) Maps like this are obtained every day by satellite and can be seen on the web (http://toms.gsfc.nasa.gov/).
(b) The graph shows how the ozone concentration varies during the year over the Amundsen–Scott South Pole Station.

Satellite measurements have also shown that ozone concentrations have gone down in other parts of the globe too. The effect is particularly dramatic in the Antarctic spring (see Figure 18) because of the special weather conditions existing then – but there is evidence of a smaller 'ozone hole' appearing above the Arctic and a general depletion of ozone stretching across northern latitudes, including Canada, the US and Europe.

CFCs: very handy compounds

In 1930 the American engineer Thomas Midgley demonstrated a new refrigerant to the American Chemical Society. He inhaled a lungful of dichlorodifluoromethane (CCl_2F_2) and used it to blow out a candle.

Midgley was flamboyantly demonstrating two important properties of CCl_2F_2: its lack of toxicity and its lack of flammability. Up to that time, ammonia had been the main refrigerant in use. Ammonia has a convenient boiling point, −33 °C, which means it can easily be liquefied by compression. Unfortunately, it is also toxic and very smelly.

Midgley had been asked to find a safe replacement for ammonia, and that meant finding a compound with a similar boiling point but which was very unreactive and therefore safe. He came up with the compound CCl_2F_2.

The reactions of some halogenoalkanes are investigated in **Activity A4.1.** By extending your results from this activity you should be able to see why Midgley was led to look at CFCs for a refrigerant gas.

CCl_2F_2 belongs to a family of compounds called chlorofluorocarbons (CFCs), which contain chlorine, fluorine and carbon. There are several members of the family, all with different boiling points. That's one of the things that made them so useful: there were CFCs with boiling points to suit different applications.

Another useful thing about CFCs was that they are very unreactive, and have low flammability and low toxicity.

CFCs seemed to be the perfect answer to many problems. You can see their main uses in Figure 19.

You can make a halogenoalkane in **Activity A4.2.**

The trouble with CFCs

When Thomas Midgley and other chemists developed CFCs, they did their job too well. They found a family of compounds that are very unreactive, and this makes them excellent for the jobs just described. The trouble is, they are *too* unreactive.

As propellants for aerosols

- Push valve
- Propellant
- Contents (insecticide, deodorant, etc)

When the valve opens, the pressure falls inside the can causing the propellant to vaporise. The propellant escapes into the atmosphere along with the other contents of the can.

As refrigerants

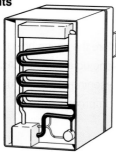

CFCs were used as refrigerants in food refrigerators and air-conditioning units. Eventually some of these refrigerants end up in the atmosphere, through leakage or when the refrigerator is scrapped.

Figure 19 CFCs were used widely.

As blowing agents for making expanded plastics

A volatile CFC was incorporated in the plastic when it was made. The heat given off during the polymerisation reaction vaporised the CFC so it 'blew' tiny bubbles in the plastic making a foam. Inevitably, some of the CFC escaped into the atmosphere during the blowing process, and more escaped when the plastic was finally disposed of.

As cleaning solvents

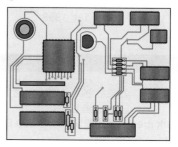

CFCs dissolve grease and were used as solvents in dry-cleaning, cleaning electronic circuits, etc. Some of the solvent escaped into the atmosphere.

The estimated lifetimes for CFCs in the troposphere are about 100 years. This gives plenty of time for them to be transported up into the stratosphere – where they are no longer unreactive.

As more became known about the effect of CFCs on ozone in the stratosphere, people began to press for their replacement. This meant that the chemical industry had an important job to do: to find compounds that would replace CFCs, but have no significant damaging effect on the ozone layer.

Progress has been very rapid. The replacement compounds are hydrochlorofluorocarbons (HCFCs), such as CH_3CFCl_2, and hydrofluorocarbons (HFCs), for example, CH_2FCF_3, or alkanes. The difference is that these molecules contain H–C bonds and are broken down in the troposphere before they have time to reach the stratosphere.

Sadly, they are not a perfect solution. Both CFCs and their current replacements are greenhouse gases and contribute to global warming (a problem we shall discuss later in this unit, in Section **A6**). Better alternatives are being sought. One being investigated is $C_3H_3F_5$.

In **Activity 4.3** you will see how the relative merits of possible refrigerants can be related to their properties and find out about the international response to the environmental impact of CFCs and HCFCs.

Not only chlorine but bromine

You have already seen that bromine atoms (produced from bromomethane given off from the oceans as well as from biomass burning) are harming the ozone layer. A significant extra amount of bromomethane reaches the stratosphere from its use as a fumigant (see **From Minerals to Elements**, Figure 11). Another group of organic bromine-containing compounds is also causing concern. These are mixed-halogen compounds called **halons**, such as $CBrClF_2$, which are used in fire-fighting systems.

Methane to the rescue

Chlorine and bromine atoms are obviously bad news in the stratosphere. They attack ozone and destroy it.

Indeed, they are so reactive that, left to themselves, they would quickly destroy most of the ozone there. Fortunately, they have not done that yet, because there are other molecules in the stratosphere that react with Cl atoms.

Methane, CH_4, is an important example of such a molecule. It is produced on Earth in large quantities by living organisms, as you will see later in Section **A6**. Most of the methane released is oxidised in the troposphere, but significant amounts of it are eventually carried up into the stratosphere.

Once in the stratosphere, methane molecules remove chlorine atoms by reacting with them like this:

$$CH_4 + Cl \rightarrow CH_3 + HCl$$

The hydrogen chloride made in this reaction may eventually be carried down into the troposphere where it can be removed in raindrops. This gets rid of some of the troublesome Cl atoms from the stratosphere, although there is not enough methane to stop concentrations of chlorine rising overall.

One reason why bromine atoms are much more dangerous to the ozone layer is that they do not react with methane in this way.

A5 *How bad is the ozone crisis?*

Ozone is a vital sunscreen gas which protects us from ultraviolet radiation. Removing it from the stratosphere may have serious consequences for the Earth. The trouble is, we do not know just what these will be.

One thing that seems clear is that cases of skin cancer and eye cataracts will increase as ozone is destroyed. It has been estimated that reducing ozone by 10% could cause a 30%–50% increase in skin cancer cases.

Figure 20 Children in New Zealand have to wear hats on school outings in summer, to protect them from harmful ultraviolet radiation.

ASSIGNMENT II

Here are three reactions involving chlorine atoms from the stratosphere:

$CH_4 + Cl \rightarrow CH_3 + HCl$	(reaction 8)
$Cl + Cl \rightarrow Cl_2$	(reaction 9)
$Cl + Cl_2 \rightarrow Cl_2 + Cl$	(reaction 10)

a For each of these reactions, answer the following questions.

  i Which bond, if any, is broken during the reaction? Use the Data Sheets to find the bond enthalpy of this bond.

  ii What new bond is made during the reaction? Use the Data Sheets to find the bond enthalpy of this bond.

  iii Use the bond enthalpies to find a value for ΔH for the reaction.

b Which of the reactions have the effect of removing chlorine atoms from the stratosphere?

c The energy required to get the bond-breaking/bond-forming processes going in a reaction is called the *activation enthalpy*. Explain why reaction 8 has a much higher activation enthalpy than reaction 9.

d Despite its higher activation enthalpy, reaction 8 normally removes Cl atoms more rapidly than reaction 9. Suggest a reason why.

But what about species other than humans? Increased ultraviolet radiation could affect species such as plankton in the oceans. That in turn could affect other organisms involved in the food chain.

And what about the weather? Changes in the amount of radiation reaching the Earth will affect the temperature of the Earth itself, which of course affects the weather.

Governments have gradually realised that it's not worth risking the global experiment that is needed to find the answers to these questions. In 1987, at an international meeting in Montreal, a procedure was agreed for restricting the production and release of CFCs into the atmosphere.

Since then, new global observations of ozone depletion have prompted amendments to strengthen the treaty: in London in 1990, Copenhagen in 1992, and in Montreal in 1997. The aim is to reduce and eventually eliminate emissions of ozone-depleting substances as a result of human activity. These now include bromomethane and other halogenoalkanes as well as CFCs.

By 1998 the developed nations had almost phased out their use of CFCs. A special fund was set up to help developing countries move away from CFC use, and many were committed to stopping their use early this century.

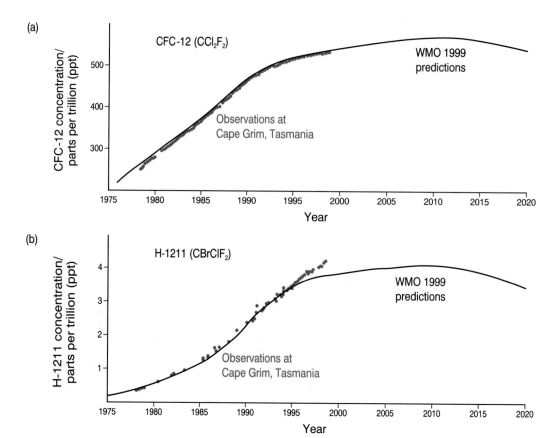

Figure 21 *Measurements on 'clean' air samples from the Cape Grim Air Pollution Station in north-west Tasmania, Australia, show the following results. (a) The build up of CFC-12 is slowing down in line with predictions from the World Meteorological Organisation (WMO) based on the Montreal Protocol. Current emissions are probably due to leakage from old refrigeration and air-conditioning systems that still contain CFC-12. (b) Emissions of the halon $CBrClF_2$ in 1998 were 60% higher than predicted under the Montreal Protocol. This compound is mainly used in fire-fighting systems. The rise is probably due to greater than expected leakage from existing systems and increased use of this type of system in the developing world, particularly China.*

World unites on ozone deal

Paul Brown
Environment Correspondent

A worldwide agreement to phase out CFCs and other ozone depleting chemicals by the year 2000 was reached in London last night.

A group of environmentally advanced countries pushed to have the date brought forward to 1997, but were blocked by the United States, the Soviet Union, and Japan. The compromise solution was a 50 per cent reduction by 1995, 85 per cent by 1997, and 100 per cent by 2000. A subclause agreed the position would be reviewed in 1992, to see if the timetable could be improved.

The agreement provided for the establishment of a new global front to help the Third World adapt to the changes.

Chris Patten, the Environment Secretary, said: "This is a major step forward in environmental diplomacy. It is a unique agreement bringing together, as it does, the establishment of environmental objectives with provision of funds and the transfer of technology."

It boded well for future global environmental agreements.

Answering criticisms that the agreement still did not go far enough, he said: "We would all like to have stopped CFCs production tomorrow but this was the best agreement that was possible, taking all considerations into account".

Negotiations had run well over the time yesterday as detailed timetables for phasing out chemicals were hammered out for inclusion in the agreement.

There was personal success for Mr Patten, who, as chairman of the conference, redrafted the final document so that both China and India felt they could pledge to sign the Montreal Protocol.

By yesterday, 59 nations had signed and most of the other 39 at the conference were expected to ratify soon.

In spite of ministerial joy at the agreement, there were doubts that the timetable would be quick enough to prevent ozone depletion being a serious problem.

Yesterday's deal meant that CFCs, the main ozone depleter, used in fridges and air conditioning, will go by 2000. Methyl chloroform, a metal cleaning agent, will be banned by 2005.

Much of the argument centred on how quickly individual chemicals could be phased out.

One of the triumphs was getting India and China, with more than one third of the world's population between them, to join. Maneka Gandhi, the Indian Environment Minister, had held out for two days for the transfer of technology from the West to be included in the agreement so India could manufacture CFCs substitutes itself.

Mrs Gandhi said yesterday the agreement now said that if technology was not transferred, that India did not have to stop the manufacture of CFCs. That placed the onus on the West to keep its promises, and on that basis she was prepared to recommend her government to ratify the protocol.

Joe Farman, the British scientist who discovered the hole in the ozone layer, said he feared the agreement was still not stringent enough.

Mr Farman, head of atmospheric dynamics at the British Antarctic Survey, said the ozone hole would go on getting bigger for some time.

Chlorine is currently 3.6 parts per billion in the atmosphere. This would grow to 4.8 parts per billion in 10 years, he said. He calculated this could lead to 18 per cent depletion in ozone in the northern hemisphere during the winter and spring by the year 2000.

Under the agreement, he calculated that it would be 2030 before the chlorine level went down to the 1986 level when the hole was first announced.

Figure 22 The Guardian *newspaper reporting in June 1990 on the international agreements reached in London about reducing CFCs. Even so, the effect of CFCs on the ozone layer will increase for many years yet.*

Figure 23 What objections can you see to Carstairs' solution to the ozone crisis?

Figure 21 shows measurements of (a) CFC-12 and (b) the halon $CBrClF_2$ taken at the Cape Grim Air Pollution Station in Tasmania, Australia. The observations are plotted along with predictions based on the Montreal Protocol.

CFCs are very stable, and they take a long time to reach the stratosphere. If the Montreal Protocol *is* observed, the amounts of chlorine and bromine atoms in the stratosphere should reach a maximum early this century, and then decline slowly. The ozone layer could return to normal by the middle of this century. But stratospheric ozone depletion will still get worse before we begin to notice any improvement.

Activity A5 should help you to summarise some of the information that has been presented in Sections **A1** to **A5**.

A6 *Trouble in the troposphere*

This story now moves a little nearer to the Earth – to the troposphere, the bottom 15 km or so of the atmosphere. But for the moment we will stick with methane, which in the stratosphere helps remove the chlorine atoms which destroy the ozone layer. In the troposphere, methane's role has a less helpful side to it. To see why, we need to look at the way the Sun keeps the Earth warm.

Radiation in, radiation out

When things get hot, they send out electromagnetic radiation. The hotter the object, the higher the energy of the radiation.

The surface of the Sun has a temperature of about 6000 K, and this means that it radiates energy in the ultraviolet, visible and infrared regions.

The Earth is heated by the Sun's radiation. The Earth's average surface temperature is about 285 K – a lot cooler than the Sun, but still hot enough to radiate electromagnetic radiation. But at this lower temperature, the energy radiated is mainly in the infrared region. The situation is illustrated in Figure 24.

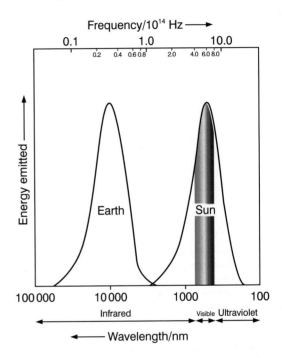

Figure 24 The radiation from the Sun which reaches the outer limits of the atmosphere and the radiation given off from the surface of the Earth (the frequencies and wavelengths are plotted here on a logarithmic scale, so each division is a factor of 10 greater than the one before).

The Sun radiates energy mainly around the visible and ultraviolet regions. Part of this energy is absorbed by the Earth and its atmosphere, and part is reflected back into space. The part that gets absorbed helps to heat the Earth, and the Earth in turn radiates energy back into space. A steady state is reached, where the Earth is radiating energy as fast as it absorbs it. Under such conditions, illustrated by Figure 26, the average temperature of the Earth remains constant.

As in all steady states, the delicate balance can be disturbed by changes to the system – in particular by changes to the quantities of various atmospheric gases. Methane is one example.

How is methane formed?

As humans, we are used to living in an airy world. We use *aerobic* respiration to oxidise carbohydrates such as glucose to carbon dioxide and water:

$$C_6H_{12}O_6 + 6O_2 \rightarrow 6CO_2 + 6H_2O$$
glucose

But some organisms live in airless places, under *anaerobic* conditions. Instead of turning carbohydrates to CO_2 and water, they turn them to other, less oxidised materials. Yeast is a good example – it converts glucose to ethanol and CO_2.

An important group of bacteria – called *methanogenic bacteria* – work in anaerobic conditions to turn materials such as carbohydrates to methane and other related substances.

Methanogenic bacteria are very common. In fact, you can be pretty sure that wherever carbohydrate is left in anaerobic conditions, methanogenic bacteria will be present, and methane will be produced. Since most biological material contains carbohydrate (or substances that can be converted to carbohydrate), it follows that methane is produced whenever biological material decays anaerobically. This may occur in

- marshes, compost heaps and waste tips, where vegetation rots without air
- rice paddy fields, where water and mud cut off air from rotting vegetation (Figure 25)
- biogas digesters where waste material is deliberately kept under anaerobic conditions
- the digestive tracts of animals, where part-digested food is acted on by bacteria (a cow releases about $500\,000\,cm^3$ of methane every day in belches!).

Baked beans are rich in carbohydrate and are notorious for causing methanogenesis in the human intestine. They contain a particular carbohydrate, called raffinose, that doesn't get broken down by normal digestive enzymes. It survives long enough to reach the intestines where it is broken down by methanogenic bacteria.

Methane is one of the most abundant of the trace gases in the troposphere (see Table 1 on page 64), and its concentration is rising.

Figure 25 Rice paddy fields cover large areas of land and are one of the biggest source of methane.

ASSIGNMENT 12

Look at the spectra in Figure 24 which show the ranges and relative intensities of radiation from the Earth and the Sun.

Methane absorbs radiation in the frequency ranges $0.39 \times 10^{14}\,Hz – 0.46 \times 10^{14}\,Hz$ and $0.85 \times 10^{14}\,Hz – 1.03 \times 10^{14}\,Hz$.

a Will methane absorb the Sun's *incoming* radiation?

b Will methane absorb the Earth's *outgoing* radiation?

c What will be the effect of methane on
  i the balance between incoming and outgoing radiation on Earth?
  ii the temperature of the Earth?

Methane is an example of a **greenhouse gas**. In effect, it traps some of the Earth's radiation that would otherwise be re-radiated into space. The effect of this is to make the Earth warmer. It's similar to the way glass traps radiation in a greenhouse, heating up the inside. Figure 27 indicates how a greenhouse works, with an attempt to show the relative wavelengths of the radiation involved.

Other gases can behave in this way too. **Activity A6** looks at the absorption characteristics of some atmospheric gases.

Do other gases have a greenhouse effect?

You should now appreciate that several atmospheric gases absorb infrared radiation, but not visible or ultraviolet light from the Sun. This means they are greenhouse gases. They let the Sun's visible radiation in, but they stop some of the Earth's infrared radiation getting out. They contribute to the **greenhouse effect**, which makes the Earth warmer.

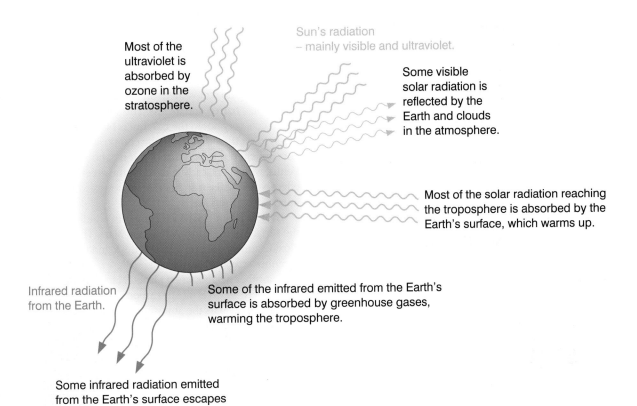

Sun's radiation
– mainly visible and ultraviolet.

Most of the ultraviolet is absorbed by ozone in the stratosphere.

Some visible solar radiation is reflected by the Earth and clouds in the atmosphere.

Most of the solar radiation reaching the troposphere is absorbed by the Earth's surface, which warms up.

Infrared radiation from the Earth.

Some of the infrared emitted from the Earth's surface is absorbed by greenhouse gases, warming the troposphere.

Some infrared radiation emitted from the Earth's surface escapes and cools it down.

Figure 26 The Earth – input and output of energy.

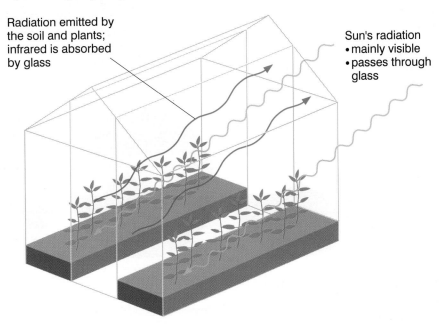

Radiation emitted by the soil and plants; infrared is absorbed by glass

Sun's radiation
• mainly visible
• passes through glass

Figure 27 Greenhouse warming.

Some gases have a more powerful greenhouse effect than others. Table 2 lists some gases, their abundance in the atmosphere and their **greenhouse factors**. The greenhouse factor is a measure of the greenhouse effect caused by the gas, relative to the same amount of carbon dioxide, which is assigned a value of 1. One molecule of methane, for example, has the same effect as about 20 molecules of carbon dioxide.

Gas	Tropospheric abundance (by volume)/%	Greenhouse factor
N_2	78	negligible
O_2	21	negligible
Ar	1	negligible
$H_2O(g)$	1*	0.1
CO_2	3.7×10^{-2}	1
CH_4	1.8×10^{-4}	20
N_2O	3.1×10^{-5}	310
CCl_3F	2.6×10^{-8}	3800
CCl_2F_2	5.3×10^{-8}	8100

Table 2 Relative contributions to the greenhouse effect of various gases in the atmosphere (averaged figure).*

ASSIGNMENT 13

Use the data in Table 2 to help with the following questions.

a Bearing in mind both their abundance and the greenhouse factor, list the gases in order of how much they contribute to the total greenhouse effect on Earth.

b Which of these gases are produced in significant amounts by human activities?

c Which gases in part **b** would it be most fruitful to try to control in order to control the greenhouse effect?

d Human activities produce a lot of water vapour, but this has very little effect on the net amount of water vapour in the atmosphere. Why?

The greenhouse effect is good for you

Without the greenhouse effect, we would not be here. By trapping some of the Sun's radiation, the atmosphere keeps the average temperature of the Earth high enough to support life.

If we had no atmosphere on Earth, it would be like the Moon – barren and lifeless. The surface of the Moon gets very hot in the day, but is bitterly cold at night. If, on the other hand, the composition of our atmosphere resembled that of our neighbouring planet Venus, with 96.4% carbon dioxide, the greenhouse effect would make it so hot that life-forms would find it impossible to survive.

Look at Figure 28, which compares conditions on the surface of the Earth, Venus and Mars. The carbon dioxide atmosphere on Venus is thick and the greenhouse effect is extreme.

Our other neighbouring planet, Mars, has an atmosphere which is mostly carbon dioxide, like that on Venus. But on Mars the atmosphere is very thin. There is only a small greenhouse effect, and so Mars is cold.

We are used to the stable, comfortable temperature on Earth and small changes in that temperature could have a dramatic effect on life.

Is the Earth getting hotter?

About 100 years ago, the Swedish chemist Arrhenius predicted that increasing amounts of carbon dioxide could lead to warming of the Earth. Average temperatures did indeed rise from 1880 to 1940, by about 0.25 °C. But then between 1940 and 1970 they fell again, by 0.2 °C.

	Venus		Earth		Mars	
Approximate composition/%	Carbon dioxide:	97	Nitrogen:	78	Carbon dioxide:	95
	Nitrogen:	3	Oxygen:	21	Nitrogen:	3
Average surface temperature/°C	427		15		−53	
Warming due to greenhouse effect/°C	446		33		3	

Figure 28 A comparison of the atmospheres, surface temperatures and warming due to greenhouse effect of the Earth and its neighbouring planets.

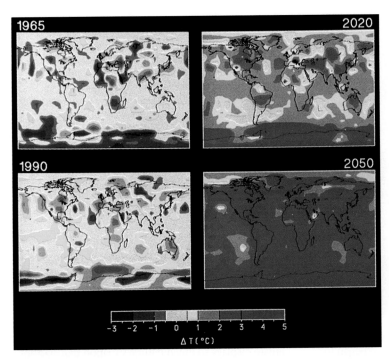

Figure 29 The global pattern of summer temperature rises as predicted by a NASA climate model, on the basis that the emission of greenhouse gases will continue to increase at current rates.

So why are we worried today? During the 1970s, measurements of carbon dioxide in the atmosphere began to show a significant increase, and new predictions began to be made about the effect of the carbon dioxide increase on the Earth's climate.

Making predictions about the climate is very difficult because it involves so many variable factors, many of which are still poorly understood.

The information available is fed into computers to give an overall **mathematical model** of the Earth's climate.

The amount of information needed is enormous. It includes, for example, the concentrations and distribution of the gases in the atmosphere, their predicted lifetimes, and details of how the concentrations are expected to change. Add to this information about variations in the intensity of the Sun's radiation, meteorological data such as air circulation patterns and cloud cover, as well as ocean temperatures and currents, and you can see why powerful computers are needed.

Models are constantly being improved as more reliable information becomes available. Figure 29 shows the global pattern of temperature rises as predicted by a computer model for a doubling in carbon dioxide concentration.

1998 is hottest year on record

By Michael McCarthy
Environment Correspondent

1998 was the warmest year for the world on record, British scientists said yesterday. It was hotter by a considerable margin than the previous record year, 1997, and means that seven of the ten hottest years in a record stretching back to 1860 have been in this decade.

It will give further credence to the belief that global warming, the overheating of the atmosphere by industrial gases such as carbon dioxide from motor vehicles and powerstations, is significantly destabilising the world's climate.

The figures were announced jointly yesterday by the Meteorological Office's Hadley Centre for Climate Prediction and Research in Bracknell, Berkshire, and the Climate Research Unit of the University of East Anglia in Norwich.

They said that the global mean temperature for 1998 is now expected to be 0.58 °C above the recent long-term average, which is based on the period 1961 to 1990. Last year, which held the previous record, was 0.43 °C above the average.

Although Britain experienced a cool and damp summer, many parts of the world, from the Mediterranean to Central America, experienced intense heatwaves in 1998.

Scientists at the Buenos Aires climate conference in November announced that the year's record sea surface temperatures, up to 2.5 °C higher than normal, had killed vast swaths of coral in the southern hemisphere.

"The top ten warmest years have all been since 1983," said David Parker, head of climate monitoring at the Hadley Centre. "This year has continued the gradual warming trend seen over the past 100 years; global temperatures in the Nineties are almost 0.7 °C above those at the end of the 19th century."

Figure 30 In 1998, the average global temperature was nearly 0.6 °C above the 1961–1990 baseline.

Figure 31 *When atmospheric scientists monitor background concentrations of atmospheric pollutants, they do so in 'clean' air away from local effects. Such a place is the Cape Grim Air Pollution Station in north-west Tasmania, Australia. The measurements for the plots in Figure 21 were made on samples of air taken here.*

At the time predictions like these were being made, field measurements of the Earth's temperature began to show increases too (Figure 30). Atmospheric scientists use an average global temperature, taken between the years 1961 and 1990, as a baseline. They then calculate an average temperature for any particular year from measurements from around the world, and calculate the difference between the temperature that year and the 1961–1990 average.

Over the last 100 years, the five hottest years, in global terms, have all been in the 1990s: in descending order 1998, 1997, 1995, 1990 and 1991.

So, is **global warming** really taking place? The climate displays so much natural variation that spotting trends is very difficult. But there is now a great deal of scientific evidence that the Earth is indeed getting warmer, and that this warming is due to human-made emissions of greenhouse gases.

A7 *Keeping the window open*

The two most significant greenhouse gases are carbon dioxide and water. Because they are so abundant in the

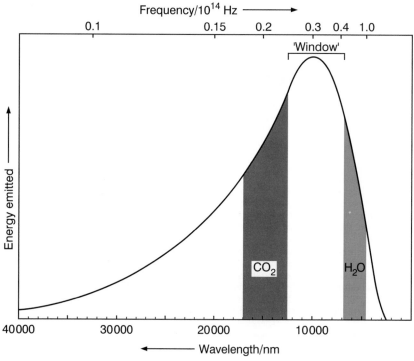

Figure 32 *The Earth's radiation spectrum, showing the regions where CO_2 and H_2O absorb strongly.*

Figure 33 It is difficult to predict accurately the effect of atmospheric water on global warming; water is a greenhouse gas, but water droplets in low clouds tend to block out the Sun and cool the surface of the Earth.

atmosphere, they absorb a lot of the infrared radiated by the Earth. Water vapour makes the larger contribution – simply because more of it is present. As Figure 32 shows, carbon dioxide and water absorb in two bands across the Earth's radiation spectrum.

Between these two bands is a 'window' where infrared radiation can escape without being absorbed. In fact, about 70% of the Earth's radiation escapes into space through this 'window'.

Gases produced by human activities can increase the natural greenhouse effect of the atmosphere. There are two types:

- Gases already naturally present in the atmosphere, which are increased in amount by human activities. Carbon dioxide is an important example. Humans burn fossil fuels, and this increases the amount of carbon dioxide in the atmosphere. This in turn increases the greenhouse effect.

- Gases which are not naturally present in the atmosphere. These gases may absorb radiation in the vital 'window' through which radiation normally escapes into space. CFCs are an example. Although CFCs are only present in the atmosphere in small amounts, they are important because they have a very large greenhouse factor – so each molecule has a big effect.

Water is different from other greenhouse gases (see **Assignment 14**). Under most conditions on Earth, water is a liquid, with some vapour associated with it. $H_2O(l)$ and $H_2O(g)$ are quickly interconverted. If human activities such as burning fuels put $H_2O(g)$ into the

atmosphere, most of it will condense to $H_2O(l)$ and eventually return to Earth. So in that sense $H_2O(g)$ isn't nearly as much of a greenhouse problem as $CO_2(g)$.

But there are two more things to consider. First, if the Earth *does* get warmer, more $H_2O(g)$ will evaporate from the oceans. That would tend to increase global warming. Second, although $H_2O(g)$ is a greenhouse gas, the droplets of $H_2O(l)$ in clouds tend to block out the Sun, as people living in the UK know well. So more water in the atmosphere *could* work in either direction – to increase or to decrease warming. This is one reason why climate modellers find it so hard to predict what will happen to the Earth's climate in the future.

At least half of the expected *increase* in greenhouse effect due to human activities is likely to be caused by carbon dioxide. So control of the greenhouse effect must focus mainly on control of the amount of carbon dioxide we produce.

ASSIGNMENT 14

Suppose the average temperature of the Earth increases due to the greenhouse effect.

a What effect will this have on the amount of water vapour in the atmosphere?

b What effect will your answer to part **a** have on further global warming?

c Explain why the role of water vapour in global warming is an example of **positive feedback**.

Figure 34 The eruption of Mt Pinatubo in the Philippines in June 1991 released enormous clouds of dust and sulphur dioxide, which led to some global cooling over the next few years; such events make predictions about climate change even more difficult.

A8 *Focus on carbon dioxide*

In this section we look more closely at the role of carbon dioxide in the greenhouse effect.

Detecting carbon dioxide

Since carbon dioxide is so crucial in the greenhouse effect, it's important to be able to make accurate measurements of its concentration in the atmosphere. Then we can keep an eye on how its concentration is changing.

The proportion of carbon dioxide in the atmosphere is fairly small – about 0.037%. In your earlier science work you probably used the limewater test to detect carbon dioxide. This test is fine when you want a **qualitative** test and when the concentration of carbon dioxide is fairly large, but it's not nearly sensitive enough to measure the small changes in atmospheric carbon dioxide concentration that occur from year to year. Later in this section you will use a **quantitative** method for measuring carbon dioxide concentration, but even this is not sensitive enough for atmospheric research.

The method most commonly used by researchers is **infrared spectroscopy**. Carbon dioxide absorbs infrared radiation – that's the reason for all the trouble. The more carbon dioxide there is, the stronger the absorption.

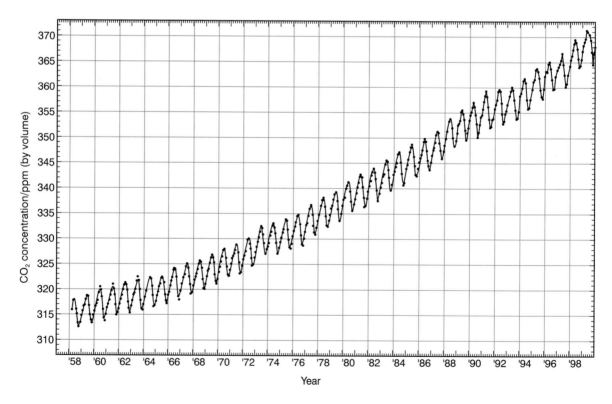

Figure 35 The build up of CO_2 in the atmosphere as recorded at Mauna Loa Observatory, Hawaii.

Using infrared measurements, it is possible to get an accurate picture of the way carbon dioxide concentration has changed over the years. The graph in Figure 35 shows measurements of atmospheric carbon dioxide concentrations made at Mauna Loa Observatory, situated at an altitude of about 3500 m in Hawaii.

ASSIGNMENT 15

a The graph in Figure 35 shows a zig-zag pattern. What time of year corresponds to
  i the peaks?
  ii the troughs?

b Suggest a reason for this zig-zag pattern.

c What was the average percentage increase in carbon dioxide concentration at Mauna Loa between 1960 and 1997?

d What problems would be involved in making a similar record of carbon dioxide concentration in London?

Where does the carbon dioxide come from – and go to?

The increase in concentration of carbon dioxide in the Earth's atmosphere is due to the increasing use of fossil fuels. Over the last 100 years, the amount of fossil fuels burned has increased by about 4% every

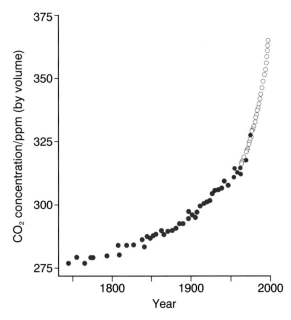

Figure 36 Carbon dioxide concentrations in the atmosphere from 1750. The data before 1958 were obtained by analysing the ice deep below the surface of Antartica, which had trapped carbon dioxide year by year (closed circles). The open circles show data from the air, obtained at Mauna Loa (see Figure 35).

year. Because of a fall in demand in the former Soviet Union, the consumption of fossil fuels levelled off during the early 1990s but is now increasing again. In 1997 the world consumption of fossil fuels for energy was almost 8.5 Gt.

$$1 \text{ Gt} = 1 \text{ gigatonne} = 1 \times 10^9 \text{ tonnes}$$

ASSIGNMENT 16

In 1994, the concentration of carbon dioxide in the atmosphere at Mauna Loa (see Figure 35) was 359 ppm (by volume). The *total mass* of carbon dioxide in the atmosphere was estimated to be about 2650 Gt at that time.

Through the four years, 1994 to the end of 1997, it is estimated that 33 Gt of fossil fuel were burned throughout the world.

In 1997 the measured percentage of carbon dioxide in the atmosphere at Mauna Loa was found to be 365 ppm (by volume).

Now answer these questions.

a Assume that the 33 Gt of fossil fuel burned in the four years were all carbon (this is a reasonable approximation since the other main element in fossil fuels, hydrogen, has a very low relative atomic mass). Calculate the mass of carbon dioxide formed by burning this mass of fuel.

b Calculate the new total mass of carbon dioxide in the atmosphere at the end of 1997, assuming that the extra carbon dioxide came only from burning fossil fuels and that it stayed in the atmosphere.

c By what percentage would you expect the concentration of carbon dioxide in the atmosphere to have increased in the four years?

d By what percentage did the concentration of carbon dioxide in the atmosphere actually increase in the four years?

e Comment on your answers to parts **c** and **d**.

Calculations of the kind you have just done suggest that – judging by the quantity of fossil fuels being burned – the increase in carbon dioxide concentration in the atmosphere *should* have been about *twice* as much as it actually has been. Atmospheric carbon dioxide concentration is not increasing as fast as we might expect. Good news – but where is all the carbon dioxide going?

Oceans soak up carbon dioxide

Oceans cover almost three-quarters of the Earth's surface. Carbon dioxide is fairly soluble in water, so large amounts of atmospheric $CO_2(g)$ dissolve in the oceans.

A8 **FOCUS ON CARBON DIOXIDE**

To understand the chemistry in this section you need an idea of dynamic equilibrium. **Chemical Ideas 7.1** will help you with this.

When carbon dioxide dissolves in water, it forms hydrated CO_2 molecules:

$$CO_2(g) + aq \rightleftharpoons CO_2(aq) \quad \text{(reaction 11)}$$

This is a reversible reaction, as you will know if you have watched what happens when you take the top off a bottle of fizzy drink. However, it is a fairly slow reaction (which is a good thing from the point of view of fizzy drink consumers), and it takes quite a long time for equilibrium to be reached.

The uptake of carbon dioxide by the oceans is quicker than this. Minute marine plants called phytoplankton use up most of the carbon dioxide which goes into the sea (Figure 37). So the concentration of 'free' $CO_2(aq)$ is small and gaseous carbon dioxide is encouraged to dissolve.

A small proportion of the $CO_2(aq)$ (about 0.4%) goes on to react with the water. It forms hydrogencarbonate ions and hydrogen ions:

$$CO_2(aq) + H_2O(l) \rightleftharpoons HCO_3^-(aq) + H^+(aq)$$
$$\text{hydrogencarbonate} \quad \text{(reaction 12)}$$
$$\text{ions}$$

Since $H^+(aq)$ ions are formed, this reaction is responsible for the acidic nature of carbon dioxide. A solution of carbon dioxide in pure water is, however, only weakly acidic. Reaction 12 does not go to completion: an equilibrium is set up with products and reactants both present in solution. Since only 0.4% of the $CO_2(aq)$ reacts, there is very much more $CO_2(aq)$ than $H^+(aq)$ present.

You can investigate some chemical equilibria in **Activity A8.1**.

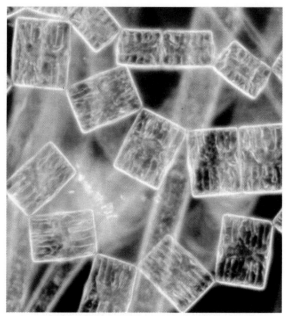

Figure 37 Phytoplankton act as a 'biological pump', removing CO_2 from the atmosphere and transporting organic carbon compounds from surface waters to deeper layers as a 'rain' of dead and decaying organisms. This is balanced by upward transport of carbon by deeper water which is richer in CO_2 than surface water.

Since the pH of a solution is related to the concentration of $H^+(aq)$ ions present, reactions 11 and 12 link together the quantity of CO_2 in the air and the pH of water which is in contact with it.

By measuring pH in a situation like this we should therefore be able to deduce from it the concentration of atmospheric carbon dioxide.

Activity 8.2 shows how the concentration of carbon dioxide in sample of air can be measured.

The global carbon cycle

The oceans are just one part – a very important part – of the carbon cycle which puts carbon dioxide into the atmosphere and takes it out again. Figure 38 shows the circulation of carbon in the *biosphere*, and the quantities involved. Have a look at the diagram and answer the questions in **Assignment 18**.

You will find out more about the role of the oceans in the global carbon cycle in **The Oceans**.

Biosphere

The biosphere comprises those parts of the Earth inhabited by living organisms or *biota*. It includes parts of the atmosphere, the oceans and the Earth's surface.

ASSIGNMENT 17

a What will happen to the position of equilibrium in reaction 11 if the concentration of atmospheric carbon dioxide increases?

b What will be the effect on reaction 12 of the change described in part **a**?

c In **Assignment 16** you found that the proportion of atmospheric carbon dioxide was not rising as fast as expected. Suggest a reason for this.

d Do you think carbon dioxide in the oceans and in the atmosphere are actually in equilibrium? Explain your answer.

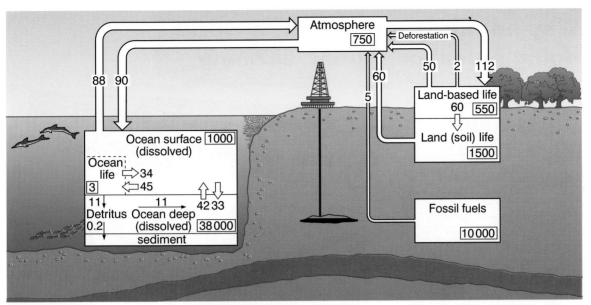

Figure 38 The global carbon cycle. The numbers in boxes are reservoirs, *showing the total mass of carbon (in Gt) in a particular part of the cycle; the numbers beside the arrows are* fluxes, *showing the rate of movement of carbon from one reservoir to another (in Gt year⁻¹). The amount of carbon in the atmosphere is growing at about 3 Gt per year, and that dissolved in the ocean surface at about 2 Gt per year.*

ASSIGNMENT 18

Look at the global carbon cycle in Figure 38.

a i What is the total mass of carbon passing *into* the atmosphere per year?

ii What is the total mass of carbon passing *out* of the atmosphere per year?

iii What is the net change in the mass of carbon in the atmosphere per year?

b Of the various processes that release carbon dioxide into and absorb carbon dioxide from the atmosphere, which do you think are likely to show significant changes over the next 50 years?

c Which single process removes carbon dioxide from the atmosphere at the greatest rate?

d The *residence time* for a reservoir is the average time that carbon spends in that reservoir. It is a measure of how fast carbon atoms are 'turned over' in the reservoir. You can calculate average residence times by dividing the size of the reservoir by the size of the flux. This gives the residence time in years. (Where the ingoing and outgoing fluxes are different, use the outgoing flux to calculate the residence time.)

Calculate the residence time for carbon in

i the atmosphere

ii land-based life

iii ocean life.

Suggest reasons for any difference between parts **ii** and **iii**. What is the significance for global change of the residence time in the atmosphere?

A9 *Coping with carbon*

It is estimated that in the middle of the 19th century, before the Industrial Revolution had really started pumping carbon dioxide into the atmosphere, the atmospheric concentration of carbon dioxide was 270 ppm.

By the early 1990s it was 353 ppm and by 2000 it was about 370 ppm. By the 2080s it will probably have doubled from its pre-Industrial Revolution value to about 540 ppm.

The increase in atmospheric carbon dioxide is a serious problem which will have far-reaching effects on the world's climate.

It is thought that the doubling of carbon dioxide concentration will cause an average temperature increase of about 2 °C. If you take into account the other greenhouse gases such as methane, which are also increasing in concentration, this 2 °C rise could be with us by the 2030s – within the lifetime of most people reading this. Although 2 °C may not sound much, it will be enough to have a dramatic effect on the global climate.

The science of the effects of carbon dioxide in an atmosphere, which is what has been presented in this unit, is fairly well known. We are far less certain about the extent to which these effects will occur, the role of other processes and the overall balance which will emerge.

One of the predicted results of global warming is a rise in the mean sea level because of the thermal expansion of water and the melting of some land ice.

Why hot air is stopping the world doing a deal on global warming

Jets causing greenhouse gas increase

AIR TRAVEL is an important and growing cause of global warming, says a report yesterday from the world's leading meteorologists.

Greenhouse gases from airliners' exhausts make up 3 per cent of global warming but by 2050 could account for 15 per cent, according to figures in the study from the UN's Intergovernmental Panel on Climate Change (IPCC).

Experts warn fossil fuels must be curbed

Spring bursts into flower a month early

The white plum blossoms at Kew confirm it; spring this year is earlier than ever.

Early blooming flowers and trees at the Royal Botanic Gardens strongly support the view of scientists who claimed last week that spring is arriving in Europe on average six days earlier than it was 30 years ago.

Figure 39 The dangers inherent in global warming are becoming an increasingly public issue and of concern to many – but not all – governments.

'So this is your response to global warming?'

Figure 40

Modelling our future climate is extremely difficult. There are people who believe we are heading for disaster from accelerating global warming fuelled by positive feedback. Others believe the Earth will develop ways of compensating for any serious departure from equilibrium.

Reducing the rate of increase of carbon dioxide in the atmosphere is a major challenge for the world at the beginning of the 21st century. In **Activity A9** you can investigate some of the possible approaches to solving the problem.

A10 *Summary*

In this unit you have looked at two problems in atmospheric chemistry: the destruction of the ozone layer and an increase in the greenhouse effect leading to global warming. These are problems that affect the whole world.

Studying the rates of chemical reactions is vital for chemists working to understand the many reactions taking place in the atmosphere. This led you to consider the factors that affect the rate of chemical reaction, in particular, the effects of concentration and temperature.

Ozone depletion is a *stratospheric problem*. At its centre is the idea of the absorption of radiation by gas molecules. Dangerous effects arise from the increased transmission of ultraviolet light through the atmosphere as ozone molecules are destroyed by radical reactions in the stratosphere.

The mechanisms of many chemical processes in the stratosphere require an understanding of radicals and their reactions. Chlorine radicals produced from CFCs act as catalysts for the depletion of ozone. This led you to study homogeneous catalysis in more detail and to study the role of catalysts in providing an alternative route for a reaction with a lower activation enthalpy.

You looked at the uses of CFCs and read about the search for suitable replacements. This led you to study the chemistry of organic halogen compounds in detail.

The greenhouse effect is a *tropospheric phenomenon*. Again it arises from the absorption of radiation by gas molecules. You have focused on the role of carbon dioxide and the need to monitor and detect its concentration. This led you to the key idea of chemical equilibrium and its importance in determining the point of balance in some chemical processes.

Activity A10 should help you to get your notes in order at the end of the unit.

THE POLYMER REVOLUTION

Why a unit on THE POLYMER REVOLUTION?

This unit has three themes. First, it tells the story of the polymer revolution – the great changes that polymers have brought to our lives. Many of the discoveries that led to important advances were made by chance, and this is the second theme of the module. The most important polymers in terms of bulk production are *addition* polymers, prepared from alkene monomers. The historical development of addition polymers is the third theme. Most of their development has taken place over the last 70 years or so.

The unit provides some essential information about polymers and the process of addition polymerisation. Through this, you are introduced to alkenes and their reactions, geometric isomerism and ideas about intermolecular forces.

Central to the unit is the relationship between the properties of a substance and its structure and bonding. This relationship helps to explain the properties of polymers.

Overview of chemical principles

In your earlier studies, you will probably have come across ideas about

- polymerisation
- how the structure of a substance determines its properties.

In this unit you will learn more about …

- the chemistry of organic molecules (**Developing Fuels** and **The Atmosphere**)
- isomerism (**Developing Fuels**)
- reaction mechanisms (**The Atmosphere**)
- radicals (**The Atmosphere**)
- electronegativity abd bond polarity (**The Elements of Life** and **The Atmosphere**)

… as well as learning new ideas about

- addition polymers and polymerisation
- reactions of alkenes
- geometric isomerism
- intermolecular forces
- the relationship between the properties of a substance and its structure and bonding.

THE POLYMER REVOLUTION

PR1 *The start of the revolution*

Polymers are produced in profusion by nature – in plants and animals and in our bodies. Synthetic polymers are so much part of our lives, both in terms of materials and culture, that it is difficult to believe that their development spans only the last 70 years or so. Indeed, they have only been used on such a massive scale over the last 50 years.

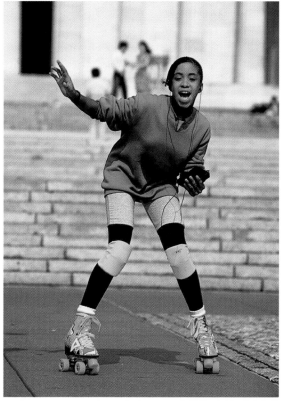

Figure 1 Tuned into plastic: earpierce with conducting plastic, polyester shirt, leggings with Lycra and roller boots with other polymers.

In the late 19th century, plastics were produced by modifying natural polymers. Celluloid, for example, was produced by reacting cellulose from plants with nitric acid (Figure 2). The first plastic to be made in significant quantities from manufactured chemicals was *Bakelite* (made from phenol and methanal). Bakelite is still used to make electrical fittings such as sockets and plugs. Although it was first made in 1872, it was not until 1910 that the process was patented and Bakelite was manufactured.

Figure 2 Hatpins and their stands, made over 100 years ago from the plastic celluloid.

The great polymers which we now know, such as nylon, Terylene, polythene, PVC, polystyrene and polyurethanes, were developed many years later, between the 1930s and the 1960s. They were not manufactured in significant quantities for the general public until the 1950s. Their growth is part of the story of the development of the chemical industry after the end of the Second World War in 1945, when crude oil began to take over from coal as the main raw material for organic compounds.

Chance has played an important part in the polymer revolution. In this unit, you will meet a number of polymers, such as poly(ethene) and poly(tetrafluoro-ethene) (Teflon), that have been discovered 'by accident'. The chance discoveries resulted from experiments that gave unexpected results, or from errors made by the chemists involved.

The key feature is that the researchers recognised the importance of what they had observed and went on to investigate further. In **Designer Polymers**, you will meet some polymers that have been specifically *designed* by chemists for particular purposes.

So this unit has three themes: the story of the great changes brought about to our lives by polymers; the importance of chance discoveries in scientific research; and the story of the historical development of addition polymers.

What is a polymer?

A **polymer** molecule is a long molecule made up from lots of small molecules called **monomers**.

If all the monomer molecules are the same, and they are represented by the letter A, an A–A polymer forms:

-- A + A + A + A -- → -- --A–A–A–A-- --

Poly(ethene) and PVC are examples of A–A polymers.

If two different monomers are used, an A–B polymer is formed, in which A and B monomers alternate along the chain:

-- A + B + A + B -- → -- --A–B–A–B-- --

Nylon-6,6 and polyesters are examples of this type of A–B polymer.

Writing out the long chain in a polymer molecule is very time consuming – we need a shorthand version. See how this is done for poly(propene):

$$—CH_2—CH—CH_2—CH—CH_2—CH—$$
$$\qquad\quad CH_3 \qquad\quad CH_3 \qquad\quad CH_3$$

or

poly(propene)

In the chain, the same basic unit is continually repeated, so the chain can be abbreviated to:

$$\left(CH_2—CH\right)_n \quad or \quad \left(\wedge\right)_n$$
$$\qquad\quad CH_3$$

where n is a very large number which can vary from a few hundred to many thousand. The part of the molecule in brackets is called the **repeating unit**.

Elastomers, plastics and fibres

Polymer properties vary widely. Polymers which are soft and springy, which can be deformed and then go back to their original shape, are called **elastomers**. Natural rubber is an elastomer.

Poly(ethene) is not so springy and when it is deformed it tends to stay out of shape, undergoing permanent or plastic deformation. Substances such as this are called **plastics**.

Stronger polymers, which do not deform easily, are just what you want for making clothing materials: some can be made into strong, thin threads which can then be woven together. These polymers, such as nylon, are called **fibres**.

Poly(propene) is on the edge of the plastic–fibre boundary – it can be used as a plastic like poly(ethene), but it can also be made into a fibre for use in carpets.

It turns out that many of the polymers which have been found by accident are the A–A polymers, known as **addition polymers**. Those that have been designed are mostly A–B polymers, known as **condensation polymers**.

Before you go any further, it will help you to become familiar with some facts, figures and applications for the more widely used plastics. You can do this in **Activity PR1**.

PR2 *The polythene story*

An accidental discovery

Imperial Chemical Industries (ICI) was formed in 1926 by the amalgamation of a number of smaller chemical companies. The prime aim of the merger was to form a strong competitor to the huge German chemical company IG Farben.

To understand this section you will need to know about a compound called **ethene** and some of its relatives in the series of compounds known as the **alkenes**. **Chemical Ideas 12.2** contains the information you will need.

In 1930 Eric Fawcett, who was working for ICI, got the go-ahead to carry out research at high pressures and temperatures aimed at producing new dyestuffs. His results were disappointing and his project was eventually abandoned.

The team then moved into the field of high-pressure gas reactions and was joined by Reginald Gibson. On Friday 24 March 1933, Gibson and Fawcett carried out a reaction between ethene and benzaldehyde using a pressure of about 2000 atm. They were hoping to make the two chemicals add together to produce a ketone:

benzaldehyde + *ethene*

They left the mixture to react over the weekend, but their apparatus leaked and at one point they had to add extra ethene.

They opened the vessel on the following Monday and found a white waxy solid. When they analysed it they found that it had an empirical formula CH_2. They were not always able to obtain the same results from their experiment: sometimes they got the white solid, on other occasions they had less success, and sometimes their mixture exploded leaving them with just soot.

The work was halted in July 1933 because of the irreproducible and dangerous nature of the reaction.

Learning to control the process

In December 1935 the work was restarted. Fawcett and Gibson found that they could control the heat given out during the reaction if they added cold ethene at the correct rate. This kept the mixture cooler and prevented an explosion. They also found that they could control the reaction rate and relative molecular mass of the solid formed by varying the pressure.

A month later they had enough of the material to show that it could be melted, moulded and used as an insulator.

Most crucial of all was the identification of the role of oxygen in the process, by Michael Perrin who took charge of the programme in 1935. When oxygen was not present, the polymerisation did not occur. Too much oxygen caused the reaction to run out of control.

The trick is to add just enough oxygen. The leak in Fawcett and Gibson's original apparatus had accidentally let in a small amount of oxygen. Without this, the discovery of poly(ethene) might not have been made. It was also Perrin who showed that even if the benzaldehyde is left out of the reaction mixture, the polymer still forms.

Poly(ethene) is an example of an addition polymer. You can find out more about the formation of polymers like poly(ethene) in the section on addition polymerisation in **Chemical Ideas 12.2**.

Poly(ethene) – or polythene as it is commonly called – is tough and durable, and has excellent electrical insulating properties. Unlike rubber which had previously been used for insulating cables, poly(ethene) is not adversely affected by weather or water.

It also has almost no tendency to absorb electrical signals. Its first important use was for insulating a telephone cable laid between the UK mainland and the Isle of Wight in 1939. Its unique electrical properties were again essential during the Second World War in the development of radar.

If you have never made an addition polymer you can do so in **Activity PR2**. You will make some poly(phenylethene) – better known as **polystyrene**.

Figure 3 Poly(ethene) was essential in the development of radar equipment in the Second World War.

The first poly(ethene) washing-up bowls appeared in the shops in 1948, and were soon followed by carrier bags, squeezy bottles and sandwich bags. Sadly, poly(ethene) and some other early polymer materials were overexploited. They were used for all manner of novelties and as cheap but poor substitutes for many natural materials. This gave *plastic* a bad name – the word is often used to describe something which looks cheap and does not last. The reputation still sticks, as an undeserved slur on many of today's excellent materials.

Figure 4 Poly(ethene) carrier bags coming off the production line.

ASSIGNMENT I

a Write an equation for the formation of poly(ethene) from ethene.

b Draw the structural formula of part of a poly(ethene) chain. Explain why poly(ethene) can be thought of as a very large alkane.

c Explain why polymerisation of ethene does not occur when there is no oxygen present, but with too much it gets out of control.

A bonus of being big

A polymer molecule is just a very big molecule. This seems obvious now, but the idea wasn't proposed until 1922 and it met with considerable criticism for the rest of that decade. Since the large molecules in a polymer are chemically similar to much smaller ones, it should be possible to predict many of a polymer's properties from the properties of substances which contain the smaller molecules.

Poly(ethene) is the simplest polymer from a chemical point of view, containing only singly bonded carbon and hydrogen atoms. It should behave like an alkane of high relative molecular mass – and in many ways it does. It burns well, and tends not to react with acids or alkalis.

Poly(ethene), like all polymer materials, is a mixture of similar molecules, rather than a pure compound, because different numbers of monomers join together in the chain-building process before polymerisation stops. Therefore, poly(ethene) does not melt sharply: it softens and melts over a temperature range. However, this happens roughly at the temperature you would expect it to for a very large alkane.

In contrast, poly(ethene)'s mechanical properties are completely unlike those of similar but smaller molecules.

At this point you need to understand about the attractions which arise between poly(ethene) molecules. These attractions (which result from a type of *induced dipole force*) arise between **all** molecules and are described in **Chemical Ideas 5.3**. Other types of intermolecular forces are also discussed here.

The first part of **Chemical Ideas 5.5** on addition polymerisation explains the factors, such as chain length, which affect the physical properties of polymers.

PR3 *Towards high density polymers*

The poly(ethene) produced by Fawcett and Gibson was what we would today call **low density poly(ethene) (ldpe)**.

Although they had some control over the product of their polymerisation process, the low density poly(ethene) they made was still quite messy at a molecular level. The polymer chains were extensively branched. This makes it impossible for the chains to fit together in an organised way: they coil around randomly taking up a lot of room, and hence lower the density of the material.

This relatively disorganised and open structure also lowers the strength of the poly(ethene). The next major advance in poly(ethene) production came with the development of **high density poly(ethene) (hdpe)**, which resulted from discoveries made by Karl Ziegler.

The German scientist Karl Ziegler was born in Helsa, near Kassel, in 1898. Encouraged to work hard by his father, the young Ziegler set up a chemical laboratory at home, where he became so advanced in his chemistry studies that he was allowed to omit the first year of his degree course.

At the age of 23 he became a professor and started on a research path which was to have a great influence on the future development of polymerisation processes.

Figure 5 Karl Ziegler (1898–1973) who, with Giulio Natta, won the Nobel Prize for Chemistry in 1963 for their work on catalysis for polymerisation.

Ziegler catalysts

Ziegler was studying the catalytic effects of **organometallic** compounds. These are compounds which contain covalent metal–carbon bonds. Strange things happened in some of his experiments when he was using an aluminium organometallic compound.

He tracked down the unexpected behaviour to tiny traces of nickel compounds left over as impurities in his apparatus after cleaning. His research group then tried putting as many different transition metal salts as possible with the alkylaluminium compound to see what would happen.

In 1953 Ziegler, with his colleagues Holzkamp and Breil, found that adding titanium compounds easily led to the production of very long chain polymers. Simply passing ethene at atmospheric pressure into a solution of a tiny amount of $TiCl_4$ and $(C_2H_5)_3Al$ in a liquid alkane caused the immediate production of poly(ethene).

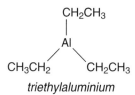

triethylaluminium

So, like Fawcett and Gibson's discovery of poly(ethene), Ziegler's achievements began accidentally with an impurity.

The poly(ethene) which Ziegler produced had an average relative molecular mass of 3 000 000 with very little branching along the polymer chain. The chains could therefore line up and pack more closely than those made by the original high-pressure process. In this form the poly(ethene) is said to be **crystalline**. This gives the polymer its higher density and greater strength.

You can simulate the packing of polymer chains using spaghetti in **Activity PR3**.

High density poly(ethene) is often used to make washing-up bowls, water tanks and piping. It is strong and can easily be moulded into complicated shapes: car petrol tanks, for example, can be made to fit neatly into the spaces under the car, something which was impossible with metal tanks. Hdpe is not as easily deformed by heat as ldpe and an early use was as Tupperware food storage containers. The ability to retain shape during heating means that hdpe articles can be heat-sterilised, making hdpe an important material for hospital equipment such as buckets and bed-pans.

ASSIGNMENT 2

Some data for low density and high density poly(ethene) are given in the table below. Use this information to answer the questions which follow.

	Density/ g cm^{-3}	Tensile strength/ MPa	Elongation at fracture/ %
ldpe	0.92	15	600
hdpe	0.96	29	350

a Will either polymer sink in water?

b Why is the tensile strength lower for ldpe than for hdpe?

c Use the data to explain the different uses of ldpe and hdpe.

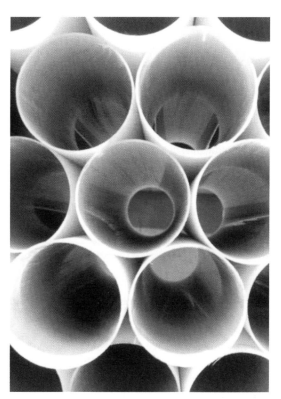

Figure 6 Piping and many containers are made from hdpe.

Ziegler patented hdpe. As a result he became a multimillionaire, and on his 70th birthday he gave $10 million to support further research at the Max Planck Institute where he worked.

Natta and stereoregular polymerisation

Giulio Natta was born in Imperia, Italy, in 1903. He studied chemical engineering at the Milan Polytechnic Institute, and after working at various universities he returned to Milan in 1938 to become Professor of Industrial Chemistry.

Figure 7 Giulio Natta (1903–79).

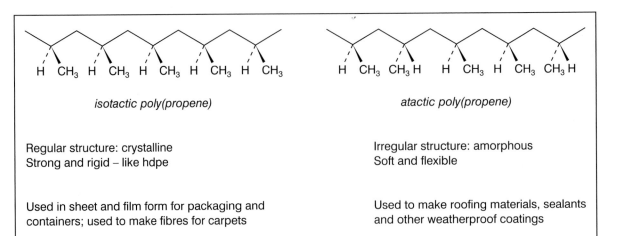

isotactic poly(propene)

Regular structure: crystalline
Strong and rigid – like hdpe

Used in sheet and film form for packaging and
containers; used to make fibres for carpets

atactic poly(propene)

Irregular structure: amorphous
Soft and flexible

Used to make roofing materials, sealants
and other weatherproof coatings

Figure 8 Isotactic and atactic poly(propene).

Natta was convinced that alkylaluminium catalysts
were the key to making **stereoregular polymers** –
polymers with a regular structure.

In March 1954 he used Ziegler's catalyst to
polymerise propene. His reaction mixture contained
two forms of poly(propene) – a **crystalline** form and
an **amorphous** (non-crystalline) form. He was able to
separate them.

In the crystalline form, the methyl groups all have
the same orientation along the polymer chain. Natta
called this the **isotactic** form. (Iso means *the same* –
as in isotope or isobar.) In the amorphous polymer,
the methyl groups are randomly orientated and this
was called the **atactic** form.

Figure 8 tells you more about these two forms of
poly(propene).

Natta went on to develop new catalysts which
allowed the polymer molecule to grow outward from
the catalyst surface like a growing hair (see Figure 10).
These catalysts, known as **Ziegler–Natta catalysts**,
have allowed chemists to tailormake specialist
polymers with precise properties. You will read more
about this in **Designer Polymers**.

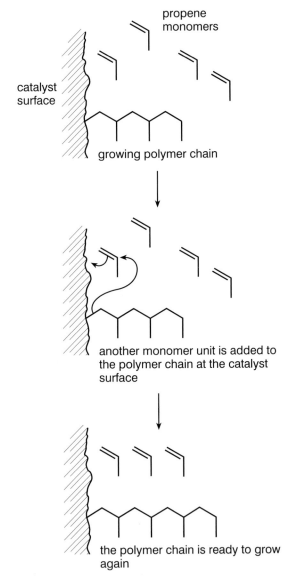

Figure 10 Poly(propene) chain growing from a catalyst surface.

*Figure 9 Isotactic poly(propene) can be drawn into fibres
and used in carpets.*

The story continues into the new millennium

Now, a new generation of catalysts for addition polymerisation is being introduced, from the research laboratory, into industry. They are called the **metallocenes**. Chemists have compared their structure to a sandwich. An example is shown below.

Here the 'filling' of the sandwich is zirconium (but it could be another transition metal, such as titanium). The two slices of 'bread' are flat organic molecules with arene ring systems.

Metallocenes are even more specific than the original Ziegler–Natta catalysts and they allow chemists to control the polymer's molecular mass as well as its structure.

Poly(ethene) and poly(propene) produced using a metallocene can be used as thin films with very interesting properties. They are even more impermeable to air and moisture than the older polymers. They are also strong and much more tear-resistant than the polymers formed by the other methods. Thus thin films of them are used to protect materials susceptible to air and moisture, such as food.

ASSIGNMENT 3

Different metallocenes produce isotactic and atactic poly(propene). It is also possible to produce a form of poly(propene) in which the orientation of the methyl groups alternates along the chain. This is called *syndiotactic* poly(propene).

a Draw the structure of the repeating unit in syndiotactic poly(propene).

b Suggest how the properties of syndiotactic poly(propene) might compare with those of the isotactic and atactic forms.

Figure 11 A very thin film of a metallocene-based polymer. It is being subjected to a test, using a ball-point pen, to show its strength and resistance to being punctured.

Another chance discovery

In 1975, Professor Walter Kaminsky was experimenting with a metallocene containing titanium as the transition metal in the centre of the sandwich. He asked one of his students to investigate the polymerisation of ethene with a modified Ziegler–Natta catalyst consisting of the metallocene and trimethylaluminium ((CH_3)$_3$Al), with the metallocene replacing the more usual $TiCl_4$.

The student found that the polymerisation was many times faster than expected. He admitted he had been lazy and had failed to bubble nitrogen through the solution to flush out the air before sealing the tube containing the reaction mixture. (Ziegler–Natta systems are normally very sensitive to oxygen.)

But why was the reaction so fast? After a six-month search, Professor Kaminsky found that for the reaction to go faster water must be present. He showed that a mixture of the metallocene, trimethylaluminium and water could increase the rate of reaction a million-fold. He found that water reacts with trimethylaluminium to form a new catalyst based on the structure:

$$\left(\begin{array}{c} Al-O \\ | \\ CH_3 \end{array}\right)_n$$

It was this substance that was responsible for the huge increase in the rate of reaction. Water vapour from the air must have been present in the student's original experiment.

Polymers formed using this method have very long chains that are uniform in length. They are more crystalline than those made using normal Ziegler–Natta catalysts and are thus stronger and more transparent.

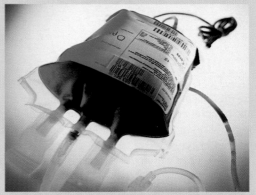

Figure 12 Polymers made using metallocenes are not easily broken down by γ-rays and so are an ideal material for medical packaging, such as blood bags, which are often sterilised using γ-radiation.

PR4 *The Teflon man*

Roy Plunkett worked in the research laboratories of DuPont, a large chemical company in the US. On April 6 1938, he went to use some tetrafluoroethene, a gas stored in a cylinder. However, the cylinder appeared empty. He decided to open the cylinder and discovered a white waxy solid. The gas had polymerised to form poly(tetrafluoroethene), PTFE, which is now marketed as 'Teflon'.

tetrafluoroethene   *poly(tetrafluoroethene)*

On examining the polymer, Plunkett discovered its now well-known anti-stick properties. But Teflon is also highly resistant to chemical attack and it is a very good electrical insulator.

Figure 13 PTFE in use. (a) The roof of the Millennium Dome is made of a woven glass fibre coated with PTFE, which is long lasting, easy to clean and non-flammable. (b) PTFE in a more familiar role as a coating for a non-stick frying pan.

Water out but not in

Poly(tetrafluoroethene), PTFE, is a hydrophobic (water-hating) material. In 1969, Bob Gore discovered that the polymer could be stretched to form a porous material which would allow water vapour but not water liquid to pass through the minute holes.

He developed a material – Gore-tex – which used this property. A layer of porous PTFE film and a layer of an oil-hating polymer act as the filling of a sandwich between an outer fabric and the inner lining (Figure 14). The oil-hating polymer allows the water vapour through, but prevents the natural oils from the skin and from cosmetics from blocking the pores in the PTFE, thus preserving its waterproofing properties.

Outer fabric

PTFE film

Oil-hating polymer

Inner lining

Figure 14 The layers that make up a Gore-tex membrane.

Figure 15 Gore-tex is ideal clothing for bad weather conditions, whether walking, climbing or, as here, working out of doors in a storm. It allows the water vapour from the body to evaporate.

PR5 *Dissolving polymers*

If soiled laundry from a hospital is mishandled, there is a risk of infection. The risk can be avoided by making the laundry bags out of a dissolving plastic. The dirty linen is safely contained until the bag is placed in the wash – then the bag dissolves and the washing is let out.

The dissolving plastic which is used is **poly(ethenol)**.

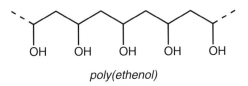

poly(ethenol)

To understand why poly(ethenol) dissolves in water, you need to find out about **hydrogen bonding**, a particularly strong type of intermolecular bonding.

Figure 16 This hospital laundry bag has a section made of dissolving plastic, which dissolves in the washing machine to release the dirty washing.

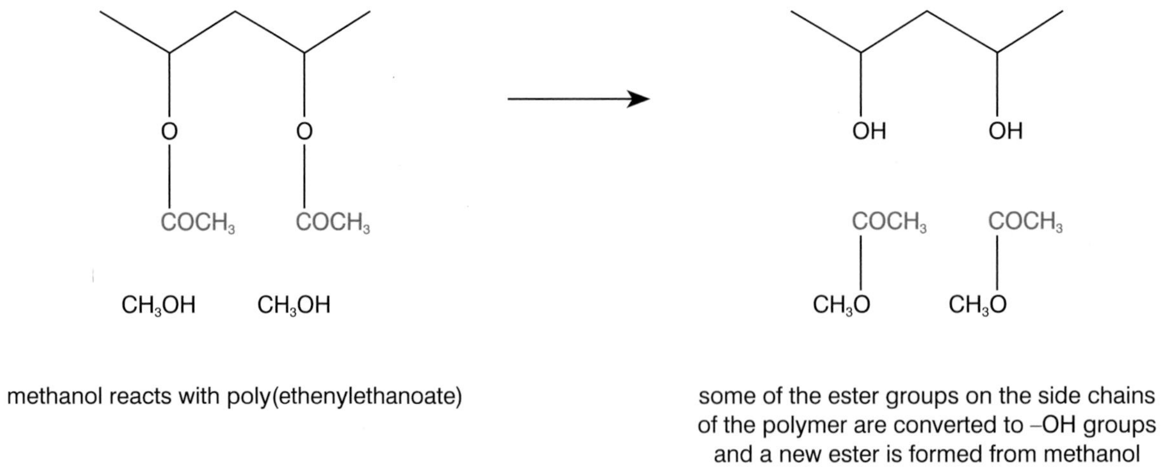

methanol reacts with poly(ethenylethanoate)

some of the ester groups on the side chains of the polymer are converted to –OH groups and a new ester is formed from methanol

Figure 17 Formation of poly(ethenol) by ester exchange.

You can read about hydrogen bonding in **Chemical Ideas 5.4**.

Activities PR5.1 and **PR5.2** will help to illustrate the ideas you are reading about.

You might think that poly(ethenol) could be made by polymerising ethenol (CH_2=CH–OH) but this compound does not exist. However, it can be made from another polymer, poly(ethenyl ethanoate), by the process illustrated in Figure 17.

The extent of reaction can be controlled by adjusting either the temperature or the reaction time. The plastic's solubility depends on the percentage of OH groups present. Table 1 shows how the two are related. Different solubilities give the plastic different uses.

% of OH groups	Solubility in water
100–99	insoluble
99–97	soluble in hot water
96–90	soluble in warm water
below 90	soluble in cold water

Table 1 Solubility of poly(ethenol).

Molecules with the structure R–O–COCH$_3$ are examples of *esters*. (You will study ester groups in more detail in **What's in a Medicine?**) Can you see why the process is called an ester exchange?

You can study the behaviour of poly(ethenol) in **Activity PR5.3**.

Activity PR5.4 provides you with a very tangible example of hydrogen bonding in action.

ASSIGNMENT 4

a What type of intermolecular bonding will there be between the chains of poly(ethenol)?

b Explain why nearly pure poly(ethenol) is insoluble even in hot water.

c How is the intermolecular bonding in this polymer affected if the ester groups are still present?

d Explain the effect on solubility of increasing the number of ester groups in the polymer.

e Hospital laundry bags are made from the form of the polymer which is soluble in only hot water. Suggest why this form is chosen.

f Suggest some other uses for poly(ethenol) film.

PR6 *Polymers that shine in the dark*

Many chemists have tried to make polymers that can conduct electricity. For example, Natta, the discoverer of isotactic poly(propene), was one such chemist. He carried out his experiments in 1955. About 15 years later, in 1971, Hideki Shirakawa and Sakuji Ikeda, working at the Tokyo Institute of Technology, found that when they directed a stream of ethyne, C_2H_2, on to the surface of a solution of a Ziegler–Natta catalyst at –78 °C, a red film was formed. When the experiment was repeated at 100 °C, the film was coloured blue. They had prepared poly(ethyne), which they later found could be made to conduct electricity.

Before you go any further you will need to have some understanding of two areas of chemistry which may be new to you. **Geometric isomerism** is a phenomenon which arises in alkenes and a number of other types of compound. **Alkynes** are a family of hydrocarbons in which there is a triple carbon to carbon bond.

Alkynes

Alkynes form a class of unsaturated hydrocarbons. They differ from the alkenes by the presence of a C≡C triple bond instead of a C=C double bond.

Ethyne, the simplest alkyne, has the structure H—C≡C—H.

Other alkynes are

propyne CH_3—C≡C—H

but-1-yne CH_3CH_2—C≡C—H

pent-2-yne CH_3CH_2—C≡C—CH_3

Alkynes are named in a similar way to alkenes, but with the suffix -yne instead of -ene.

Ethyne is a linear molecule. There are two groups of electrons around each carbon atom and these are furthest apart when the bond angle is 180°.

You may understand the shapes of these molecules better if you make models of ethyne and the other alkynes shown in this box.

You can read about geometric isomerism in **Chemical Ideas 3.5**.

ASSIGNMENT 5

a Name the alkynes with the following structures:

i CH_3—CH_2—C≡C—CH_2—CH_3

ii CH_3—CH—C≡C—CH_2—CH_3
            |
           CH_3

iii ⬡—C≡C—CH_3

b Draw structures for the following alkynes:
  i but-2-yne
  ii diphenylethyne
  iii dimethylpent-2-yne.

When ethyne molecules polymerise, the product, poly(ethyne), is a molecule which contains alternating single and double bonds.

H—C≡C—H   H—C≡C—H   H—C≡C—H
ethyne

↓

—CH=CH—CH=CH—CH=CH—
poly(ethyne)

There is more than one form of poly(ethyne) because geometric isomerism causes there to be two ways in which the double bonds can be arranged relative to one another. The structures of two forms of poly(ethyne) are shown in Figure 18; however, of course, mixed *cis* and *trans* arrangements are also possible.

cis-poly(ethyne)

red

(copper-coloured when thicker)

trans-poly(ethyne)

blue

(silver-coloured when thicker)

Figure 18 The structures of two forms of poly(ethyne).

Molecules with alternate single and double bonds have **conjugated systems**. Benzene has a conjugated system arranged in a six-membered ring. However, it is better to regard the electrons forming the second bond in each double bond as being *delocalised* – spread around the whole benzene ring – rather than localised between adjacent carbon atoms. This is why we represent benzene as

 rather than

The *trans* isomer of poly(ethyne) is shown below:

with a dotted line to represent the delocalised electrons. This isomer conducts electricity better than the *cis* isomer.

One of Shirakawa's students repeated the experiment to prepare poly(ethyne) by polymerising ethyne using a Ziegler–Natta catalyst. By mistake, he used 1000 times too much catalyst. He produced a silvery, metallic-looking film which was crystalline, which was later found to be the *trans* form of the polymer (the colour changing from blue, as the film is thicker).

Alan MacDiarmid, of the University of Pennsylvania, saw the film on a visit to the laboratory. MacDiarmid invited Shirakawa to spend some time at his university, to work together to improve the electrical properties of the polymer.

Success came a year later. If the silvery material is 'doped' with iodine, it conducts electricity 1 000 000 000 times better than previous forms of poly(ethyne).

Conducting polymers of this type are now used as antistatics (Figure 19).

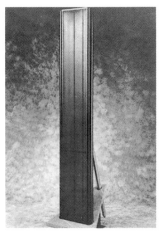

Figure 19 A conducting polymer is used as the membrane in this high-performance loudspeaker.

Other monomers can be used to produce conducting polymers. You can make and test your own conducting polymer in **Activity PR6**.

Conjugated polymers continue to be of great interest. In 1988, Andrew Holmes, at the University of Cambridge, was working on some polymers produced from alkynes. He had formed a collaboration with a physicist, Richard Friend, to study these polymers and a related one called PPV (from its old name poly(*p*-phenylene vinylene)). The repeating unit of PPV is shown below.

Then came the fateful day in 1989, when one of Andrew Holmes's students, Paul Burn, took along some PPVs he had made to Jeremy Burrough at the physics laboratory so that the polymers could be tested as insulating materials for transistors. The test involved applying a voltage to a thin film of the polymer. On applying the voltage, there on the bench was the polymer glowing with an eerie green light. The first light-emitting polymer had been discovered.

Since then, other conjugated polymers have been produced, including some based on the structure of the original PPV.

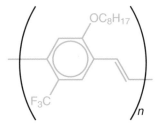

Emits orange-red light

Emits yellow-orange light

From that chance discovery, chemists have developed polymers that, on applying a voltage, will emit a range of colours from red to blue.

The possible uses are enormous. They include video and computer screens that are 2 mm thick, televisions the size of a credit card and televisions you can roll up into a tube when not in use. All these possibilities were stimulated by the original discovery of poly(ethyne) as a conducting polymer.

Figure 20 Polymers that shine in the dark are revolutionising the design of TV sets.

PR7 *Summary*

In this unit you have seen how our ideas about addition polymers have developed. As a result of this development, chemists have learned how to build up simple molecules into very long polymer chains. The polymers they have produced are new materials with unique sets of properties not possessed by any natural substances.

You began by reading about the accidental discovery of poly(ethene), which led you to a study of alkenes and their reactions. You saw how the properties of poly(ethene) depend on its structure and the intermolecular forces between the polymer chains. The instantaneous dipole–induced dipole forces that are present in all substances are important here. For other addition polymers with polar groups on the chain (such as PVC), permanent dipole–permanent dipole forces are also present, and the attractive forces between the chains are stronger.

You went on to find out about the discovery of Ziegler–Natta catalysts and the use of metallocenes to control the polymerisation process. These developments have allowed chemists to control the structure of the polymer formed – and hence control its properties.

Dissolving polymers have hydroxyl (–OH) groups on the polymer chain. To explain why they dissolve in water, you needed to find out about hydrogen bonding, a particularly strong type of intermolecular attraction.

Finally, you were introduced to some exciting work taking place in the development of conducting polymers and polymers that emit light when a voltage is applied. Through this you met another family of hydrocarbon compounds, the alkynes, and learned about geometric isomerism in alkenes.

One of the themes of this unit was the accidental nature of many of the important discoveries in polymer chemistry. They came about because the scientists recognised that they had observed something unusual and interesting, and went on to investigate further.

Activity PR7 allows you to check that your notes cover the important chemistry contained in this unit.

WHAT'S IN A MEDICINE?

Why a unit on WHAT'S IN A MEDICINE?

This unit introduces the importance of the pharmaceutical industry, both to the personal health of the individual and the financial health of the nation. Through a study of aspirin it illustrates many of the principal activities involved in the development of a medicine. Finally, it considers the problems of development and safety testing.

During the unit, you will study the application of instrumental methods for determining the structure of molecules, and practise extraction of a natural product, organic synthesis and the use of test-tube reactions to identify functional groups.

The chemistry of alcohols, phenols and carboxylic acids is studied in some detail. You will see how alcohols and phenols can react with carboxylic acids to produce esters.

Overview of chemical principles

In this unit you will learn more about …

ideas introduced in earlier units in this course
- the electromagnetic spectrum (**The Elements of Life**)
- the interaction of radiation with matter (**The Elements of Life** and **The Atmosphere**)
- mass spectrometry (**The Elements of Life**)
- alcohols (**Developing Fuels**)

… as well as learning new ideas about
- molecular structure determination
- infrared spectroscopy
- phenols
- carboxylic acids
- esters.

WM1 *The development of modern ideas about medicines*

This unit is about medicines and the pharmaceutical industry. Many pharmaceuticals are complex compounds, but in this unit we focus on the chemistry of a simple and familiar substance – aspirin.

The active ingredients of **medicines** are **drugs** – substances which alter the way your body works. If your body is already working normally the drug will not be beneficial, and if the drug throws the body a long way off balance it may even be a **poison**. When your body is working wrongly a medicine prevents things getting worse and can help bring about a cure, for example when you take aspirin or penicillin. Not all drugs are medicines: alcohol and nicotine are not medicines but they certainly are drugs. Some drugs, eg opium, may or may not be medicines depending on your state of health.

The study of drugs and their action is called **pharmacology**; the art and science of making and dispensing medicines is called **pharmacy**.

People have been using medicines for thousands of years – most of that time with no idea how they worked. Their effectiveness was discovered by trial and error, and sometimes there were disastrous mistakes.

Today's medicines are increasingly designed to have specific effects, something which is becoming easier as we learn more about the body's chemistry and begin to understand the intricate detail of the complex molecules from which we are made.

Work at this level comes into the field of **molecular pharmacology** and you will gain some insight into this in a later unit, **Medicines by Design**.

In **Activity WM1** you will find out about the importance of the UK pharmaceutical industry.

WM2 *Medicines from nature*

Modern pharmacy has its origins in folklore, and the history of medicine abounds with herbal and folk remedies. Many of these can be explained in present-day terms and the modern pharmaceutical industry investigates 'old wive's tales' to see if they lead to important new medicines.

One such tale is the 'Doctrine of Signatures' which proposes that illnesses can often be cured by plants which are associated with them. A simple example is the use of dock leaves (which grow near nettles) to attempt to cure nettle stings. However, in this case, no scientific validity has been found.

Figure 1 Feverfew has been used since ancient times for the treatment of migraine; research in the 1970s confirmed that it was an effective medicine for this disorder.

Medicines from willow bark

Marshy ground was thought to breed fevers, and so the bark and leaves of willow trees which often grew there were tried as a remedy against fever.

In 400 BC, Hippocrates recommended a brew of willow leaves to ease the pain of childbirth, and in 1763 The Reverend Edward Stone, an English clergyman living in Chipping Norton, Oxfordshire, used a willow bark brew to reduce fevers.

He argued by the 'Doctrine of Signatures': 'As this tree delights in a moist or wet soil, where agues chiefly abound, I could not help applying the general maxim, that many remedies lie not far off from their causes.'

The 'Doctrine of Signatures' is no longer a current pharmaceutical theory, but it is now known that there is a compound in willow bark and leaves which does have an effect in curing fevers.

Figure 2 Extracts from willow trees have been used in medicine for thousands of years.

'Culpeper's Herbal'

Nicholas Culpeper (1616–1654) rose to fame in the wave of enthusiasm for astrological botany which swept England in the 17th century. He believed that plants were 'owned' by certain planets, stars, etc. He also believed that the celestial bodies were the causes of diseases, and that illness could be cured by administering a plant 'owned' by an opposing body, or sometimes by a sympathetic body.

His *Herbal* has been published since 1640 under many titles and contains many references like the following one for the willow.

Willow tree

The leaves bruised with pepper, and drank in wine, help in the wind-colic.

It grows 60–70 feet (18–21 m) high, has a rough bark and narrow, sharp-pointed leaves on its whitish grey branches. It produces yellow male and green female catkins. Also called the white willow.

Where to find it: *Beside running streams and in other moist places.*

Flowering time: *Spring*

Astrology: *The Moon owns it.*

Medicinal virtues: *The leaves, bark and seeds are used to staunch the bleeding of wounds and other fluxes of blood in man or woman. The decoction helps to stay vomiting and also thin, hot, sharp salt distillations from the head upon the lungs, causing consumption.*

Water gathered from the willow when it flowers, by slitting the bark, is good for dimness of sight or films that grow over the eyes. If drank it provokes the urine, and clears the face and skin from spots and discolourings. The decoction of the leaves, or bark in wine, takes away scurf and dandruff, if used as a wash.

Modern uses: *The bark of the willow contains salicin from which aspirin is derived. Herbalists use the bark and leaves as an astringent tonic and as a preventive treatment against diseases that are apt to recur, such as malaria.*

The decoction, made by boiling 1 oz (28 g) of bark in 1 pt (568 ml) of water until the mixture measures 1 pt (568 ml) is given in doses of 1–2 fl oz (28–56 ml) for fevers, diarrhoea and dysentery. The powdered root can be taken in sweetened water in doses of one teaspoonful. An infusion of the leaves – 1 oz (28 g) to 1 pt (568 ml) of boiling water – is a useful digestive tonic.

Figure 3 Nicholas Culpeper the 17th-century herbalist.

The substance extracted in the recipe you have just read has no pharmacological effect by itself. The body converts it by hydrolysis and oxidation into the active chemical, *salicylic acid*. (The acid is named after the Latin name for willow: **salix**.)

Figure 4 Sixteenth-century herb gathering.

In **Activity WM2** you can produce some of the fever-curing chemical from a naturally occurring substance.

WM3 *Identifying the active chemical in willow bark*

In **Activity WM2** you saw how thin-layer chromatography could be used to show that the substance you obtained was possibly salicylic acid.

How can we find out the chemical structure of compounds like salicylic acid? One way is to use chemical reactions, and in this section you will learn how chemical tests reveal the presence of particular functional groups in salicylic acid.

Some —OH group chemistry

A knowledge of some relatively simple test-tube experiments can often be used effectively in the identification of unknown substances. For example, you may already know about the use of bromine solutions for detecting double bonds between carbon atoms in alkenes.

Three chemical tests are particularly helpful in providing clues about the structure of salicylic acid.

1. An aqueous solution of the compound is weakly acidic.
2. Salicylic acid reacts with alcohols (such as ethanol) to produce compounds called esters. Esters have strong odours, often of fruit or flowers.
3. A neutral solution of iron(III) chloride turns an intense pink colour when salicylic acid is added.

Tests 1 and 2 are characteristic of **carboxylic acids** (compounds containing the —COOH functional group); test 3 indicates the presence of a **phenol** group (an —OH group attached to a benzene ring).

Before you read any further, you need to find out about the chemistry of compounds containing these functional groups.

You can remind yourself about the structure of alcohols by reading **Chemical Ideas 13.2**.

Chemical Ideas 13.3 tells you about the structure of carboxylic acids and some compounds related to them.

In **Chemical Ideas 13.4** you can compare the behaviour of the —OH group in an alcohol, a phenol and a carboxylic acid.

Activity WM3 allows you to investigate the behaviour of —OH groups in alcohols, phenols and carboxylic acids.

WM4 *Instrumental analysis*

Although chemical tests provide evidence for the presence of carboxylic acid and phenol groups in salicylic acid, instrumental techniques are today's most efficient research tools. In this section you will learn about three frequently used instrumental techniques:

- mass spectrometry (m.s.)
- infrared (i.r.) spectroscopy
- nuclear magnetic resonance (n.m.r.) spectroscopy.

Making use of infrared spectroscopy

One of the very first things which would be done with any unidentified, new substance is to record its **infrared (i.r.) spectrum**. Figure 6 shows the i.r. spectrum of salicylic acid.

Chemical Ideas 6.4 tells you about infrared spectroscopy.

An i.r. spectrum measures the extent to which electromagnetic radiation in part of the i.r. region is transmitted through a sample of a substance. The frequency ranges which are absorbed provide important clues about the **functional groups** which are present. The functional groups absorb at similar frequencies in many different compounds so an absorption pattern provides a kind of fingerprint of the molecule.

The i.r. spectrum of salicylic acid shows clear evidence of the presence of the C=O and —OH groups.

Figure 5 An i.r. spectrophotometer: the sample is inserted in the chamber on the right-hand side of the machine; the spectrum can be viewed on the screen, or a trace can be drawn out by the chart recorder on the left.

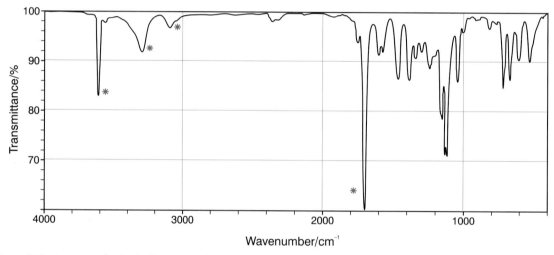

Figure 6 The i.r. spectrum of salicylic acid (in the gas phase).

ASSIGNMENT I

Examine the i.r. spectrum of salicylic acid shown in Figure 6. Compare the absorptions marked by an asterisk (*) with the characteristic absorption bands of the different functional groups listed in **Chemical Ideas 6.4**, and suggest which groupings could be responsible. (Information about i.r. absorptions is also listed in the **Data Sheets: Table 22**.)

Do these groupings correspond to what you know of the formula for this compound?

Evidence from n.m.r. spectroscopy

A second instrumental technique which would be applied to an unidentified compound is **nuclear magnetic resonance (n.m.r.) spectroscopy**.

This investigates the different chemical environments in which the nuclei of one particular element are situated.

Often this element is hydrogen. The nucleus of a hydrogen atom consists of just one proton, and the **proton n.m.r. spectrum** for salicylic acid is shown in Figure 7. The spectrum shows that salicylic acid contains

- one proton in a —COOH environment
- one proton in a phenolic —OH environment
- four protons attached to a benzene ring.

In this case, the n.m.r. spectrum is quite complicated because all six hydrogen atom nuclei are in different environments in the molecule, and five of them give signals which are close together. Although n.m.r. spectroscopy is of limited value in this case, it provides a powerful technique for determination of the structure of many organic compounds.

(a)

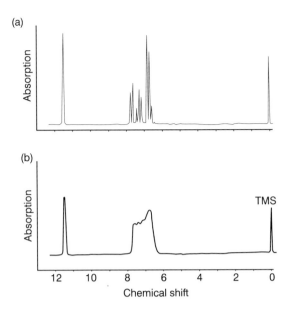

(b)

TMS

12 10 8 6 4 2 0

Chemical shift

Figure 7 The proton n.m.r. spectrum of salicylic acid (the signal labelled TMS is made by a reference compound called tetramethylsilane).

The n.m.r. spectrum in Figure 7(a) is a high-resolution spectrum: to explain this in full would be beyond the scope of this course. The spectrum in Figure 7(b) represents a low-resolution spectrum: information is lost in such a spectrum, but it is much simpler and the positions of the signals still tell us about the environments of the hydrogen atoms. You will learn more about n.m.r. spectroscopy in **Engineering Proteins**.

_____ **ASSIGNMENT 2** _____

Examine the structure of 2-hydroxybenzoic acid shown below. Explain why the following hydrogen atom nuclei are in different environments within the molecule:

a hydrogen atoms 1 and 6

b hydrogen atoms 3 and 4.

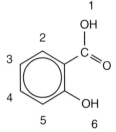

2-hydroxybenzoic acid

The mass spectrum of salicylic acid

A combination of i.r. and n.m.r. spectroscopy shows that salicylic acid has an —OH group and a —COOH group both attached to a benzene ring; in other words, a better name for salicylic acid is hydroxybenzoic acid.

However, there are three possible isomeric hydroxybenzoic acids: 2-hydroxybenzoic acid, 3-hydroxybenzoic acid and 4-hydroxybenzoic acid. A decision about which isomer salicylic acid is can be made by analysis of the **mass spectrum** of salicylic acid.

Figure 8 Mass spectrometers are now small enough to sit on a benchtop. This one is linked to a gas chromatograph which is used to separate the components of a mixture. The mass spectrum of each component is recorded.

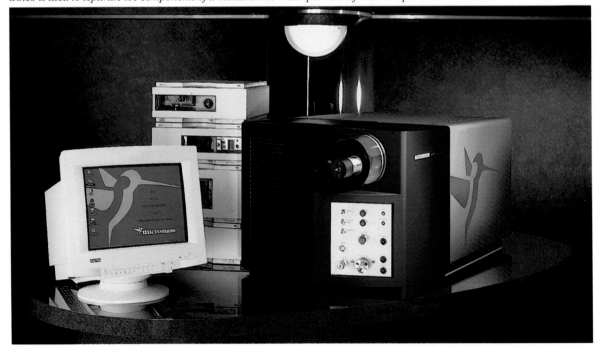

III

In **Chemical Ideas 2.1** you saw how information about the relative abundance of isotopes of an element can be obtained from a mass spectrum.

Chemical Ideas 6.5 tells you how mass spectrometry is used to find the structure of compounds.

A mass spectrum shows signals which correspond to positively charged ions formed from the parent compound, and fragment ions into which the parent compound has broken down.

Figure 9 shows the mass spectrum of salicylic acid. The signal at mass = 138 is from the parent ion, called the **molecular ion**. Modern machines, such as the one in Figure 8, give very accurate mass values for the signals in the spectrum. For salicylic acid, the high-resolution spectrum (Figure 9a) gives a mass of 138.0317 for the molecular ion. This confirms that the substance has an empirical formula of $C_7H_6O_3$.

The way in which a parent ion breaks down, its **fragmentation pattern**, is characteristic of that compound. In this case, comparison with a database of known mass spectra identifies salicylic acid as 2-hydroxybenzoic acid. Thinking about the way that the 3-hydroxybenzoic acid and the 4-hydroxybenzoic acid isomers would break down also leads to the conclusion that these isomers could not form some of the fragments observed in the mass spectrum of salicylic acid. For example, the signal at mass = 120 could only be formed from breakdown of 2-hydroxybenzoic acid.

Activity WM4 uses accurate M_r values, isotope peaks and a database to lead you to the formula of salicylic acid. It shows you how chemists can use fragmentation patterns to deduce or confirm a molecular structure.

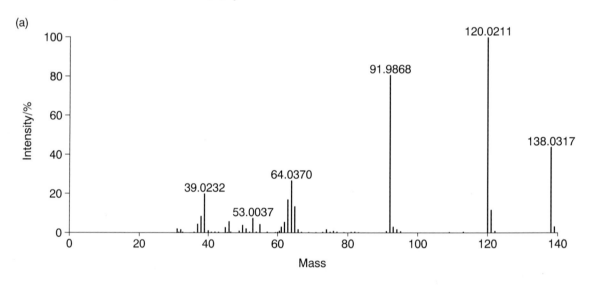

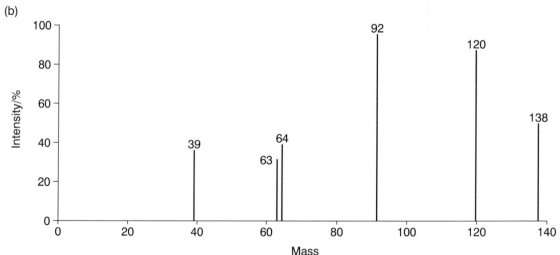

Figure 9 The mass spectrum of salicylic acid (a) high-resolution spectrum, (b) low-resolution spectrum. The relative heights are not the same as different experimental conditions have been used.

Drawing the evidence together

1. Chemical tests showed the presence of a phenolic —OH group and a carboxylic acid —COOH group.
2. Infrared spectroscopy showed that the —OH and —C=O groups were certainly present.
3. Nuclear magnetic resonance spectroscopy confirmed that there were hydrogen atoms in three types of environment: —COOH, —OH attached directly to a benzene ring and —H attached directly to a benzene ring.
4. Mass spectrometry showed that salicylic acid was the same compound as that stored in the database as 2-hydroxybenzoic acid. The structure of 2-hydroxybenzoic acid is

5. The mass spectrum fragmentation pattern showed that the structure could not be

or

WM5 The synthesis of salicylic acid and aspirin

Medicines which are 'natural products', ie those which come directly from plants, may be difficult to obtain when needed. The supply may be seasonal, may depend on weather conditions and may be liable to contamination. Collecting plants from their natural habitat is not environmentally acceptable.

Chemists, therefore, do not want to rely on willow trees as their source of 2-hydroxybenzoic acid. Once the chemical structure of the active compound in a plant is known, chemists can instead begin to search for ways of producing it artificially.

Simple inorganic substances, such as aluminium chloride, can be synthesised directly from their elements, but larger, more complex molecules cannot be made directly in this way. Instead, chemists search for a compound which is already known, and which has a similar structure to the required compound and can be modified.

At the end of the 19th century, the compound phenol was already well known in the pharmaceutical industry – it has germicidal properties. It was also readily available as a product from heating coal in gas-works. Its molecular structure differs from that of 2-hydroxybenzoic acid by only one functional group. The problem in synthesis is to introduce this extra group in the right position and without disrupting the rest of the molecule.

ASSIGNMENT 3

Compare the structural formulae of the starting and finishing compounds:

phenol

and

2-hydroxybenzoic acid

What extra atoms have to be added?

Refer to an organic chemistry textbook to find out the conditions needed to bring about this change.

In this particular case, by careful control of the conditions, carbon dioxide can be combined directly with phenol to give 2-hydroxybenzoic acid. This general method is known as the Kolbe synthesis (details can be found in many organic chemistry textbooks) and an industrial version of it was developed by the German chemist Felix Hoffmann. Thus synthetic 2-hydroxybenzoic acid of reliable purity became available and it was marketed by the chemical company Bayer.

Synthetic 2-hydroxybenzoic acid was widely used for curing fevers and suppressing pain, but reports began to accumulate of irritating effects on the mouth, gullet and stomach. Clearly the new wonder medicine had unpleasant side-effects. Chemists had a new problem – could they modify the structure to reduce the irritating effects, whilst still retaining the beneficial ones?

Figure 10 Meadowsweet (Spiraea ulmaria), from which salicylic acid was first extracted in 1835; aspirin got its name from 'a' for acetyl (an older word for ethanoyl) and 'spirin' for spirsaüre (the German word for salicylic acid).

Figure 11 Felix Hoffmann who first synthesised aspirin in a chemically pure and stable form in 1897.

Hoffmann prepared a range of compounds by making slight modifications to the structure of 2-hydroxybenzoic acid. His father was a sufferer from chronic rheumatism and Hoffmann tried out each of the new preparations on him to test the effects. This was a bit more primitive than modern testing of medicines. It is not recorded what Hoffmann senior thought of all this, but he survived long enough for his son to prepare, in 1898, a derivative which was as effective as 2-hydroxybenzoic acid and much less unpleasant to use.

The effective product was 2-ethanoylhydroxybenzoic acid (or acetylsalicylic acid). This is known as *aspirin*.

aspirin

The problem was that aspirin is not very soluble in water. It was at first available as a powder in sachets. Bayer then decided to pellet the powder and aspirin became the first medicine to be sold as tablets.

Aspirin belongs to a class of compounds known as **esters**, and Hoffmann used the process of **esterification** to produce aspirin. **Chemical Ideas 13.4** introduces you to the structure of esters. You will find out more about esters and esterification later in the course in **Designer Polymers**.

In **Activity WM5.1** you can convert 2-hydroxybenzoic acid into aspirin and in **Activity WM5.2** you will use infrared and mass spectra to identify some of the molecules used in the synthesis

WM6 *Delivering the product*

Protecting the discovery

To develop a new medicine costs an enormous amount of money. The price charged by the pharmaceutical company must be sufficient not only to cover the costs of production and marketing, but also to recover the development costs. If other companies could simply copy the medicine, they would be able to sell it at much lower prices. This is where **patents** become important.

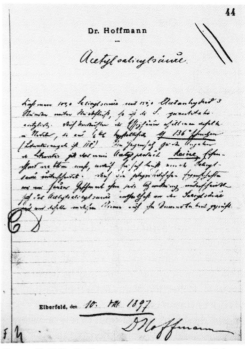

Figure 12 Felix Hoffmann's lab notes (top) and the trademark certificate (bottom) for aspirin, dated March 6, 1899.

When a pharmaceutical company discovers a new medicine, it takes out patents to protect the discovery. Patents only apply to one country, so several patents must be taken out to prevent companies in other countries manufacturing the medicine. When the medicine is approved, the pharmaceutical company markets it under a trade name or *brand name*. Patents only last for a specific amount of time, but while the patent is in force, no other company can manufacture the medicine in that country.

Pharmaceuticals are usually complex compounds with long and unwieldy chemical names. For convenience they are known by shorter, trivial names. These are called 'generic' names in the pharmaceutical business.

So most pharmaceuticals have three names:

- their chemical name
- their generic name
- their brand name.

An example is the compound 2-(4-(2-methylpropyl)phenyl)propanoic acid, known by the generic name *ibuprofen*, and marketed, for example, as Brufen (by Knoll) or simply supplied as Ibuprofen.

By the time a patent runs out, the company which discovered the medicine will hopefully have sold enough of it to cover its development costs. Afterwards, any company can produce and sell the medicine – usually under its own, new brand name. That's why there are so many ibuprofen tablets around.

In 1899, when the Bayer company in Germany first decided to market 2-ethanoylhydroxybenzoic acid, it was not a new compound, so the compound itself could not be protected by patents. However, the company patented the process by which it was made and also sought copyright for the trade name *Aspirin* in as many countries as possible.

The trade protection lasted until the First World War, when other countries were no longer able to obtain aspirin from Germany. American firms were restrained by the patent agreements from producing it and UK chemists were busy with the war effort, but an Australian pharmacist, George Nicholas, developed a process for producing a 'soluble' form of the medicine and marketed this under the new trade name *Aspro*. As part of war reparations, the German rights to the trade name Aspirin were given up in the UK and Commonwealth countries.

Figure 14 Early advertising for Aspirin in The Netherlands; the slogan on the vehicle means "Aspirin conquers every pain".

Because there are no development costs, the new brands can be made more cheaply than the original one. Currently, the Government encourages doctors to prescribe some medicines by their generic name. The pharmacist is then free to dispense the cheapest brand – which helps to keep health service costs down.

Figure 13 On March 6, 1999, 30 professional mountaineers draped the 120 metre high building which is Bayer's headquarters in fabric made of a woven polyester coated with PVC, using 32 zip fasteners. It was in celebration of the 100th anniversary of Bayer registering the trademark Aspirin®. The Guiness Book of Records *has this as the biggest aspirin box in the world!*

Figure 15 Aspirin was on board when the American spacecraft Apollo 11 landed on the Moon in 1969. The aspirin tablets produced each year would make a path to the Moon and back.

Different ways of buying aspirin

There are some 200 analgesic (pain-relieving) formulations worldwide which contain aspirin.

There are so many because:

- the medicine may come in different forms, eg solids, soluble substances and syrups
- other compounds may be present to help relieve other symptoms which occur along with the one being treated
- there may be other substances present to help the action of the principal compound
- for all these different formulations, and aspirin itself, there are many companies each producing their own brand-name equivalent.

Figure 16 Tablet production today: batches of 6–10 tonnes are processed at a time.

Manufacturers need to analyse samples from each batch of a medicine to ensure that it has been properly blended and each tablet contains exactly the stated amount of active ingredient.

More critically, because these medicines are so easily available and widely used, many households keep some permanently in stock. These are not always safely stored and are sometimes kept unlabelled. Each year many thousands of cases of accidental poisoning occur. Hospital analysts need techniques for establishing quickly which compounds – and how much – are present in tablets, so that the correct treatment can be given.

In **Activity WM6** you can perform an aspirin assay, in other words an experiment to find out the amount of aspirin in a medicine.

The safety of aspirin

People tend to think of aspirin as a safe medicine, because it is so familiar. But like all medicines, it is only safe if taken in the recommended dose. The lethal dose of aspirin is 30 g for an adult of average size. A typical aspirin tablet contains 0.3 g (300 mg) of aspirin, so 100 tablets could be a lethal dose. Unpleasant symptoms would be experienced with far fewer tablets than this. The recommended dose of aspirin is no more than 12 tablets a day, and it is not recommended at all for children under 12 years old.

Since 1998, aspirin has not been able to be sold in packets containing more than 16 tablets (each of 300 mg), unless you consult a pharmacist.

WM7 The miraculous medicine

The life cycle of modern medicines is often short because medical and pharmaceutical research offers better remedies all the time. However, there is a drug that is so cheap that it is generally bought by the patient rather than by an NHS prescription. The drug has such miraculous properties that over 1 trillion tablets have been consumed since it was first launched on the market. Every year about 50 billion tablets are consumed in the UK. The drug? Aspirin.

Aspirin is now is over 100 years old and new uses continue to emerge. Nearly 40% of the aspirin now sold is used to prevent heart disease by inhibiting blood clotting, and hence to reduce the incidence of problems such as strokes. About 25% is used to treat the symptoms of arthritis and another 15% to relieve headaches.

The discoveries of its potential in, for example, heart disease, came from observations by doctors on their patients. By careful analysis, Dr Laurence Craven, in the US, noticed that his male patients who took aspirin suffered fewer heart attacks. Similar work is being carried out to observe its effects on certain cancers, diabetes and Alzheimer's disease, amongst others.

Sir John Vane

Sir John Vane, of the Wellcome Research Laboratories in Beckenham, shared the Nobel Prize in Medicine in 1982 for his work on a group of hormones called *prostaglandins*.

Prostaglandins are formed in the body from unsaturated carboxylic acids. They are released from tissue when it suffers disease or stress in order to maintain its normal function, and so they act to defend the cells against sudden change.

Different prostaglandins help to control reactions in different parts of the body. For example, one type of prostaglandin is involved in the prevention of peptic ulcers, whereas another keeps the kidneys healthy. However, the presence of prostaglandins can cause inflammation, fever and pain and this can be relieved by aspirin. Sir John's work has helped to show how aspirin is able to act as a painkiller.

Another prostaglandin induces thickening of the blood. Aspirin greatly reduces the 'stickiness' of the platelets in the blood so that they do not clump together and form clots. It is now the most effective drug in the treatment of strokes. In fact, aspirin has an effect wherever prostaglandins are being produced.

Figure 18 Sir John Vane.

Sir John has written about his start in chemistry:

"At the age of 12, my parents gave me a chemistry set for Christmas and experimentation soon became a consuming passion in my life. At first, I was able to use a Bunsen burner attached to my mother's gas stove, but the use of the kitchen as a laboratory came to an abrupt end when a minor explosion involving hydrogen sulphide spattered the newly painted decor and changed the colour from blue to dirty green!

"Shortly afterwards, my father, who ran a small company making portable buildings, erected a wooden shed for me in the garden, fitted with bench, gas and water. This became my first real laboratory, and my chemical experimentation rapidly expanded into new fields."

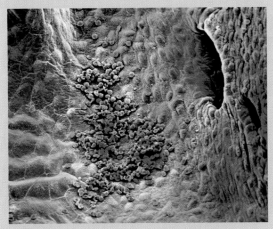

Figure 17 Platelets sticking together in the early stages of blood clot formation (magnification about × 550).

WM8 *Development and safety testing of medicines*

A great deal of time, money and effort is spent by the pharmaceutical industry in discovering and developing new medicines. The following account comes from *An A to Z of British Medicines Research*, published by the Association of the British Pharmaceutical Industry (ABPI).

The stages in the medicines research and development process

... terms such as 'discovery research', 'development' and 'phase I, II, or III clinical trials' are used [by the pharmaceutical industry]. It is important that the reader understands these terms, so as not to be misled into believing that progress is more advanced than it really is. The time-scale for developing a new medicine is surprisingly long and the average for all medicines is about 12 years (Figure 19). The many stages involved in developing a conventional medicine (Figure 20) are

described below. However, the procedures for biological products (cytokines, growth factors, gene therapy, etc) differ in a number of respects and are covered by their own requirements and regulations.

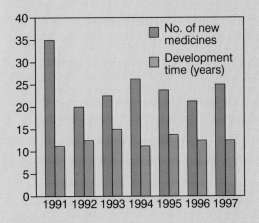

Figure 19 Number of new medicines marketed in the UK and the average development time from 1991 to 1997.

Figure 20 *The stages in the discovery and development of a new medicine.*

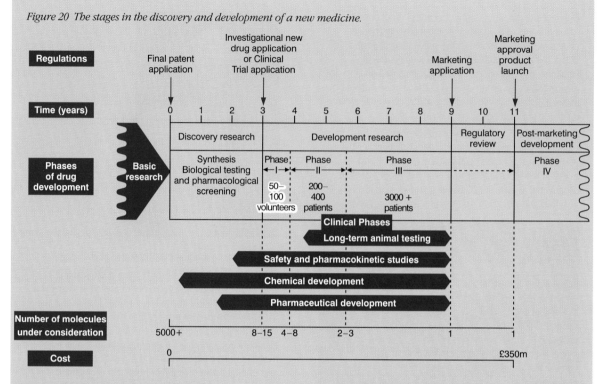

Discovery research

Discovery research relates to the activities of chemists, biologists and pharmacologists who extract, synthesise and test new molecules. For every new medicine that reaches the patient, many thousands of other molecules fall by the wayside. When a useful activity for a molecule has been identified, chemists optimise its structure by making many close variations (called *analogues*), to try and maximise the desired effects. Once this is done, the improved molecule enters the development stage.

Development research

Before a new potential medicine can be given to humans, much work has to be done to determine whether it is acceptably safe, whether it is sufficiently stable, and how it is likely to be absorbed and excreted by the body. It is also necessary to prepare a dosage form that suits specific medical needs, such as an injection, a capsule, a tablet, an aerosol, or a suppository. Once all this has been completed, a long and complex process of clinical studies begins. These are generally divided into three distinct phases.

Phase I trials

Phase I trials are the first time the new substance is administered to humans, usually in studies of healthy, informed volunteers conducted under the close supervision of a qualified doctor. The purpose is to determine if the new compound is tolerated and behaves in the way predicted by all the previous experimental investigations. Initial doses will be the lowest possible consistent with obtaining the required

information, but may gradually be raised to the expected therapeutic dose level. If the compounds under investigation are particularly powerful, as in cancer treatments, for example, it may be that people who actually have the condition will take part in these trials.

Once the data from volunteers are available, an application has to be made to the UK Medicines Control Agency for a certificate to conduct clinical trials. The information is reviewed by independent medical and scientific experts, who make their recommendation on whether further trials can start or whether more information is required. If a certificate is granted, a new medicine will then pass through two further phases of clinical trials before the company can seek a licence for its more widespread use.

Phase II trials

Phase II trials are the first in which the illness is actually treated. Different dose levels may be given to different patient groups to establish whether the compound is suitable for further study or should be abandoned. Patient numbers in these trials are usually small.

Phase III trials

Phase III trials only follow encouraging results in the Phase II study. As this stage, the new medicine is likely to be compared with a 'dummy' medication, called a *placebo*, and possibly with another medicine already used for the disease under investigation, to provide a reference standard. Patients are allocated randomly to one of the groups and, during the trials, neither the doctor nor the patient knows which preparation is being given. When the code is broken, a positive result would be indicated by an improvement in those patients who received the real medication compared to those on placebo. Phase III trials usually involve much larger patient groups, so that the results can be analysed statistically. If the medicine proves successful and well tolerated at this stage, the way is open for a product licence application to be made (see Figure 20) which includes all aspects of the data generated on the new medicine and runs to many volumes.

Later stages of development

Of 10 to 15 compounds reaching Phase I studies, only one is likely to survive through to licensing. Also, the time-scales for the above studies are very variable. Thus, if a new compound is an antibiotic for urinary tract infections, a positive result will be apparent in each patient within a few days as the infection is eradicated. However, for chronic diseases such as multiple sclerosis, AIDS, arthritis, or for some forms of cancer, the trial may last for more than a year in each patient and involve long-term follow-up to ensure that any observed positive effects are of lasting value.

Despite these complexities, the number of entirely new medicines reaching the British public has remained steady for the last five to six years (see Figure 19) after a fall at the end of the 1980s, though the time from discovery to marketing has slowly drifted upwards.

Development of aspirin

Note that the medicine you have been looking at closely in this unit, aspirin, did not go through all the safety testing described here, because it was developed long before all these safety procedures were established. But aspirin has been known for so long and so widely that its use is accepted by most people – provided the recommended dosage isn't exceeded.

Which medicine to develop?

You have seen how expensive it is to develop a medicine from discovery right through to marketing (see Figure 20). A firm will have wasted millions of pounds if it produces a medicine which does not meet a perceived need by the medical professionals or the public. Very careful decisions have to be made at several stages to ensure that the medicine is both safe and commercially viable.

Activity WM8 gives you a chance to experience decision-making like that involved when a pharmaceutical company considers which medicine should be developed.

Figure 21 Pharmaceutical research has to be carried out in conditions of safety and cleanliness.

WM9 *Summary*

In this unit you have learned about some of the chemistry associated with the pharmaceutical industry using one familiar, important medicine – aspirin – as an example. You saw how the analytical techniques of mass spectrometry and infrared spectroscopy allow us to identify the compound which is responsible for its pharmacological activity.

You were introduced briefly to another analytical technique, called nuclear magnetic resonance spectroscopy, which is also useful in determining the structure of molecules. You will find out more about this technique in **Engineering Proteins**.

A study of aspirin led you to find out about a new series of organic compounds called esters. You saw how esters can be formed by the reaction of an alcohol or a phenol with a carboxylic acid. This led to a more general study of compounds that contain the –OH group.

This unit also introduced you to some new experimental techniques. You learned how to heat volatile liquids under reflux and how to use thin-layer chromatography to identify the components of a mixture.

Knowledge of the chemical reactions of organic functional groups gives us the power to construct molecules of compounds which are essential for our wellbeing from readily available starting materials. You experienced something of the scale, complexity and costs involved in the production of a medicine for mass use.

Activity WM9 will help you to check your notes on this unit.

DESIGNER POLYMERS

Why a unit on DESIGNER POLYMERS?

This unit picks up the thread from **The Polymer Revolution** and continues the story of polymers and polymerisation. This time the main theme is the development of condensation polymers and the ways that chemists have found to produce polymers with specific properties to meet particular needs.

An understanding of the relationship between the properties of polymers and their structure and bonding is the key to designing new polymers. This will require you to consider in more detail the factors that affect the properties of polymers and will allow you to revise earlier work on intermolecular forces.

The story begins with the invention of nylons and polyesters, and then moves on to more recent condensation polymers, such as Kevlar and PEEK, and polymers designed to degrade under certain conditions.

During the unit you will revisit and extend some of the organic chemistry you have met in earlier units – the chemistry of alcohols, carboxylic acids and esters – as well as learning about two new series of compounds – amines and amides.

Disposing of polymers in landfill sites is no longer acceptable on environmental and economic grounds. There are three possible solutions, involving recycling, incineration or the use of degradable polymers. The advantages and disadvantages of each are discussed.

Overview of chemical principles

In this unit you will learn more about …

ideas introduced in earlier units in this course

- polymers and polymerisation (**The Polymer Revolution**)
- alcohols (**Developing Fuels** and **What's in a Medicine?**)
- carboxylic acids and esters (**What's in a Medicine?**)
- intermolecular forces (**The Polymer Revolution**)
- the relationship between structure and bonding, and properties (**The Polymer Revolution**)

… as well as learning new ideas about

- condensation polymerisation
- reactions of amines and amides
- use of acyl chlorides to make esters and amides
- effect of temperature changes on polymers
- disposal of polymers.

DP1 *Designer polymers*

Producing a perfect copy

If you had lived in medieval times and visited a monastery, you might have found that one of the most important rooms was the one in which monks made copies of religious documents, painstakingly copying out the original by hand.

Fortunately, today we don't have to copy out by hand everything we want duplicates of: we can use a photocopier. Unlike some of the medieval bibles, the results are not works of art, but they are very good copies and it is almost impossible to tell the difference between the copy and the original. You probably accept this as a normal, everyday fact. But do you know how the photocopier works?

At the heart of the photocopying process is a photoreceptor surface which is sensitive to light. In many machines, the photoreceptor is in the form of a metal drum coated with a very thin layer (about 10^{-5} m thick) of a special polymer called *polyvinyl carbazole*.

The polymer is made from a compound, vinyl carbazole, with the chemical structure

vinyl carbazole

A remarkable property of polyvinyl carbazole is that it exhibits **photoconductivity** – it conducts electricity much better when light shines on it than when it is in the dark.

Figure 1 Part of a decorative initial letter in the Lindisfarne Gospels, written by Northumbrian monks in the late 7th or 8th century; religious documents like this were laboriously copied by hand.

In a photocopier, the surface of the drum is first given an electrostatic charge so that the polymer becomes charged. The interesting behaviour starts when light shines on the drum. If it is in the light, the polymer becomes conducting and so its charge disappears. But if it is in the dark, the polymer retains most of its charge.

A piece of paper with a picture on it is laid on a glass plate and illuminated, so that an *image* of the picture is projected onto the drum.

Can you imagine what happens? The polymer stays charged where there were dark areas in the picture, but loses its charge where the picture was white. So an electrically charged image of the picture is held on the drum.

You can follow the stages involved in producing a photocopy by looking at Figure 2.

Something new

There are many other 'designer polymers' with useful properties. Take *poly(1,1-difluoroethene)*: it is **piezoelectric**, that is it generates electricity when it is bent or twisted. If a sheet of poly(1,1-difluoroethene) is made to shake and wobble by a sound wave, the electrical signals can be processed and the sound wave detected. In reverse, electrical signals sent to a piece of this polymer can make it wobble and act as a type of loudspeaker.

In this unit you will learn about a polymer – *Kevlar* – which is about as strong as steel but five times lighter, and *PEEK* – a heat-resistant polymer. You will also learn about a polyester designed to hydrolyse slowly in the body to release non-toxic substances, so that it can be used for surgical stitches, and other polymers designed to degrade under certain specific conditions.

Figure 2 How a photocopier works.

1 Charge drum

The surface of the drum is given a positive electrical charge.

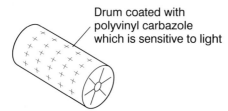

Drum coated with polyvinyl carbazole which is sensitive to light

2 Project image onto drum

The document to be photocopied is placed face-down on the document glass. When the start button is pressed, the document is exposed to a light which scans across its surface, and an image of the document is projected through a system of lenses and mirrors onto the surface of the drum.

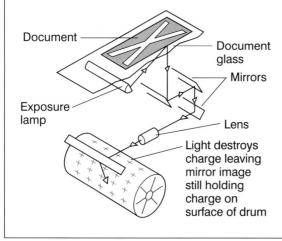

Document

Document glass

Mirrors

Exposure lamp

Lens

Light destroys charge leaving mirror image still holding charge on surface of drum

3 Develop image

Negatively charged ink powder or toner is dusted over the drum. It sticks to the areas where there is a charge, so an exact copy of the document is held on the drum.

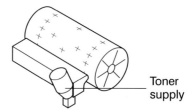

Toner supply

4 Transfer image to paper

A sheet of ordinary paper is now passed across the surface of the drum. A charge below the paper attracts toner from the drum to the paper.

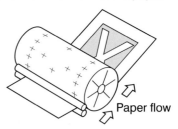

Paper flow

5 Fuse image to paper

Hot rollers fuse the image to the paper to produce an exact photocopy of the original document.

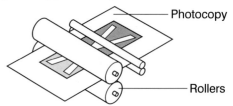

Photocopy

Rollers

All these polymers have been designed with a particular application in mind. Making them poses many chemical and technological problems. Chemists have to develop ways of making the required quantities of starting compounds at a reasonable cost, and have to combine them together in the right way to produce a usable material; engineers have to understand how to use and process the polymers.

As you found out in **The Polymer Revolution**, some early polymers were discovered almost by accident. But as our knowledge and understanding of the polymerisation process and the properties of polymers have grown, the creation of new polymers has become more systematic.

Many of the polymers in this unit are **condensation polymers**. Most of these are A–B type polymers (see p. 94) formed from a condensation reaction between two different monomers. The story starts with the invention of an important condensation polymer, nylon.

DP2 *The invention of nylon*

Wallace Carothers joined the US chemical company DuPont in 1928. He led a team investigating the production of polymers that might be used as fibres. This was around the time when scientists were beginning to understand more about the structure of polymers, so Carothers had a scientific basis for his work.

Before you read this section, it will be helpful for you to revise your previous work on carboxylic acids in **Chemical Ideas 13.3** and **13.4**.

You will also need to know about **amines** and **amides**, two important families of nitrogen-containing organic compounds. You can read about these in **Chemical Ideas 13.8**.

It was already known that wool and silk have protein structures, and are polymers involving the **peptide linkage** —CONH—. Chemists had also begun to discover that many natural fibres are composed of molecules which are very long and narrow – like the fibres themselves. Figure 3 shows part of a protein chain in a silk fibre.

Carothers did not make his discoveries by accident; he set about systematically trying to create new polymers. In one series of experiments he decided to try to make synthetic polymers in which the polymer molecules were built up in a similar way to the protein chains in silk and wool. Instead of using amino acids

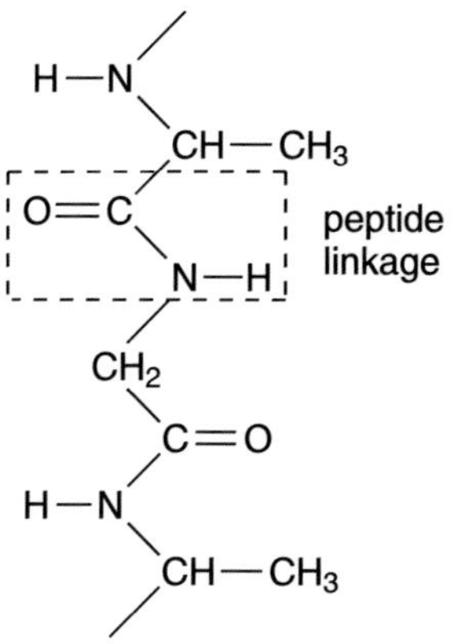

Figure 3 Part of a protein chain in silk.

(the starting materials for proteins), Carothers began with **amines** and **carboxylic acids**.

Amines are organic compounds which contain the —NH₂ functional group. When an —NH₂ group reacts with the —COOH group in a carboxylic acid, an **amide** group —CONH— is formed. In the process, a molecule of water is eliminated. Reactions of this type are known as **condensation reactions**.

carboxylic acid   amine   amide group

+ H_2O

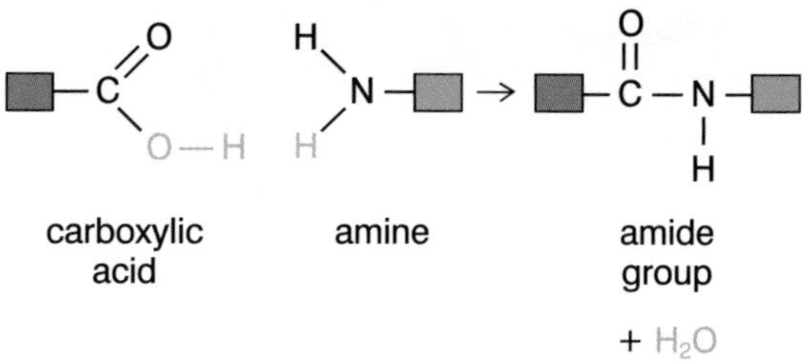

Carothers used *di*amines and *di*carboxylic acids which contained reactive groups in *two* places in their molecules, so they could link together to form a chain. In this way he was able to make polymers in which monomer units were linked together by amide groups. The process is called **condensation polymerisation** because the individual steps are condensation reactions (Figure 4).

Examples of a diamine and a dicarboxylic acid that can be made to polymerise in this way are

$H_2NCH_2CH_2CH_2CH_2CH_2CH_2NH_2$ 1,6-diaminohexane
$HOOCCH_2CH_2CH_2CH_2COOH$ hexanedioic acid

Because the group linking the monomer groups together is an amide group, these polymers are called **polyamides**. More usually, though, they are known as **nylons**.

Figure 4 Formation of a polyamide.

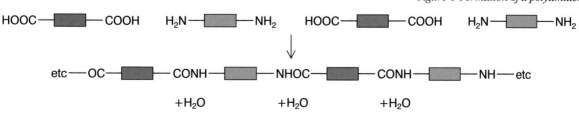

The industrial preparation of nylon from a diamine and a dicarboxylic acid is quite slow. It is easier to demonstrate the process in the laboratory if an **acyl chloride** derivative of the acid is used: 1,6-diaminohexane and decanedioyl dichloride react readily and the equation is

$$n\text{H}_2\text{N(CH}_2)_6\text{NH}_2 \quad + \quad n\text{ClCO(CH}_2)_8\text{COCl} \quad \rightarrow$$
$$\text{1,6-diaminohexane} \qquad \text{decanedioyl dichloride}$$

$$-(\text{NH(CH}_2)_6-\text{NH}-\text{CO}-(\text{CH}_2)_8\text{CO})_n- \quad + \quad 2n\text{HCl}$$

Figure 5 Advertising nylon stocking in Los Angeles when the polymer was first being used, 1939–40.

The —CONH— group is also found in proteins where it joins amino acids together. You can see this in the structure of the silk protein chain in Figure 3. In this case the secondary amide group is given the special name **peptide group**.

Figure 6 In the Second World War (1939–45), the production of nylon for stockings took second place to its use for parachutes.

Wallace Carothers

When the American Chemical Society celebrated its 75th birthday in 1998, it asked its members to vote on the most distinguished American chemist of the twentieth century.

The four who came top were Linus Pauling, Robert Woodward, Glen Seaborg – all Nobel Prize winners – and Wallace Carothers. The first three spent their careers in universities. Carothers was different. His career was largely spent with DuPont, the giant US chemical company. Between 1928 and 1937 he and his colleagues at DuPont created much of the foundation of modern polymer chemistry.

Sadly, Wallace Carothers was not to see the development of his invention of nylon. He had been troubled with periods of mental depression since his youth. Despite his success with nylon, and other inventions such as neoprene (the first commercially successful rubber), he felt that he had not accomplished much and had run out of new ideas.

His unhappiness was compounded by the death of his favourite sister, and on 29th April 1937 he checked into a Philadelphia hotel room and died after drinking a cocktail of lemon juice laced with potassium cyanide.

He was 41. His daughter Jane was born 7 months later.

Figure 7 Wallace Carothers holding Louisa, the daughter of Julian Hill, who carried out the original experiments that led to nylon.

Carothers discovered nylon in the spring of 1935 and by 1938 the first product using nylon, 'Dr West's Miracle Toothbrush', appeared. Nylon stockings were seen for the first time in 1939 (Figure 5). Most of the nylon produced at this time was used in place of silk for parachute material (Figure 6), so nylon stockings did not become generally available in Britain until the end of the Second World War.

Naming nylons

A nylon is named according to the number of carbon atoms in the monomers. If two monomers are used, then the first digit indicates the number of carbon atoms in the diamine and the second digit indicates the number of carbon atoms in the acid. So nylon-6,6 is made from 1,6-diaminohexane and hexanedioic acid. Nylon-6,10 is made from 1,6-diaminohexane and decanedioic acid. This is illustrated below:

1,6-diamino*hexane* + *hex*anedioic acid → nylon-6,6

1,6-diamino*hexane* + *dec*anedioic acid → nylon-6,10

It is also possible to make nylon from a single monomer containing an amine group at one end and an acid group at the other, eg nylon-6 is —(NH—$(CH_2)_5$—CO)$_n$— and is made from molecules of $H_2N(CH_2)_5COOH$.

ASSIGNMENT I

a Name the nylons which contain the following repeating units:

  i —HN—$(CH_2)_6$—NHCO—$(CH_2)_6$—CO—

  ii —HN—$(CH_2)_9$—NHCO—$(CH_2)_7$—CO—

  iii —HN—$(CH_2)_4$—NHCO—$(CH_2)_2$—CO—

b Write out the repeating units and give the names for the polymers formed from the following molecules:

  i $HOOC(CH_2)_5NH_2$

  ii $H_2N(CH_2)_5NH_2$ and $HOOC(CH_2)_5COOH$.

c Nylon-5,10 can be formed by the reaction of a diamine with a diacyl dichloride.

  i Write down the structures of the two monomers.

  ii What small molecule is lost when the monomers react?

  iii Draw the structure of the repeating unit in nylon-5,10.

If you did not make a nylon in an earlier chemistry or science course, **Activity DP2.1** allows you to make some nylon-6,10.

In **Activity DP2.2** you can take some nylon molecules apart again.

Hard choices; right decisions

The development of polyamides, with their potential to replace silk as well as to open up entirely new markets, required some hard choices. Many different types of nylon had been prepared. Wallace Carothers favoured nylon-5,10, because the material was easy to process and would be appropriate for detailed studies of the relationships between polymer structures and properties. A principal drawback would have been the high cost of the decanedioic acid needed and this could have limited the widespread use of the polymer.

Elmer Bolton, the director of chemical research at DuPont, held a different view. He believed that a better choice would be nylon-6,6. Its physical properties would allow more applications than nylon-5,10. In addition, the starting materials (1,6-diaminohexane and hexanedioic acid), which have six carbon atoms each, could both be made from readily available and cheap chemicals, for example, benzene. Even so, a synthesis of the diamine from benzene had first to be developed, as did techniques to process the polymer, which has a higher metling point than nylon-5,10.

Despite these initial difficulties, Bolton foresaw greater commercial possibilities for nylon-6,6. The success of nylon depended on the combination of Carother's pioneering research and Bolton's brilliant ideas about its development. This coming together of basic science and applied technology must occur in industry if the product is to be successful.

Nylon machine parts

You know that nylon is widely used as a fibre, but it is also a very important **engineering plastic** – a material which can be used in place of a metal in things like machine parts.

Its usefulness arises from its excellent combination of strength, toughness, rigidity and abrasion resistance, as well as its chemical unreactivity in many environments. As an engineering material it is far superior to poly(ethene) or poly(propene). The polymer chains in nylon need only be about half as long as hdpe chains to show the same strength. The more powerful intermolecular forces which act between nylon chains are the source of this increased strength.

In nylon there is hydrogen bonding between adjacent polymer chains. In poly(ethene) and poly(propene) there are only the weaker instantaneous dipole–induced dipole attractive forces between adjacent chains.

At this point it will be helpful to look again at **Chemical Ideas 5.4** to check your understanding of the intermolecular forces responsible for nylon's strength.

In and out of fashion

Towards the end of the 1970s, people's tastes were moving away from nylon clothes and back towards the look and softer feel of natural fibres.

One of the problems with nylon fibres is that they are **hydrophobic** – they repel water. The nylon fabrics produced did not absorb moisture, and did not allow water vapour to escape through the weave. This made them rather sweaty and uncomfortable to wear.

Chemical companies were facing a big downturn in the demand for their nylon, bringing major financial losses, but the high cost of developing a completely new polymer on the scale necessary to replace nylon was too high. To make things worse, the machines which had been specifically designed for making nylon fabrics could not be used for cotton or other natural fibres.

ICI's answer to this problem was to redevelop their nylon so that it bore a much closer resemblance to natural fibres. The first steps were to slim down the thickness of the nylon filament to the equivalent of cotton, and to add a delustrant to the fibre to reduce the shiny appearance. The major breakthrough came when they developed a process for changing the shape and texture of the nylon yarn. They were able to create a large number of loops along the nylon filaments by blowing bundles of them apart with high-pressure air.

The new fibre was called *Tactel*. When the fibre is woven, the loops give the material a softness and texture similar to cotton.

Further refinement of the Tactel family of yarns has led to fabrics which are waterproof but which 'breathe'. For example, very fine yarns have been developed which allow water *vapour* to escape through the weave but do not allow *liquid* water in. Material made from these yarns is ideal for lightweight raincoats and ski-wear.

The solution to the problem of falling demand for nylon was not chemical – it did not involve the creation of a new polymer to suit the new fashion – but *technological*. It involved discovering new ways of handling the existing polymer to produce materials of the type people wanted.

Figure 8 Tactel fabrics 'breathe' but remain waterproof.

DP3 Polyesters: from clothes to bottles

The development of **polyesters** follows a similar story to that of nylon. A brilliant idea of UK chemist Rex Whinfield, in the 1940s, led to an incredible journey, developing materials undreamt of by their originators.

Whinfield knew that Carothers had tried to produce polyesters, but felt that Carothers had not used the most likely acid to give the necessary properties for the polymer. Whinfield used 1,4-benzenedicarboxylic acid (terephthalic acid) and reacted it with ethane-1,2-diol. The structure of the polyester formed is

$$\left[\!\!\begin{array}{c} \overset{O}{\underset{\|}{C}}\!\!-\!\!\bigcirc\!\!-\!\!\overset{O}{\underset{\|}{C}}\!\!-\!\!O\!\!-\!\!CH_2\!\!-\!\!CH_2\!\!-\!\!O \end{array}\!\!\right]_n$$

The 'old' name for this polyester is polyethylene terephthalate, often known as PET.

You can read about the formation and reactions of esters in **Chemical Ideas 13.5**.

The polyester is produced as small granules, which are melted and squeezed through fine holes. The resulting filaments are spun to form a fibre (Figure 9). The fibre, known as Terylene and Dacron, is widely used in clothes, such as suits, shirts and skirts, and for filling anoraks and duvets, as it gives good heat insulation. The polyester can also be made into sheets for audio tapes and X-ray films.

Figure 9 The birth of a continuous fibre.

A newer use is for packaging. The granules of the polyester are heated to about 240 °C and further polymerisation takes place – a process known as *curing*. When the polymer is then stretched and moulded, the molecules are orientated in three dimensions. The plastic has great strength and is impermeable to gases. No wonder PET is widely used for bottling carbonated drinks (Figure 10).

Figure 10 Granules of PET which are heated and moulded into bottles.

The versatility of the polymer arises because the molecules can be aligned in one, two or three dimensions (Figure 11).

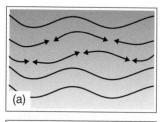

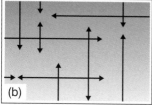

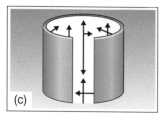

Figure 11 (a) Yarn: the molecules are mostly in one dimension (Terylene and Dacron). (b) Film: the molecules are in two dimensions. (c) For bottles: the polyester molecules are aligned in three dimensions.

Figure 12 From clothes to bottles.

ASSIGNMENT 2

a Draw out full structural formulae for the two monomers used to make the polyester, PET.

b What types of intermolecular attractions will exist between the polymer chains in PET?

Disappearing in the body

Over the last 20 years, threads made from a special polyester have been used by surgeons to stitch together the sides of wounds. What is special about the thread is that it disappears as the wound heals.

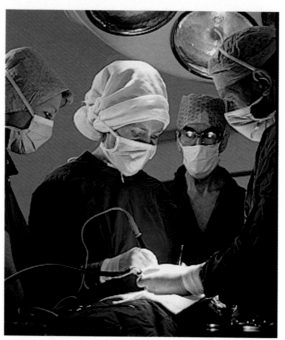

Figure 13 Surgeons using polyester thread, which is hydrolysed slowly in the body to harmless compounds after the wound has healed.

The polyester is made from monomers that contain both a hydroxyl group (–OH) and a carboxylic acid group (–COOH).

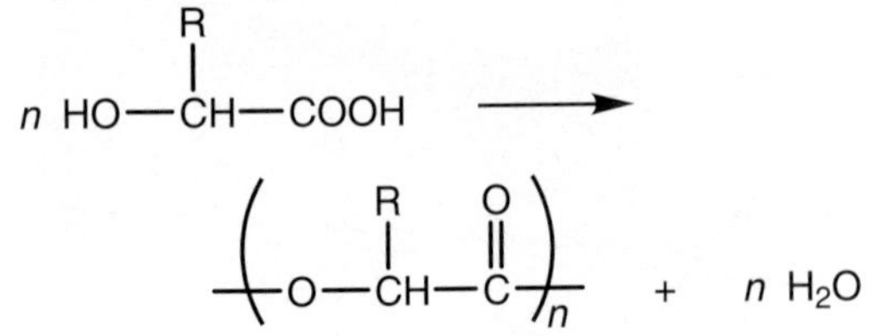

$$n \; HO\!-\!\underset{\underset{\displaystyle R}{|}}{C}H\!-\!COOH \longrightarrow$$

$$\left(\!\!\!\begin{array}{c} \\ O\!-\!\underset{\underset{\displaystyle R}{|}}{C}H\!-\!\underset{\overset{\displaystyle O}{\|}}{C} \end{array}\!\!\!\right)_{\!\!n} + \; n \; H_2O$$

The monomer used may be lactic acid ($R = CH_3$) or glycolic acid ($R = H$) (or a mixture of the two).

The polymers form strong threads, but water in the body slowly hydrolyses the ester bonds. The hydrolysis products (lactic acid or glycolic acid) are non-toxic. Indeed, lactic acid is a normal breakdown product of glucose in the body.

ASSIGNMENT 3

In **The Polymer Revolution** you met a dissolving polymer, poly(ethenol), which is used to make laundry bags in hospitals. It can be used to make soluble threads for surgical stitches.

Polyester thread made from lactic acid is also used for stitches.

Compare the way the two polymers operate and explain why the polyester is more suited to use inside the body.

Figure 14 Tablets have coatings that dissolve slowly. For example, aspirin tablets contain fine crystals of aspirin coated in a polymer (ethyl cellulose). This allows the aspirin to pass through the stomach and into the gut, where it is released.

Another use for the degradable polyesters is the controlled delivery of a medicine. The medicine is dispersed throughout a tablet of the polymer, which is then implanted in a suitable part of the body. The medicine is released at a rate determined by the rate of hydrolysis of the polyester.

The rate of hydrolysis of the polymer is very important. It is affected by the relative molecular mass of the polymer and its crystallinity. By tuning and balancing these features in polyesters made from lactic acid and glycolic acid, a precise rate of hydrolysis can be achieved.

DP4 *Kevlar*

The first aramids

After the invention of nylon, chemists began to make sense of the relationship between a polymer's structure and its properties. They were able to predict strengths for particular structures, and research was directed at inventing a 'super fibre'. In the early 1960s, DuPont were looking for a fibre with the 'heat resistance of asbestos and the stiffness of glass'.

The aromatic polyamides seemed promising candidates: the planar aromatic rings should result in rigid polymer chains and, because the ratio of carbon to hydrogen is high, they require relatively large concentrations of oxygen before they burn.

The first polymeric aromatic amide – an **aramid** – was made from 3-aminobenzoic acid. The polymer could be made into fibres and was fire-resistant, but it was not particularly strong. The zig-zag nature of the chains prevented the molecules from aligning themselves properly.

A polymer was needed which had straighter chains, and which could be made from readily available and reasonably cheap starting materials. An example is shown in Figure 15.

This substance turned out to have all the right properties except one. The problem was its insolubility, which made it precipitate out of solution before long polymer chains had been able to grow.

The only suitable solvent seemed to be concentrated sulphuric acid. The company engineers were not impressed! However, the remarkable properties of the 'super fibre' were enough to encourage investment in a plant which uses concentrated sulphuric acid as a solvent. The polymeric material produced was called *Kevlar*.

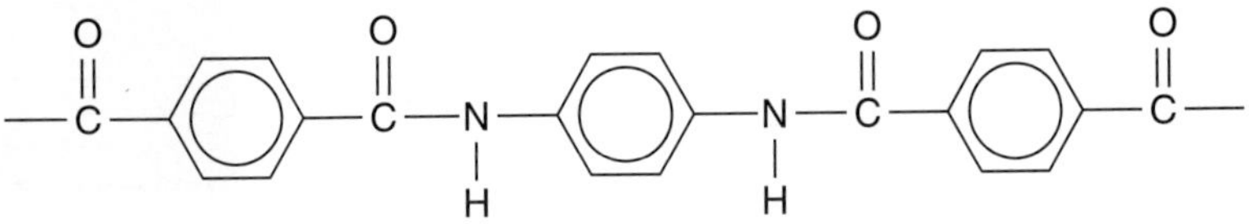

Figure 15 The molecular structure of Kevlar.

ASSIGNMENT 4

a Draw out a small section of the structure of the polymer which would be made from 3-aminobenzoic acid.

b Kevlar is made from two monomers, a diamine and a dicarboxylic acid. Look carefully at Figure 15 and then draw out the structural formulae of the two monomers used.

c The intermolecular forces in Kevlar are disrupted by concentrated sulphuric acid. That is why it dissolves. How do you think the forces are affected?

d Kevlar fibres are produced by squirting the solution in concentrated sulphuric acid into water. Suggest why the polymer precipitates out when the solution is diluted in this way.

Why is Kevlar so strong?

Kevlar is a fibre which is fire-resistant, extremely strong and flexible. It also has a low density because it is made from light atoms: carbon, hydrogen, oxygen and nitrogen. Weight for weight, Kevlar is around five times stronger than steel! One of its early uses was to replace steel cords in car tyres: the Kevlar tyres are lighter and last longer than steel-reinforced tyres.

Kevlar is strong because of the way the rigid, linear molecules are packed together. The chains line up parallel to one another, held together by hydrogen bonds. This leads to sheets of molecules. The sheets then stack together regularly around the fibre axis to give an almost perfectly ordered structure. This is illustrated in Figure 16.

The important thing to realise about Kevlar is that it adopts this crystalline structure because of the way the polymer is processed to produce the fibre. And this is a consequence of the work put in by the DuPont team.

Developing a market for Kevlar

A full-scale commercial plant for the production of Kevlar required an investment of $400 million. It was therefore essential that there would be a market for the product. Although Kevlar has some remarkable properties it could not simply replace existing materials. Detailed market development had to take place alongside technical development. In other words, new uses had to be found for the new polymer.

The flat molecules are held together by hydrogen bonds to form a sheet

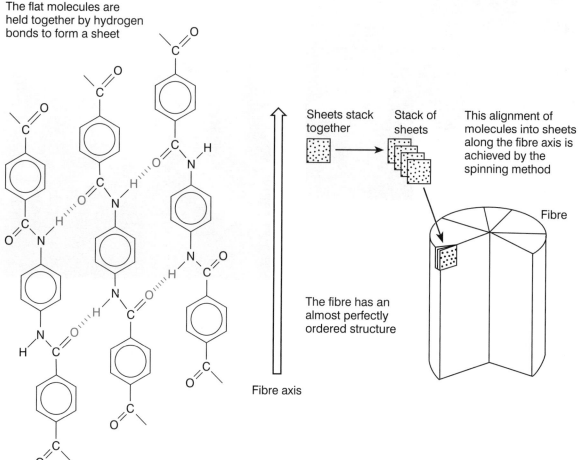

Sheets stack together

Stack of sheets

This alignment of molecules into sheets along the fibre axis is achieved by the spinning method

Fibre

The fibre has an almost perfectly ordered structure

Fibre axis

Figure 16 An illustration of the crystalline structure of Kevlar.

Many uses have now been developed in addition to replacing steel in tyres. Kevlar ropes have 20 times the strength of steel ropes of the same weight. They last longer too. A stiffer form of Kevlar is used in aircraft wings, where its strength combined with its low density is important. And it's ideal for making bullet-proof vests and jackets for fencers.

You can compare models and structures for Kevlar and a nylon in **Activity DP4.**

Figure 17 Kevlar is also used to strengthen the rigging of ocean-going yachts and tennis racquet frames.

DP5 *Taking temperature into account*

PEEK

The story of PEEK begins in the early 1960s when John Rose, a chemist at University College London, moved to the Plastics Division of ICI. He was put in charge of a team which had a brief to develop new polymers.

Rose decided to investigate high temperature materials. He knew that these would need to have high melting points and also be resistant to oxidation. For these reasons, he decided that his new polymers would have to be based on aromatic compounds.

The team tried to join aromatic units together in as many ways as possible. They had to develop new types of reactions and solve many other problems along the way. They also had to look for monomers which were reasonably cheap to make.

Out of all this came PEEK. Its structure is shown in Figure 18.

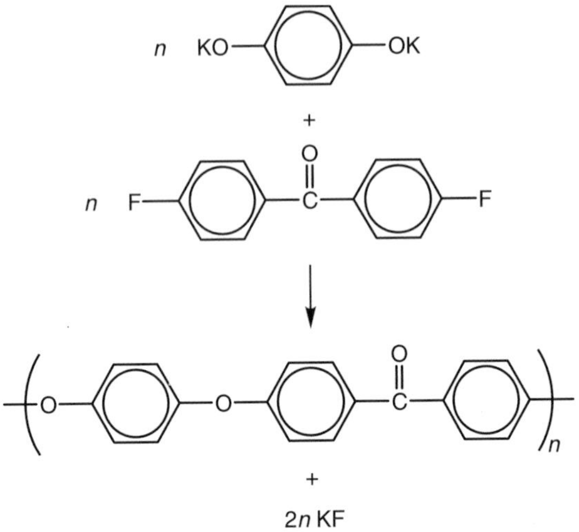

$$n \quad KO - \bigcirc - OK$$
$$+$$
$$n \quad F - \bigcirc - \overset{\overset{\displaystyle O}{\|}}{C} - \bigcirc - F$$
$$\downarrow$$
$$\left[-O - \bigcirc - O - \bigcirc - \overset{\overset{\displaystyle O}{\|}}{C} - \bigcirc - \right]_n$$
$$+$$
$$2n \text{ KF}$$

Figure 18 The equation for the formation of poly(ether-ether-ketone), PEEK.

Although it is expensive, the very wide range of applications of PEEK made it worth developing.

Figure 19 PEEK is a polymer which can be used to make precision articles, for example cogs, which can withstand very high temperatures. Polymers have been used for some time in the passenger compartments of cars and airliners, but PEEK and related compounds can be used inside the engines where it is really hot.

You can find out more about the way polymers are affected by heat in **Chemical Ideas 5.5**.

Activity DP5 looks at the effect of temperature on an everyday substance – bubble gum.

A clever idea

Although aeroplane crashes are infrequent, they are catastrophic when they occur. Even if you survive the impact, you are still in danger from fire. To help reduce the risk, seats, overhead lockers and wall panels can be made from PHA (polyhydroxyamide). When heated strongly, PHA loses water and rearranges to form another polymer, PBO, which is fire-resistant. No smoke is produced – very important in an aeroplane.

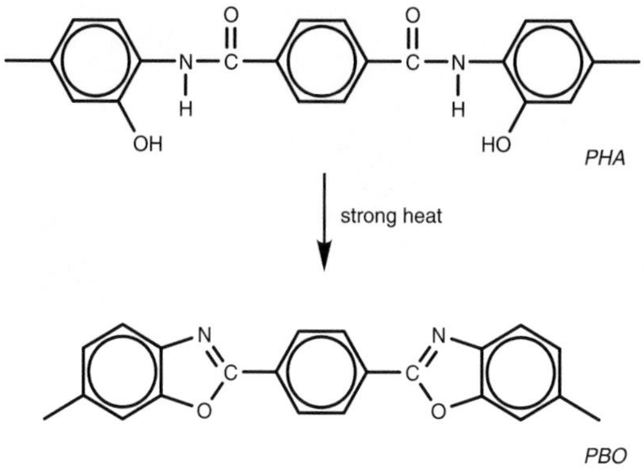

PHA

strong heat

PBO

So why not make the seats and lockers from PBO in the first place? Because it is very difficult to fabricate – whereas PHA is easy.

Figure 20 There are stringent international regulations to ensure passengers are as safe as possible in the event of fire. The use of PHA is one way chemists are contributing to the solution to this problem.

Mixing it

In this course, you have by no means come across all the polymers which are available to manufacturers. But suppose you have a particular application in mind.

Even a full list would not contain enough substances to allow you to be sure of finding one with just the properties you were seeking. You can bet that at some point you would say, "What we need is something like X but which behaves a bit like Y."

These days, it is far less expensive to modify existing polymer materials than to develop new ones. So, often, well-known polymers are combined to produce new materials which show some of the properties of the individual components. For example, sheets of different polymers can be stacked together to form *laminates*, or polymers can be mixed to produce *composites*.

It is also possible to mix things at a molecular level and to make **polymer alloys**, in which polymers are mixed when molten to give a new material with the desired properties, or to make **copolymers**, or to add **plasticisers**.

You can read more about copolymers and plasticisers in **Chemical Ideas 5.5**.

DP6 *Poly(ethene) by design*

In **The Polymer Revolution**, you met two forms of poly(ethene) that were the results of accidental discoveries. One is the low density variety, ldpe, produced using high pressures. The other, the high density variety, hdpe, does not need high pressures, but needs instead an organometallic compound as catalyst.

Both of the forms have distinctive properties which lend themselves to different applications. The demand for low density poly(ethene) was growing, but, considering building new plants to withstand high pressures was a daunting prospect for companies because of the very high capital costs.

The problem was solved by using a Ziegler–Natta catalyst, the route to hdpe, but with one significant change, the feedstock. Instead of using pure ethene, small amounts of another alkene, hex-1-ene, are added and small side-chains are produced along the polymer chain (Figure 21).

The resulting polymer molecules do not pack together as regularly as they do in hdpe, where the chains are linear with few branches. The new polymer is known as **linear low density poly(ethene), lldpe**. It is lower melting than hdpe and is more flexible. By using different alkenes and varying the amount added, chemists are able to change the structure, and hence the properties, of the poly(ethene) produced. The polymers are widely used and are in great demand.

$$\mathrm{CH_2}-\underset{\underset{C_4H_9}{|}}{\mathrm{CH}}-\mathrm{CH_2}-\mathrm{CH_2}-\mathrm{CH_2}-\mathrm{CH_2}-\mathrm{CH_2}-\underset{\underset{C_4H_9}{|}}{\mathrm{CH}}-\mathrm{CH_2}-$$

Figure 21 A section of an lldpe chain.

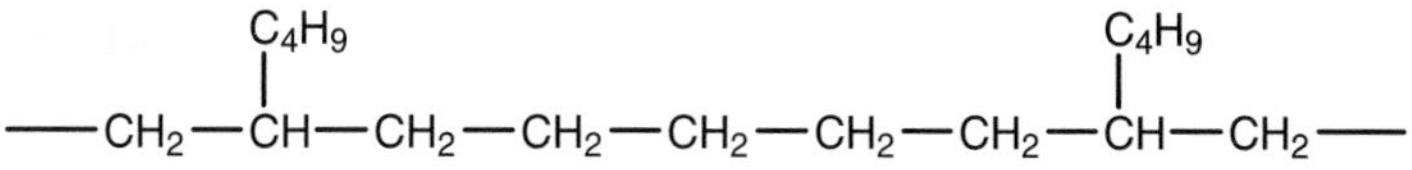

_____ **ASSIGNMENT 5** _____

a Draw out the skeletal formula for the section of a lldpe chain shown.

b Explain why

   i lldpe has a lower density than hdpe

   ii lldpe has a higher tensile strength than ldpe.

Figure 22 An extrusion machine producing poly(ethene) sheet tubing. Small granules of poly(ethene) are fed into the extrusion machine and softened by being heated. An extrusion machine can produce sheets, hollow pipes and guttering.

DP7 *Throwing it away ... or not?*

Is there a problem?

In Europe, over 3000 million tonnes of waste are produced each year. While plastics only account for about 0.6% by mass, this is still a vast amount and most of it is household waste. In the UK, about 11% of household waste is plastic and this amounts to over 2 million tonnes per year.

Much of our waste is disposed of by dumping in landfill sites. Sites are getting harder to find and plastics are notorious in not being readily decomposed when buried. Waste disposal this way is becoming more expensive and there is a search for other means. Recycling sounds attractive, degradable plastics even more so.

Figure 23 Much of our plastic waste still goes into landfills.

Recycling plastics

The obvious answer is to **recycle** the polymers, for most of them are thermoplastics and can be reworked without decomposition. However, there are many problems in trying to reuse domestic waste. For one thing, the plastics have to be sorted (Figure 24) and generally this is a very expensive process. A better source of plastics for recycling is the waste from factories where plastic articles are being fabricated.

Figure 24 Plastics being sorted manually for recycling.

Figure 25 These articles have been produced from fizzy drinks bottles made from PET.

Another approach is to recycle plastics chemically by converting them back to their monomers and repolymerising. This is only practicable where there is enough high-quality single material waste available. Some polyesters and polyamides are being recycled in this way.

A third, perhaps more fruitful, possibility is to crack the polymer and break it into smaller molecules. These small molecules can then be used as feedstocks in the chemical industry (Figure 27). Small trial cracking plants are being built which can be used on a relatively small scale to prepare feedstock for existing chemical processes.

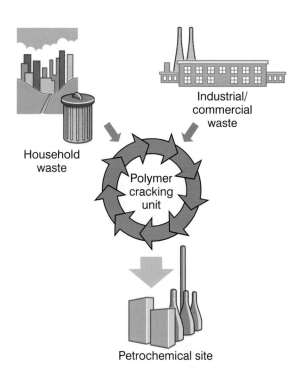

Figure 27 Increasingly, polymer waste will be recycled to produce feedstock (hydrocarbons and synthesis gas) for the chemical industry.

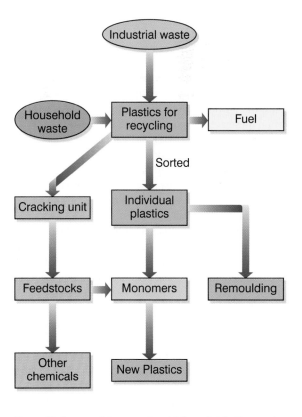

Figure 28 Summarising ways in which waste plastics can be used to save energy and oil.

Figure 26 The sails of this yacht are made from recycled poly(ethene).

Is burning an option?

Simply burning plastics prior to disposing of the residual ash is becoming unacceptable in terms of the environment. It is also a waste of valuable energy. But **incinerators** can be effective and a great effort is being made to develop furnaces which trap harmful emissions, such as toxic gases and smoke, and then use the energy to generate heat and power.

Degradable plastics

Most plastics are not degradable because decomposer organisms (bacteria and fungi) do not have the enzymes needed to break them down. There are three important categories of degradable plastic:

- **biopolymers** – made by living organisms and can be broken down by bacteria

- **synthetic biodegradable plastics** – which are broken down by bacteria

- **photodegradable plastics** – which break down in sunlight.

Biopolymers

Poly(hydroxybutanoate), PHB, is a natural polyester made by certain bacteria.

PHB

When nutrient glucose is in short supply, the bacteria break down the PHB, eating the polymer to survive. Thus the polymer can be made into plastic articles and when these are no longer needed, the plastic can be broken down by the bacteria which helped to make it.

The plastic has superior properties to other plastics such as poly(propene) and polystyrene, but it is costly, about 10 times the price of these conventional polymers. This has prevented the material from being successful commercially.

However, the story does not finish here. PHB is a member of a series of related compounds known as the polyhydroxyalkanoates (sometimes called PHAs – but don't confuse with the PHA, polyhydroxamide, on p. 131), which all have similar structures. PHB has a methyl side group on its polymer chain; other members of the series have different side groups.

New work is in progress in which plants, such as oil seed crops and cotton, are being genetically modified to produce a range of PHAs. Think of that! Plastics not being made from oil but by photosynthesis of carbon dioxide and water. Then, when they are not wanted, being readily decomposed back to carbon dioxide and water. A truly renewable plastic.

Figure 29 Protecting plants from frost is one use for biodegradable plastics, seen in close up above (× 400). Starch granules, coloured orange, are embedded in the plastic.

Synthetic biodegradable plastics

Some plastic bags are made of poly(ethene) which has starch granules encapsulated in it (Figure 29). The starch is digested by microorganisms in the soil when the plastic bag is buried. The bag then breaks up into very small pieces of leftover poly(ethene) which have a large surface area and biodegrade more quickly.

Photodegradable plastics

Carbonyl groups ($C{=}O$) absorb radiation in the wavelength range 270 nm–360 nm (about 10^{15} Hz frequency). This corresponds to light in the near ultraviolet region of the spectrum. These groups can be incorporated into polymer chains to act as energy traps. The trapped energy causes fission of bonds in the neighbourhood of the carbonyl group, and the polymer chain breaks down into short fragments which can then biodegrade.

For example, carbonyl groups have been

incorporated in the process of manufacture into poly(ethene) chains.

DP8 *Summary*

In this unit you have seen how chemists have learnt to use their knowledge of polymers and polymerisation to make new polymeric materials with specific properties. They can do this by modifying existing polymers (either chemically or physically) or by designing new polymers.

You began by reading about the discovery of nylon, a polymer designed to imitate the structure of silk. You learned about condensation reactions and the properties and reactions of amines and amides. The story then moved on to polyesters, another type of condensation polymer. This allowed you to revisit and extend earlier work on alcohols, carboxylic acids and esters. You met two modern condensation polymers, Kevlar and PEEK, which led you to think about the effect of temperature on polymers.

You saw how chemists are now able to manipulate the structure of poly(ethene), an addition polymer, so it too can be 'designed' to suit particular needs.

As in **The Polymer Revolution**, the link between structure and bonding, and properties is central to the unit and this allowed you to apply your knowledge and understanding of intermolecular forces in new situations.

Finally, you considered some environmental and economic issues connected with the disposal and recycling of plastics.

Activity DP8 allows you to check that your notes cover the important chemistry contained in this unit.

ENGINEERING PROTEINS

Why a unit on ENGINEERING PROTEINS?

This unit introduces you to proteins – one of the most versatile classes of chemicals found in all living things. Through the example of insulin – a hormone crucially important for life – you learn about the structures of proteins, their synthesis in cells, and how chemists are able to modify their structure and function: the technique of protein engineering. The unit ends by looking at enzymes – another vitally important class of proteins.

To understand how proteins are formed, you need to know about amino acids and you need to revisit earlier work on carboxylic acids, amines and amides. In that sense the unit carries forward your study of organic chemistry. You will also meet another analytical technique for structure determination, nuclear magnetic resonance spectroscopy.

But explanations of the structures and behaviour of proteins and other macromolecules found in cells are based on physical chemical ideas: in particular, molecular shape and intermolecular bonding, and chemical equilibrium. Studying the behaviour of enzymes also provides an opportunity to extend your chemical knowledge, this time about the rates of chemical reactions.

Overview of chemical principles

In this unit you will learn more about …

ideas introduced in earlier units in this course
- amines and amides (**Designer Polymers**)
- carboxylic acids (**What's in a Medicine?** and **Designer Polymers**)
- condensation reactions (**Designer Polymers**)
- stereoisomerism (**The Polymer Revolution**)
- intermolecular forces (**The Polymer Revolution** and **Designer Polymers**)
- spectroscopic techniques (**What's in a Medicine?**)
- chemical equilibrium (**The Atmosphere**)
- the rates of chemical reactions (**The Atmosphere**)
- catalysts (**Developing Fuels** and **The Atmosphere**)

… as well as learning new ideas about
- amino acids
- optical isomerism
- nuclear magnetic resonance spectroscopy
- protein structure
- protein biosynthesis, DNA and RNA
- the shapes of molecules and molecular recognition
- genetic engineering
- the effect of changes in concentration on a chemical equilibrium, and the use of K_c
- enzymes
- rate equations, reaction orders and half-lives.

ENGINEERING PROTEINS

EP1 *Christopher's story*

Christopher is 11 years old. Just over 2 years ago he suddenly became very ill; he was taken into hospital where tests showed that he had developed *diabetes*.

In the UK, there are about 20 000 people under the age of 20 who have been diagnosed with diabetes.

For some reason, possibly as a result of an earlier viral infection, Christopher's pancreas had stopped producing a *hormone:* **insulin**. Hormones are chemicals which regulate the rate at which we carry out our body activities. Insulin controls the uptake of glucose and some other sugars by our cells. Without insulin, Christopher's cells could not absorb glucose, and its concentration in his blood was building up. Untreated, he would have died within a few weeks.

Today, Christopher looks like any other 11-year-old boy. He goes to school; he rides his bike; he swims. In fact, he's very fit because he takes regular exercise and eats a healthy diet. However, regular insulin injections are essential.

Christopher's morning regime goes like this. The first thing he does is to test his blood-glucose level. He pricks his finger, puts a drop of blood onto a strip of test paper and places the test paper into a machine. After a few seconds he reads the digital display and writes the figure into a notebook. This isn't part of the treatment; it's part of the monitoring process which helps him to keep his blood-glucose concentration within normal limits. The reading tells him how successfully he is balancing his insulin dose, his food

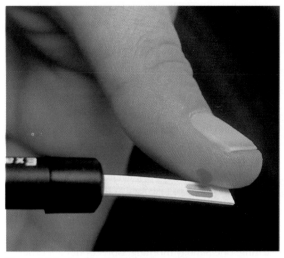

Figure 2 Monitoring blood-glucose concentration.

intake and his exercise. It is a way of enabling Christopher to understand more about his diabetes control and manage his diabetes more effectively.

After the blood test, Christopher goes to the fridge and takes out two bottles of insulin, mixes his dose and injects the insulin into a fold of skin below his ribs; from there it will act more quickly than from, say, an arm. Then, however hungry he might be, he must wait half an hour before he eats his breakfast.

Figure 1 Christopher needs regular injections of insulin so that the cells in his body can absorb glucose.

Figure 3 Christopher's typical breakfast.

Christopher's breakfast looks like anyone else's. Although it is important for him to limit sugar by cutting out sugary drinks for everyday use, he does not have to avoid sugar completely. He can still enjoy family meals, including some cakes, biscuits and puddings. However, he must base his meals and snacks mainly on starchy carbohydrate foods (such as bread, potatoes, rice, pasta and cereals) and eat plenty of fruit and vegetables.

Perhaps the most unusual thing about Christopher's day is the frequency and regularity of his meals, particularly as he must keep his weight at a reasonable level. He eats breakfast, lunch and dinner, and those meals are interspersed with snacks. Whilst he is growing, this suits him fine: he says he's always hungry anyway! It may not be quite so easy when he's grown up, and it's certainly difficult to keep eating when he's ill. During illness he still needs to eat regular amounts of carbohydrate foods or take regular drinks, including some sugary drinks, to maintain his blood-glucose level.

Half an hour before his evening meal, he gives himself another injection of insulin. Life can get exciting: one Sunday morning, Christopher's mother dropped one of the insulin bottles on the kitchen floor and it broke. She phoned the local pharmacist who came out specially to open the shop. As long as Christopher gets his twice-daily dose of insulin, he remains fit and well. Without insulin, he would die. The pharmacist took the situation very seriously.

More about insulin

Insulin is a hormone, and a **protein**. Many hormones are proteins (for example human growth hormone) but there are others, like the sex hormones, which are not.

Figure 5 Protein in the form of collagen is an important part of animal skins – leather is made of animal skin, so consists partly of protein.

The name *protein* (meaning 'first thing') was coined by Berzelius in 1838. He had little idea what proteins were, but he recognised their importance because they were so widespread in living things.

Proteins are big molecules; they are *natural* polymers with relative molecular masses up to about 100 000. They play a key role in almost every structure and activity of a living organism (Figure 4). That's why they are regarded as among the most important constituents of our bodies.

Insulin was first isolated in 1921 by two young scientists, F G Banting and C H Best, working in

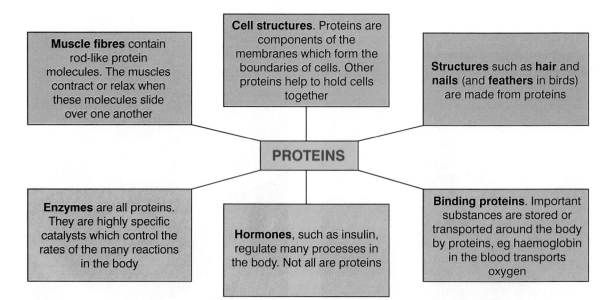

Muscle fibres contain rod-like protein molecules. The muscles contract or relax when these molecules slide over one another

Cell structures. Proteins are components of the membranes which form the boundaries of cells. Other proteins help to hold cells together

Structures such as **hair** and **nails** (and **feathers** in birds) are made from proteins

PROTEINS

Enzymes are all proteins. They are highly specific catalysts which control the rates of the many reactions in the body

Hormones, such as insulin, regulate many processes in the body. Not all are proteins

Binding proteins. Important substances are stored or transported around the body by proteins, eg haemoglobin in the blood transports oxygen

*Figure 4 Proteins perform other functions in our bodies in addition to acting as hormones. **Fibrous proteins** are the major structural materials. **Globular proteins** are involved in maintenance and regulation of processes, and include hormones and enzymes.*

Toronto. Only 1 year later it became available for the treatment of diabetes in the form of a preparation extracted from the pancreas of pigs or cows. Before then, diabetes was usually fatal, and diabetics spent their numbered days living miserable lives on starvation diets which were designed to contain as little sugar or starch as possible. Vegetables which had been boiled three or four times and agar jelly (a protein) were standard fare.

Injections of insulin have transformed the situation. Today we don't even need to use pig or beef insulin – we have synthetic human insulin available from the biotechnology industry. But the treatment still doesn't mimic natural insulin production as in non-diabetic people.

Splitting the six-pack

What happens to insulin concentrations when we eat? Figure 6 shows how the concentration of insulin in the blood of a non-diabetic person would be expected to behave after a meal.

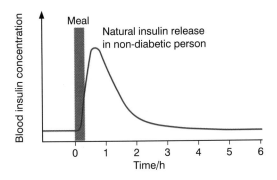

Figure 6 Insulin concentrations in the blood of a non-diabetic person. The maximum concentration may be about 1 × 10⁻¹⁰ mol dm⁻³.

Notice four points about the graph in Figure 6:

- a concentration as low as 1×10^{-10} mol dm⁻³ of insulin in the blood can have profound physiological effects
- there is a low-level, 'background' concentration of insulin in the body, even when food is not being digested
- insulin release follows rapidly after eating
- the insulin concentration peaks soon after eating and then falls off, eventually reaching the 'background' level.

We don't just make insulin when we eat: we make it all the time, and it is stored in special cells in the pancreas. It is not stored in the form of individual molecules, but as **hexamers** – six molecules clustered together because of interactions between their surfaces.

When the concentration of glucose in the blood rises, insulin hexamers are released into the bloodstream. As the solution of insulin becomes more dilute the hexamers burst apart to form dimers and then, on further dilution, monomers. The monomers are quickly carried to where they are needed.

Injecting insulin

Why does Christopher have to wait for half an hour between his insulin injection and his meal? Figure 7 shows how his blood-insulin concentration might vary. (Also shown, for easy comparison, is the curve from Figure 6 for a non-diabetic person.)

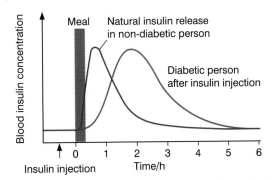

Figure 7 Insulin concentrations in the blood of a diabetic person following an insulin injection (injections also contain a slow-acting form of insulin which produces an effect for up to 12 hours; after that the insulin level falls to zero).

The solution Christopher injects into the muscle tissue just under his skin contains insulin as hexamers. However, the hexamers are too big to pass immediately through the capillary membranes into his blood. They have to spread out at the injection site and become diluted before they can break up into dimers and then monomers. The monomers can pass into his bloodstream (Figures 8 and 9). This takes time. That's why Christopher has to inject his insulin half an hour before he starts to eat.

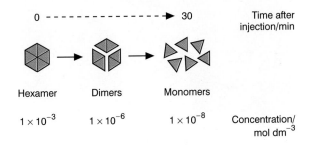

Figure 8 Insulin hexamers break up to form dimers and eventually monomers, as the solution becomes more dilute.

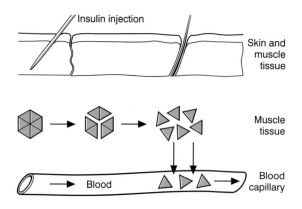

Figure 9 Injecting insulin. Although the insulin hexamers are too large to pass through the blood capillary membrane, the monomers are able to do so.

Christopher uses two bottles to prepare his dose because his injection is a mixture of two types of insulin. One has been prepared so that it breaks down easily and produces an effect close to the rapid response of normally produced insulin; the other breaks down slowly and gives him a low-level insulin concentration which is not permanent but which does last for up to 12 hours.

Christopher's injection is an improvement on the early preparations but, even so, the two curves in Figure 7 are clearly different. Because the insulin concentrations are different, diabetic people will still have unusual blood-glucose concentrations, and this can lead to complications and illness later in life. That's why Christopher also has to be so careful about his diet.

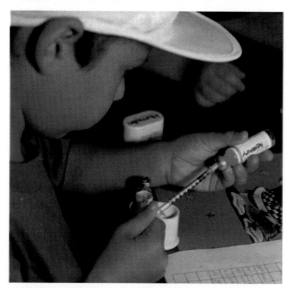

Figure 10 Christopher injects insulin hexamers into the tissues just below his skin; it takes about half an hour for the hexamers to diffuse apart, split into monomers and enter his bloodstream.

But the big difference, of course, is that normal insulin release switches on and off every time we eat – our insulin levels fit in with our food intake. It's the other way round for Christopher: his meals have to be planned around his levels of injected insulin.

To help Christopher further and make his injections mimic natural insulin release more closely, the half-hour time delay during which hexamers begin to break down needs to be eliminated. That would mean injecting insulin monomers which can get straight into the bloodstream: Christopher could then have injections along with his meal. Life would be closer to normal – and so would his blood-glucose concentrations. The problem is that insulin only exists as a monomer at very low concentrations. Thus it would mean injecting a much larger volume of liquid.

During the 1980s, scientists developed methods of **protein engineering** that can be used to modify the structure of a protein and so alter its properties. One possibility would be to modify the structure of insulin so that the monomers are stable and do not combine to form hexamers.

However, before the modified monomeric insulin could be designed, scientists needed a thorough understanding of the protein's structure and function. In particular, scientists needed to know

- the composition and structure of insulin
- the shape of the monomer and the regions of the molecule which interact to form dimers and hexamers
- the types of intermolecular interaction which occur in the hexamer
- how to modify the monomer proteins to make them more stable and so prevent them sticking together.

In this unit you will find out about one way of making and modifying proteins using **genetic engineering** – but first you need to know more about proteins and how they are made in the body.

EP2 *Protein building*

Amino acids: the building blocks of proteins

Figure 11 illustrates the composition of a molecule of human insulin.

The abbreviations in circles in Figure 11 represent the α-**amino acids** which have combined to form insulin. There are two short chains of these **amino acid residues** – the parts of the original amino acid molecules which are joined together to form the protein. All the proteins in the living world are made from just 20 α-amino acids, all of which have the same general structure as shown in Figure 12.

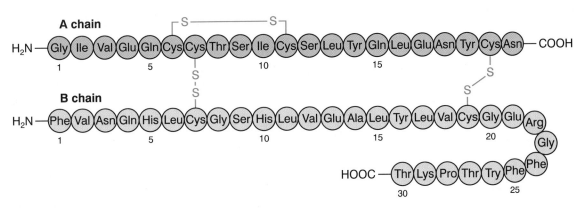

Figure 11 Human insulin. The two chains are held together by –S–S– links.

This is a good time to remind yourself of the chemistry of some important functional groups. You can read about carboxylic acids and their reactions in **Chemical Ideas 13.3** and **13.4**, and about amines and amides in **Chemical Ideas 13.8**.

You can learn more about amino acids in **Chemical Ideas 13.9**.

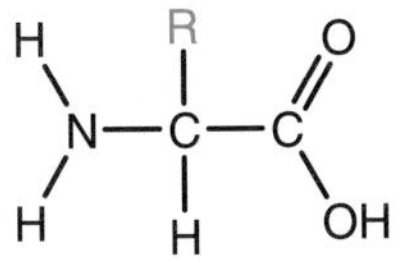

The amino group (–NH₂) and the carboxylic acid group (–COOH) are both attached to the same carbon atom: the α-carbon. That is why they are called α-amino acids.

Figure 12 General structure of α-amino acids. An α-carbon atom is one adjacent to a functional group. The R group is different in different amino acids (see Table 1).

Each amino acid has a different side chain, labelled R. Table 1 shows the structures of these R groups and the names of the amino acids, together with their abbreviated symbols.

In all of us, even adults who have stopped growing, proteins need to be continuously replaced. This is obvious when your hair and your nails grow after they have been cut. It is also important that hormones and some enzymes (such as the digestive enzymes) are made when needed, then destroyed once they have done their job, so that they do not go on producing their effects after the need for them has passed.

We replace our proteins from the food we eat, but we do not need to eat proteins which are identical to the ones being replaced. We do not need to eat human hair or finger nails to grow our own! In fact, we don't even have to eat animal protein; we can supply our needs by eating plants.

This is because our bodies break proteins down into their constituent 'building blocks' – amino acids. Eight

of these are usually classified as *essential amino acids* – our bodies cannot synthesise them: we must take them in as food. The other 12 can be made from carbohydrate and other amino acids in the body.

The amino acids are then reassembled to make our own collection of proteins. For each of us, this collection is unique: most of your proteins will be identical to those found in other humans, but some will be different. Human proteins are also different from the proteins of other animals. However, in some cases very similar proteins are found not just throughout the animal kingdom but in plants and microorganisms as well: for example, the enzymes which oxidise glucose in cells' metabolism.

All these millions of proteins are built up from the same small number of amino acids. What makes each protein different is the *order* in which the amino acids are joined to one another. This is called the **primary structure** of the protein. You have already seen the primary structure of human insulin in Figure 11.

Figure 13 The sweetener aspartame *is made from a combination of two amino acids; it is not a sugar and so it can be used by diabetics.*

Amino acid	Abbreviation	R group	Amino acid	Abbreviation	R group
glycine	Gly	—H	cysteine	Cys	—CH_2—SH
alanine	Ala	—CH_3	methionine	Met	—CH_2—CH_2—S—CH_3
valine	Val	—CH(CH_3)(CH_3)	aspartic acid	Asp	—CH_2—C(=O)—OH
leucine	Leu	—CH_2—CH(CH_3)(CH_3)	glutamic acid	Glu	—CH_2—CH_2—C(=O)—OH
isoleucine	Ile	—CH(CH_2—CH_3)(CH_3)	asparagine	Asn	—CH_2—C(=O)—NH_2
phenylalanine	Phe	—CH_2— ⟨benzene ring⟩	glutamine	Gln	—CH_2—CH_2—C(=O)—NH_2
proline	Pro	HN—⟨ring⟩—COOH	tyrosine	Tyr	—CH_2— ⟨benzene ring⟩—OH
tryptophan	Trp	—CH_2— ⟨indole ring⟩	histidine	His	—CH_2— ⟨imidazole ring⟩
serine	Ser	—CH_2—OH	lysine	Lys	—CH_2—CH_2—CH_2—CH_2—NH_2
threonine	Thr	—CH(CH_3)(OH)	arginine	Arg	—CH_2—CH_2—CH_2—NH—C(=NH)—NH_2

Table 1 The 20 amino acids which make up proteins (for clarity, the whole amino acid has been drawn out in the case of proline).

Activity EP2.1 allows you to investigate the reactions of amines and amino acids

In **Activity EP2.2** you can break down aspartame and identify its amino acid components.

ASSIGNMENT 1

The amino acids in Table 1 can be grouped according to key features of their R groups: for example, whether their side chains are polar and will interact strongly with water, or whether they are non-polar and will disrupt water's intermolecular bonding. Use the abbreviated symbols for the amino acids to answer the questions below.

a List four amino acids in each case which you think have
  i non-polar side chains
  ii polar side chains
  iii ionisable groups on their side chain.

b Look at the structures of leucine and isoleucine. Explain why isoleucine is so named.

c List one amino acid in each case in which the R group contains
  i a primary alcohol group
  ii a secondary alcohol group
  iii a phenol group
  iv a carboxylic acid group.

Chemists use spectroscopic techniques to give information about the structures of molecules. In **What's in a Medicine?**, you met infrared spectroscopy and mass spectrometry. Another important analytical technique involves **nuclear magnetic resonance (n.m.r.) spectroscopy**.

In **Activity EP2.3** you can use n.m.r. spectroscopy to distinguish between some amines and amides.

You can read more about n.m.r. spectroscopy in **Chemical Ideas 6.6**.

Making peptides

When amino acids combine to form a protein such as insulin, the carboxylic acid group on one amino acid joins on to the amino group on the next, and a molecule of water is lost from between them. This type of process is called a **condensation**, and the –CONH– (secondary amide) group which links the amino acid residues in the product is called a **peptide link**. Two amino acids joined in this way make a *dipeptide*.

two amino acids
(R and R' represent different side chains)
produce a dipeptide

Representing amino acid sequences

The dipeptide obtained by condensing the carboxylic acid group of glycine with the amino group of alanine has the structure

This dipeptide would be abbreviated to **Gly Ala**. The convention of reading peptide groups in the direction with the free NH_2 group on the left is very important if the amino acid sequence is to be read unambiguously.

ASSIGNMENT 2

a Draw the structure of the dipeptide Ala Gly.

b In the tripeptide Ser Gly Ala, which amino acid has an unreacted
  i NH_2 group?
  ii COOH group?

You met amides and condensation reactions in **Designer Polymers** storyline, Section **DP2**.

Chemists cannot make amino acids react together directly. They have to make the –COOH group more reactive: for example, by turning it into an acyl chloride (also known as an acid chloride). To make proteins, they also need to take into account another property of amino acids.

Amino acid molecules are not flat: they have a three-dimensional shape based upon the tetrahedral arrangement of the four bonds around the α-carbon atom. When we look at them properly in this way, all the amino acids in Table 1, with the exception of glycine, exist in *two* isomeric forms known as **D** or **L optical isomers**. Proteins are built up from only the L isomers.

Chemical Ideas 3.6 tells you about optical isomerism.

You can remind yourself about the shapes of molecules by reading **Chemical Ideas 3.3** and can revise earlier work on stereoisomerism and geometric isomerism in **Chemical Ideas 3.5**.

Cells are able to build up proteins directly from L-amino acids. That's why chemists are learning how to use bacterial and yeast cells to make proteins – in many ways it's better than using traditional techniques.

As a chemist you would need three things before you could synthesise a protein in the laboratory:

- a set of instructions for the protein – in other words, something which told you its primary structure
- supplies of the pure amino acids ready for you to use in the appropriate steps
- a way of making the amino and carboxylic acid groups react with one another more easily.

If we look at how a cell makes its proteins we can see a close parallel with what the chemist would do.

In **Activity EP2.4** you can build models for some amino acids and investigate optical isomerism further.

Activity EP2.5 illustrates one way your body can recognise the different D and L forms of a molecule.

Activity EP2.6 helps you to summarise what you have read so far in Section **EP2**.

How cells make proteins

The instructions specifying the primary structure of insulin are carried by molecules of a **ribonucleic acid** (**RNA**). There are many different types of RNA molecules. **Messenger RNA (or mRNA)** molecules provide a code which tells the cell which amino acids to put together, in which order, to make a protein. Just as there are many different proteins, so there are many different mRNA molecules.

The amino acids which the cell has to use are not in separate containers as they would be in the laboratory: they are dissolved and mixed together in the fluid within the cell. **Transfer RNA (or tRNA)** molecules play a similar role to the person who separates the amino acids for you at the chemical supplier. They both select and separate the amino acids needed for protein synthesis.

The cell has a different tRNA for each different amino acid. It also has a set of enzymes which recognise the tRNA and its corresponding amino acid. The enzymes catalyse the formation of an ester bond between the carboxylic acid group of the amino acid and an –OH group on the tRNA. Once the enzyme has joined the amino acid to the tRNA, the resulting complex can diffuse to the place where the protein is being made.

Figure 14 shows that RNA molecules consist of strands formed from **ribose** sugar molecules and **phosphate** groups. One of four **bases** is attached to each ribose unit. The order of bases shown in Figure 14 is for illustration only: the sequence is different in different RNA molecules.

It is the bases which form the code for protein synthesis and we can regard the sugar–phosphate strand as just a 'backbone' on which they are held. Two simpler ways of depicting RNA are also shown in Figure 14.

The cell's catalyst for protein production is a small particle called a **ribosome**. This contains a third type of ribonucleic acid, **ribosomal RNA (or rRNA)** bound to protein molecules. When the mRNA and the ribosome have collided in the fluid in the cell, the ribosome moves along the mRNA, rather like a bead on a chain, reading the code, and catalysing the reactions which join the amino acids together.

ASSIGNMENT 3

a In the formation of RNA, what type of reaction is responsible for the linking of
  i the ribose and phosphate?
  ii the ribose and base?

b The skeletal formula of ribose is shown in the illustration of RNA (Figure 14). Draw a full structural formula for ribose.

Cracking the code

How can *four* bases code for *20* amino acids? If a single base told the cell to build an amino acid into a protein, there would need to be 20 different bases. But there are only four bases, so protein building can't be coded by single bases.

Writing the code in pairs of bases isn't enough either – it would only cover 16 amino acids (you should be able to write down 16 different pairs of the four letters U, C, A and G).

In fact, a **triplet code** is needed, where a combination of three bases tells the cell which amino acid to use. There are now more combinations than there are amino acids, so some amino acids are defined by several base triplets.

The combinations, or **codons**, are shown in Table 2 on p. 146. Notice that there are also codons which *stop* the protein chain building.

ASSIGNMENT 4

Table 2 shows that for some amino acids only the first *two* bases of the RNA codon are important. The identity of the third base does not matter. Make a list of these amino acids.

For other amino acids, it is important that all *three* bases are correct. Make a list of these amino acids.

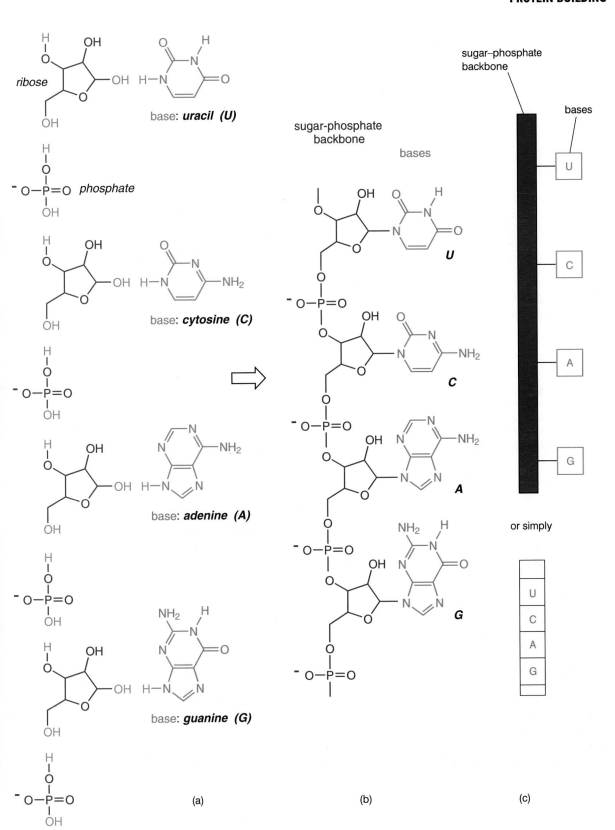

Figure 14 Representations of the structure of RNA: (a) how groups join together; (b) a skeletal formula; and (c) two simpler ways of showing the structure.

First base	Second base				Third base
	U	C	A	G	
U	UUU Phe	UCU Ser	UAU Tyr	UGU Cys	U
	UUC Phe	UCC Ser	UAC Tyr	UGC Cys	C
	UUA Leu	UCA Ser	UAA Stop	UGA Stop	A
	UUG Leu	UCG Ser	UAG Stop	UGG Trp	G
C	CUU Leu	CCU Pro	CAU His	CGU Arg	U
	CUC Leu	CCC Pro	CAC His	CGC Arg	C
	CUA Leu	CCA Pro	CAA Gln	CGA Arg	A
	CUG Leu	CCG Pro	CAG Gln	CGG Arg	G
A	AUU Ile	ACU Thr	AAU Asn	AGU Ser	U
	AUC Ile	ACC Thr	AAC Asn	AGC Ser	C
	AUA Ile	ACA Thr	AAA Lys	AGA Arg	A
	AUG Met	ACG Thr	AAG Lys	AGG Arg	G
G	GUU Val	GCU Ala	GAU Asp	GGU Gly	U
	GUC Val	GCC Ala	GAC Asp	GGC Gly	C
	GUA Val	GCA Ala	GAA Glu	GGA Gly	A
	GUG Val	GCG Ala	GAG Glu	GGG Gly	G

Table 2 The triplet base codes (codons) for each amino acid used in mRNA.

The tRNA molecules can recognise and bind to the codons on mRNA through **anti-codons**. Base G in the anti-codon specifically recognises base C in the codon (and vice versa). Bases U and A recognise one another similarly. So, for example, in Figure 15, anti-codon CGG binds to codon GCC. Similarly, the anti-codon GUA will bind to codon CAU, a codon for histidine.

Figure 16 summarises the roles of the different types of RNA in protein synthesis.

Figure 17 shows how the codons on a mRNA molecule are read as they are threaded through the ribosome and 'translated' into a protein chain.

You can think of the ribosome as sliding along the mRNA chain rather like a bead. tRNA molecules, each carrying an amino acid, feed into the front of the ribosome and the protein chain grows from the back.

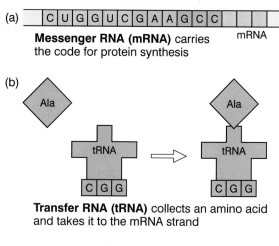

(a) Messenger RNA (mRNA) carries the code for protein synthesis

(b) Transfer RNA (tRNA) collects an amino acid and takes it to the mRNA strand

(c) Ribosomes contain ribosomal RNA (rRNA) They move along the mRNA chain, reading the code and catalysing protein synthesis

Figure 16 The roles of the different types of RNA.

–OH group forms an ester link to the amino acid – in this case alanine (Ala)

Anti-codon for binding to a codon on mRNA: in this example it would bind to GCC, the codon for alanine

Figure 15 Schematic representation of a tRNA molecule showing the three bases which form the anti-codon.

ASSIGNMENT 5

a Use the codons from Table 2 to predict the peptides obtained if RNA molecules with the following patterns of bases were used:
   i AAAAAA ...
   ii CGCGCGCGC ...
   iii UACCUAACU
b Predict the anti-codons for these amino acids
   i Trp   ii Asp.

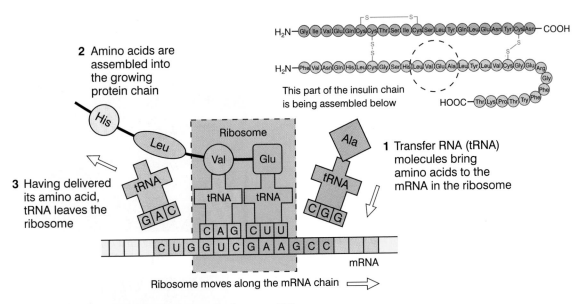

2 Amino acids are assembled into the growing protein chain

This part of the insulin chain is being assembled below

1 Transfer RNA (tRNA) molecules bring amino acids to the mRNA in the ribosome

3 Having delivered its amino acid, tRNA leaves the ribosome

Ribosome

Ribosome moves along the mRNA chain ⟹

mRNA

Figure 17 Protein synthesis and the reading of codons on mRNA.

In recent years, chemists have learned a lot about how molecules recognise one another. Molecules must have shapes which fit neatly together so that groups can easily form intermolecular bonds.

The bases in RNA all have flat shapes, and Figure 18 shows how the correct pairs of bases (known as **base pairs**) fit neatly together and place groups in just the right positions for hydrogen bonds to form.

hydrogen bonding between uracil and adenine

RNA chain

or U ⫤ A

RNA chain

hydrogen bonding between cytosine and guanine

RNA chain

or C ⫥ G

RNA chain

Figure 18 Molecular recognition and bases on RNA (the symbol ⫤ is used to represent two hydrogen bonds; ⫥ represents three hydrogen bonds).

A permanent record

mRNA is destroyed when production of its corresponding protein is no longer required. It represents a temporary set of instructions for producing a protein. The cell keeps the permanent record in the nucleus in the form of other strand-like molecules known as **deoxyribonucleic acid** (**DNA**) molecules. DNA and RNA both consist of sugar–phosphate strands with attached bases, but there are important differences (Table 3):

- the sugar, ribose, in RNA is replaced by deoxyribose in DNA, hence the change of letters from R to D
- the base uracil (U) in RNA is replaced by the base *thymine* (T) in DNA
- two strands of DNA are normally paired off together in the *double helix* arrangement proposed in 1953 at Cambridge by Francis Crick and James Watson.

RNA	DNA
ribose	deoxyribose - note the absence of an OH group
uracil	thymine

Table 3 Differences between the structures of RNA and DNA.

Crucial to Crick and Watson's double helix model was the understanding that pairs of bases in DNA can form hydrogen bonds together: A with T, and C with G (Figure 19).

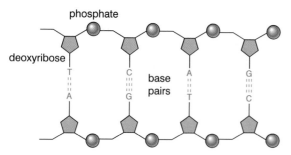

Figure 19 Two DNA strands held together by hydrogen bonds between pairs of bases (again ≡≡≡ and ≡≡≡ are used to represent sets of two and three hydrogen bonds).

The hydrogen bonding interactions between C and G are identical to those which occur between RNA codons and anti-codons. The A ≡≡≡ T interaction is similar to that between A and U in RNA: the bases thymine and uracil differ only by the presence of a methyl group in thymine.

Figure 20 shows how the double helix of DNA molecules is held together by hydrogen bonding between the bases.

Crick and Watson, together with the physicist Maurice Wilkins at King's College, London, received the 1962 Nobel Prize in Physiology and Medicine 'For their discoveries concerning the molecular structure of nucleic acids and its significance for information transfer in living materials'.

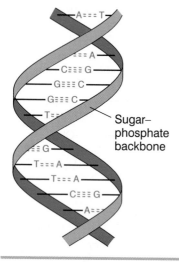

Figure 20 An illustration of the DNA double helix.

Sugar–phosphate backbone

In **Activity EP2.7** you can make a model of the DNA double helix.

You can find out about the events leading up to Crick and Watson's breakthrough and read their original paper in **Activity EP2.8**.

Figure 21 Francis Crick (right) and James Watson (left) in 1953 with their model of part of a DNA molecule.

Figure 22 Rosalind Franklin (1920–58). She worked at King's College, London, with Maurice Wilkins, using X-ray crystallography to study DNA. She was one of the four people closest to the discovery of the structure of DNA, but she could not be considered for the Nobel Prize as she had died of cancer in 1958. She was 37.

When a cell starts protein production, the record in the DNA has to be turned into an RNA message carried by mRNA, which is then 'read' to form the protein, as shown in Figure 17.

The full process is summarised in Figure 23.

With only very few exceptions, all cells use the same system, which we can sum up as:

DNA codes for RNA
RNA codes for proteins

Another important contrast between DNA and RNA is that each DNA molecule contains the information for the production of many different mRNA molecules, but, in general, each mRNA molecule is a set of instructions for just one protein. A DNA segment responsible for a particular protein is called a **gene**.

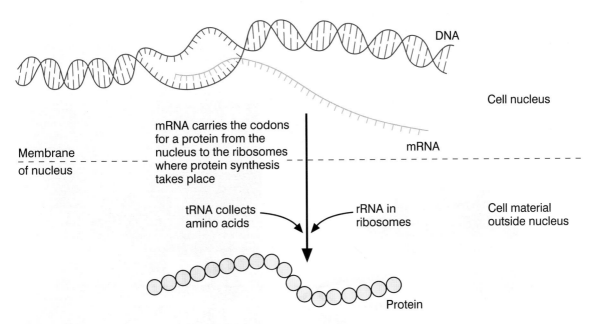

Figure 23 *A summary of protein synthesis in higher organisms.*

The full set of all the genes of an organism its called its **genome**. We now know the complete sequence of bases in the genomes of many organisms. For example, the genome of the bacterium *Escherichia coli* consists of about 4.7×10^6 base pairs, which account for over 4000 genes. The bacterium can therefore produce over 4000 different proteins.

The human genome consists of about 3.5×10^9 base pairs. Mapping the human genome is an ambitious, and costly, undertaking. The *Human Genome Project* was set up in 1988 to achieve this. It is a vast international venture which has taken 15 years to complete the first stage, the sequencing of the bases. The next stage is to identify exactly how they are divided into genes, another colossal task, as there are expected to be over 80 000 genes. The way will then be open to understand what the resulting proteins actually do.

Figure 24 *The Human Genome Project has a database that contains records of all the information collected so far about the human genome.*

Every cell in your body contains a full set of genes in its nucleus and so has a DNA molecule which carries, for example, the gene for insulin production. But the gene is only 'switched on' in the special pancreas cells which make insulin.

In addition to very many genes, DNA molecules contain base combinations which start or stop RNA production, as well as regions of 'junk DNA' which appear to have no function – perhaps relics of earlier predecessor organisms or with a function yet to be discovered.

ASSIGNMENT 6

One strand of DNA in a cell nucleus carries the following sequence of bases: CAGT.

a Write down the corresponding sequence of bases on the mRNA strand which copies from it.

b Write down the corresponding sequence of bases on the other DNA strand in the double helix.

DNA finger-printing

DNA finger-printing is based on the fact that no two people (apart from genetically identical twins) share the same DNA sequence. It is now used regularly to help investigate very serious crimes such as murder and rape.

A trace of blood, semen or skin can be used. A solution of an enzyme (a restriction endonuclease) is added to the sample, which cuts the DNA at particular sites into a specific pattern of fragments.

The resulting solution is applied to a gel which is subjected to an electric field (Figure 25). The DNA fragments are electrically charged due to the negative charges carried by the phosphate groups and so the different-sized DNA fragments move at different speeds through the gel towards the positive electrode. The process is known as *gel electrophoresis*.

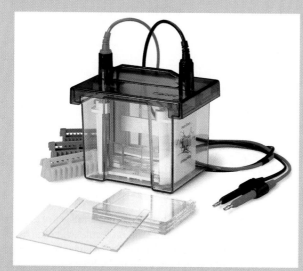

Figure 25 Electrophoresis equipment to analyse the composition of DNA in the cell nucleus.

To see the pattern of a DNA profile test, radioactive tracers are added which bind to DNA fragments. The plate on which the gel is spread is then exposed to a photographic film. A series of bands are seen which are compared with the DNA sample from the suspect (Figure 26).

DNA profiling is also used in paternity disputes, to prove who is the father of a child, in medical analysis of genetic relationships between different people and populations, and in the identification of decayed remains (Figure 27).

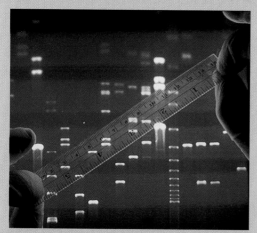

Figure 26 The DNA in a cell is unique to that individual. Fragments of the DNA are shown in the band pattern.

LA FAMILLE IMPÉRIALE DE RUSSIE

Figure 27 The identity of the bones thought to be those of the last Tsar of Russia, Nicholas II, who was killed during the Revolution in 1918, was confirmed by comparing DNA extracted from the bones with DNA from some of the Tsar's known living relatives, including the Duke of Edinburgh.

EP3 *Genetic engineering*

Changing genes

How can we make human insulin? You might think it could be done by removing some human pancreas cells, growing them by supplying them with the appropriate amino acid nutrients, and then extracting the insulin. Unfortunately this can't be done: human cells do not grow like this.

But the technology for handling other types of more robust cells, particularly bacterial and yeast cells, is well established. Bacteria can be used to produce chemicals like lactic acid. Brewers have used yeasts for thousands of years to convert sugars into ethanol; more recently, pharmaceuticals like penicillin and oxytetracycline have been made in fermenters using moulds.

In recent years, scientists have learnt how to build human genes into bacterial or yeast cells. These cells can then be used to generate proteins. The insulin gene can be used in this way. The first human insulin was made using the bacterium *E. coli* and became available in 1982. More recently, yeast cells – which more closely resemble human cells – are being used.

Figure 28 Genetically engineered yeast cells are fermented under carefully controlled conditions; the cells multiply rapidly and produce a precursor of insulin, which is converted to human insulin using enzymes.

Taking genes from the cells of one type of organism and putting them into the cells of another type of organism is called **genetic engineering** or *recombinant DNA technology*. The details of the method vary according to the gene and the microorganism: no two processes are identical. Figure 29 represents a generalised approach using bacterial cells.

New genes go out to work

Proteins such as insulin, human growth hormone and factor 8 (the blood clotting agent given to haemophiliacs) are products of genetic engineering. An important advantage of such products is that they are purer than the ones prepared by traditional means. For example, genetically engineered factor 8 removes the risk of haemophiliacs contracting AIDS, hepatitis, new variant CJD or other hazards, through blood products from infected donors.

A single tobacco plant can produce the same yield of factor 8 as a 1000-dm^3 fermentation vessel of mammalian cells. The entire world demand for this protein could be satisfied by 1000 tobacco plants into which human genes have been implanted.

But the applications and potential of genetic engineering reach wider than pharmaceuticals.

The body's immune system defends it against virus infection by recognising the protein coating on the outside of the virus. If the coat protein is produced via genetic engineering – minus its dangerous contents – it can be injected into the body to act as a vaccine. Hepatitis B vaccine is produced and works in this way.

Modified bacteria and fungi, which turn potentially harmful materials into harmless forms, could be used widely by industry and environmental protection agencies. A new oil-digesting 'superbug' has been created which contains a collection of genes from several bacteria. The bacteria have been chosen because they each metabolise different components of crude oil. When the appropriate genes from all the bacteria are combined, the resultant bacterium can break down all the chemicals found in crude oil.

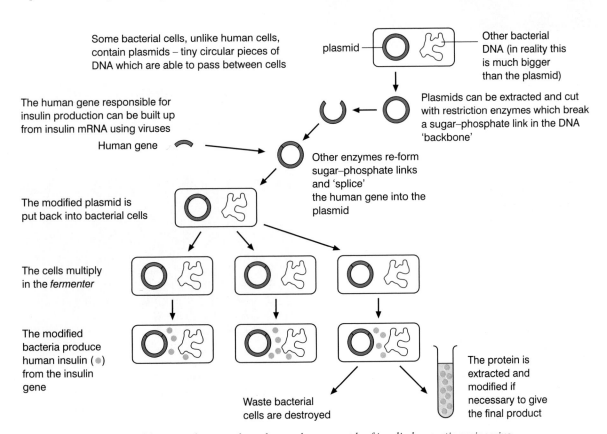

Figure 29 An illustration of the general approach used to produce a sample of insulin by genetic engineering.

PHA polymers (polyhydroxyalkanoates) are made by bacteria and lead to biodegradable plastics. You might have read about them in the **Designer Polymers** storyline, Section **DP7**. The gene which leads to the production of the polymers has been placed into another bacterium which bursts open when it is heated. The polymer can then be extracted more easily.

Cystic fibrosis affects 1 in every 2000 European children. The disease attacks the lungs and the pancreas and is difficult to treat (Figure 30).

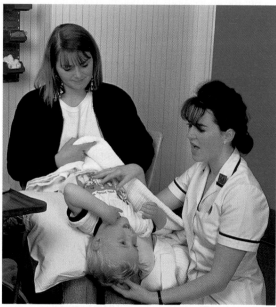

Figure 30 People with the inherited disease cystic fibrosis have to undergo regular physiotherapy to loosen mucus and ensure that their lungs remain clear and disease-free. Gene therapy, now being trialled, offers hopes of a cure.

It was discovered in 1985 that cystic fibrosis results from a defect in a single gene. The gene codes for a protein that acts as a channel, controlling the passage of chloride ions across cell membranes, particularly those in the lining of the airways in the lungs. The chloride ions carry water molecules with them, which keep the airway surface wet and clean. With the defective gene, the correct protein is not produced. The channels are blocked and the airways are drier and susceptible to infection.

The big breakthrough came in 1989 when the defective gene was isolated. This opened up the possibility of *gene therapy* and trials are now in progress. The aim of gene therapy is to smuggle healthy copies of the gene into the cells lining the lungs. The healthy gene is inserted into a *bacterial plasmid* and *liposomes* (tiny spheres of fat molecules which pass easily through cell membranes) are being used to transfer the gene into the cells. In the first stage of the trial, the gene and liposome carrier are being sprayed onto the cells lining the nose. If successful, the next stage will be to spray directly into the lungs.

The techniques of genetic engineering can be applied to produce plants with modified properties. Examples include introducing a gene which enhances flavour (for example, more flavorsome tomatoes) and others which give plants resistance to pesticides and herbicides. There is also the prospect of using plants to produce pharmaceutically important proteins, drugs, or even high-value polymers.

Figure 31 Commercially produced tomatoes are usually picked before they are ripe and so will have not developed any flavour. They can be genetically modified so that, once ripe, they deteriorate more slowly. This means that they can be picked when they are ripe and have their full flavour, but then remain fresh long enough for them to be transported to shops and markets.

Figure 32 Maize can be genetically modified to make its own insecticide.

The corn-borer is a particularly destructive pest on crops of maize.

It is well known that the bacterium *Bacillus thuringiensis* produces a δ-endotoxin protein. When eaten by the corn-borer, the protein hydrolyses to form a deadly poison for the insect. Recent work is helping to make the protein stable so that maize needs only one application. This knowledge provided another idea, to genetically modify maize to produce its own δ-endotoxin and this approach is being developed.

Salt-loving plants

A plant has been genetically modified so that it can flourish in salt solutions. If the new technology can be applied to wheat, corn or rice, the yields of these crops on irrigated land (which tends to have high concentrations of salt) will increase. This would be particularly important for people living in developing countries, many of which tend to be arid.

The plants in Figure 33 are being watered with a 0.2 mol dm⁻³ solution of sodium chloride.

Figure 33 The plant on the right has been genetically modified to tolerate salt solutions.

These advances in science and technology have raised concerns about the environmental impact of genetically modified organisms. It is possible, with careful regulation and specialised equipment, to contain the bacteria or yeast that produce a modified protein such as insulin. However, plants grow in open fields and there are concerns about whether the modified genes will transer to other species and have an, as yet unknown, impact on biodiversity and the environment. Carefully designed and controlled experiments are needed to see just what the impact will be.

ASSIGNMENT 7

Genetic engineering could bring great benefits to medicine, agriculture, industry and to pure research. But it could also be a dangerous experiment with nature that we might live to regret.

Draw up a list of the pros and cons of using genetic engineering techniques. You might like to extend this to a class discussion.

EP4 *Proteins in 3D*

Folded chains

Cow, pig and human insulin all have different primary structures, but they are equally effective at treating diabetes. So scientists need to know about more than just primary structure before they understand how insulin works. One thing they now know is that insulin, like most proteins, has a precise shape which arises from the **folding** together of the chains. The action of insulin is critically dependent on this shape.

As long as different molecules fold to the same shape, they may have the same action. Chain folding gives proteins their three-dimensional shape; it also places chemical groupings in positions where they can interact most effectively.

Four types of interactions are important in chain folding:

- **instantaneous dipole–induced dipole attractive forces** between non-polar side chains on amino acids such as phenylalanine and leucine. The centres of protein molecules tend to contain amino acids like these so that the non-polar groups do not interfere with the hydrogen bonding between the surrounding water molecules.
- **hydrogen bonding** between polar side chains (eg –CH₂OH in serine and –CH₂CONH₂ in asparagine), *and* between the peptide groups which link the chain together. If amino acids with polar side chains are situated on the outside of proteins, hydrogen bonds can also form to water molecules surrounding the protein.
- **ionic attractions** between ionisable side chains, such as –CH₂COO⁻ in aspartic acid and –CH₂CH₂CH₂CH₂NH₃⁺ in lysine.
- **covalent bonding**. The –SH groups on neighbouring cysteine residues can be oxidised to form –S–S– links: look back at Figure 11 on p. 141. There are three such links in human insulin; two hold the A and B chains together.

You can revise earlier work on instantaneous dipole–induced dipole forces and hydrogen bonding by reading **Chemical Ideas 5.3** and **5.4**.

The chains in a protein are often folded or twisted in a regular manner as a result of hydrogen bonding. Two arrangements of the protein chain are common:

- tightly *coiled* into a **helix** where the C=O group of one peptide link forms a hydrogen bond to an N–H group *four* peptide links along the chain
- stretched out into regions of *extended* chain, which lie alongside one another and hydrogen bond to form a **sheet**.

The two components of a protein's shape – helix and sheet – are sometimes referred to as its **secondary structure** (Figure 34).

(a)

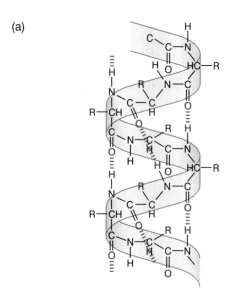

(b)

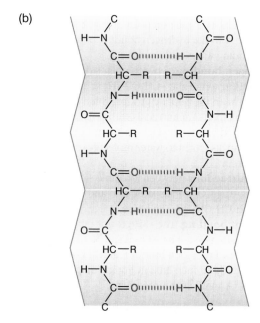

Figure 34 The secondary structure of a protein involves folding as a result of hydrogen bonding. This figure shows the protein chain folded into (a) a helix and (b) a sheet.

The chains may then fold up further. The overall shape is stabilised by instantaneous dipole–induced dipole attractive forces, by hydrogen bonding, by ionic attractions and by covalent bonding. The overall shape of a protein is sometimes called its **tertiary structure**.

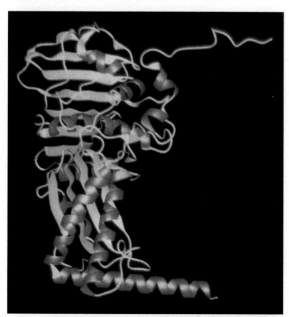

*Figure 35 The tertiary structure of a protein; here the enzyme gyrase is illustrated as a **ribbon diagram**. The helices (in red) and the sheets (in yellow) show up clearly in this representation. The thin white lines represent irregular regions. (Gyrase is involved in the DNA unwinding process.)*

Compare the ribbon diagram of insulin in Figure 36 with Figure 11 on p. 141, which shows its primary structure.

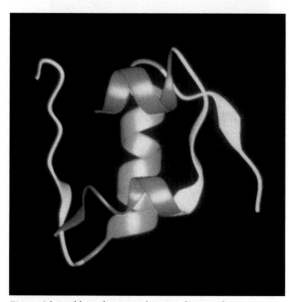

Figure 36 A ribbon diagram showing the two chains in insulin. There are two short helical sections (in red) in the A chain. The B chain contains a helical section (in blue) between amino acids 9 (Ser) and 19 (Cys).

Insulin hexamers

The same ideas about intermolecular interactions help to explain why insulin molecules stick together to form dimers and hexamers. The **space filling model** of the insulin monomer illustrated in Figure 37 shows that one side of its structure consists mainly of amino acids with non-polar side chains.

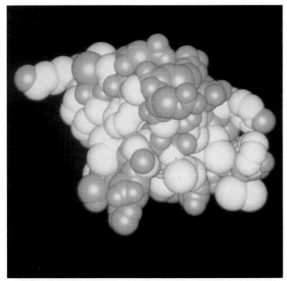

Figure 37 Space filling representation of an insulin monomer: the white spheres represent amino acid residues with non-polar side chains.

Left exposed to water, this non-polar side of the insulin monomer would disrupt water's hydrogen bonding. The non-polar sides of two insulin monomers come together when a dimer is formed, so the disruption is reduced. Figure 38 shows that a similar thing happens when hexamers form.

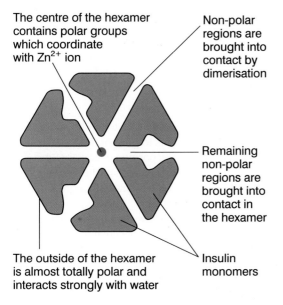

The centre of the hexamer contains polar groups which coordinate with Zn^{2+} ion

Non-polar regions are brought into contact by dimerisation

Remaining non-polar regions are brought into contact in the hexamer

The outside of the hexamer is almost totally polar and interacts strongly with water

Insulin monomers

Figure 38 An insulin hexamer.

The arrangement of several protein sub-units into a bigger unit is sometimes referred to as **quaternary structure**.

So, for insulin, we have

- **primary structure**: the order of amino acid residues
- **secondary structure**: coiling of parts of the chain into a helix or the formation of a region of sheet
- **tertiary structure**: folding of the secondary structure into an insulin monomer
- **quaternary structure**: joining of monomers into a hexamer.

Figures 39 and 40 show a space filling diagram and a ribbon diagram of the insulin hexamer.

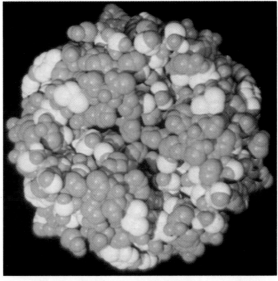

Figure 39 A space filling diagram of the quaternary structure of an insulin hexamer: the white spheres represent amino acid residues with non-polar side chains.

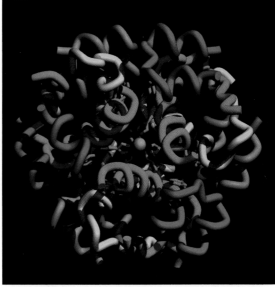

Figure 40 A ribbon diagram of the insulin hexamer.

Working out the structure of insulin

Insulin was recognised as a protein in 1928. Its primary structure was revealed in 1955 by Professor Fred Sanger and his team at Cambridge. They broke the protein into shorter bits by hydrolysis, then used chromatography to identify these bits. By piecing together the bits, they eventually worked out the amino acid sequence. Professor Sanger received a Nobel Prize for his work on the amino acid sequencing of proteins and a second Nobel Prize for more recent work on sequencing DNA.

The three-dimensional structure of insulin has been worked out from X-ray diffraction studies carried out by Professor Dorothy Hodgkin's group at Oxford. Professor Hodgkin received a Nobel Prize in 1964 for earlier work, particularly work on the structure of vitamin B_{12} and penicillin. She is only the third woman, after Marie Curie and her daughter Irène Joliot-Curie, to receive a Nobel Prize for Chemistry.

Figure 41 Professor Dorothy Hodgkin with a model of insulin.

Figure 42 A Nobel Prize medal, presented by The Royal Swedish Academy of Sciences. The Latin inscription 'Inventas vitam juvat excoluisse per artes' loosely translates as 'And they who bettered life on Earth by new-found mastery'. It would be difficult to find a phrase more fitting for Dorothy Hodgkin.

EP5 *Giving evolution a push*

Insulins that stay single

Insulin monomers are better for diabetics than insulin hexamers because they work faster. So is there a way of injecting monomers rather than hexamers?

The trouble is that the hexamer and its monomers are in dynamic equilibrium:

insulin hexamer	$\rightleftharpoons$	3 insulin dimers	$\rightleftharpoons$	6 insulin monomers
Ins_6	$\rightleftharpoons$	$3Ins_2$	$\rightleftharpoons$	$6Ins$

and the positions of these equilibria depend on the concentration of the insulin in solution. Look back at Figure 8 on p. 139. As hexamers spread out from the injection site, they become diluted and the equilibria shift towards the right.

Insulin monomers are the main form present when the concentration of the solution has fallen from 1×10^{-3} mol dm^{-3} to 1×10^{-8} mol dm^{-3}. But injecting a dilute monomer solution just would not be practicable: the volume of solution needed would be enormous!

You met the idea of chemical equilibrium in **Chemical Ideas 7.1**. You can find out more about the effect of concentration on the position of an equilibrium in **Chemical Ideas 7.2**.

ASSIGNMENT 8

The equilibrium between the insulin hexamer and insulin dimer can be represented as

$Ins_6 \rightleftharpoons 3Ins_2$

Write an expression for the equilibrium constant K_c for this reaction.

What are the units of K_c?

A possible solution might be to modify the structure of the insulin monomers to prevent them sticking together so readily. The position of equilibrium for the modified insulin would be much further to the right and the monomers would be stable in more concentrated solutions.

Figure 38 on p. 155 shows that insulin monomers fit together to form dimers so that chemically identical faces – faces which possess the same amino acid sequences – are in contact. A closer look at the amino acid residues on the dimerising surfaces shows that most of them have non-polar side chains, but some are polar or contain ionised groups.

For example, residue B13 (amino acid 13 in the B chain) is a glutamic acid with a negatively charged side chain containing the $-COO^-$ group. This is placed next to residue B9 – serine with a polar side chain containing the $-CH_2OH$ group – in the dimer (Figure 43).

Figure 43 Space filling representation of an insulin dimer showing the positions of amino acids serine B9 (in yellow) and glutamic acid B13 (in red).

What would happen if B9 were changed to aspartic acid – another amino acid with a negatively charged side chain? Four negatively charged COO^- groups – two aspartic acid/glutamic acid pairs – would be placed together and the dimer would be blown apart.

Changes like this can be studied before doing experiments, by making use of **computer graphics**. Powerful programs can be used to generate views of the protein structure, like the one in Figure 43, which can be rotated through any angle to study the shape in detail.

Chemists need to know the precise position of every atom in a molecule before they can generate a computer graphics representation of its structure. Information like this is available from **X-ray diffraction studies** on insulin crystals (Figure 44).

Figure 44 Insulin crystals. Protein crystals are often hard to grow; fortunately, insulin produces beautiful crystals.

Designer genes

The next step was to prepare a sample of the modified insulin and compare its action with that of human insulin in diabetic patients. But how can the new form of insulin be made?

There is no need to join 51 amino acids together by chemical synthesis. We know how cells make proteins, and we can insert insulin genes into yeast or bacteria. All that needs to be done is to alter the human insulin gene so that a codon for serine (eg AGG) is changed to a codon for aspartic acid (eg CTG). Then the yeast or bacterial cells will make the modified insulin for us. (See Table 2, p. 146, for information about codons and amino acids.)

This really isn't as difficult as it seems – thanks to the remarkable ability of molecules to recognise one another.

Figure 45 shows the part of the insulin gene which tells the cell to join together the six amino acids of the B chain around B9 (serine). Using genetic engineering techniques this gene is built into a bacterial plasmid. The plasmid can then be 'unzipped' by certain viruses to form two single-stranded DNA molecules.

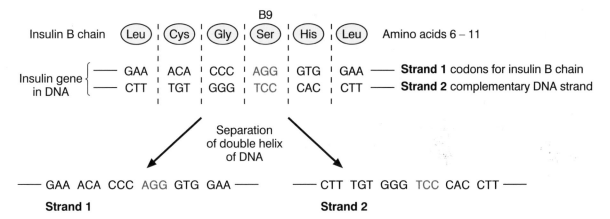

Figure 45 Part of the human insulin gene which codes for the B chain around B9 (serine).

If chemists synthesise a small piece of DNA with 18 bases in the same sequence as that shown for Strand 1, it will recognise and stick to Strand 2 at precisely the part shown in Figure 45. It will do this because T sticks to A and G sticks to C, *and* because the chance of finding the same sequence of 18 bases anywhere else in the plasmid is, for all practical purposes, zero. The small piece of DNA will be like a 'chemical magnet', only sticking to the right section on the complementary DNA strand.

Normal processes within the cell build bases onto the ends of the small section of synthetic DNA to complete the double helix, producing a plasmid which is indistinguishable from the one at the start (Figure 46).

1 Synthetic DNA (shown in green) sticks to correct sequence of bases on Strand 2

Strand 2

    CTT TGT GGG TCC CAC CTT
    GAA ACA CCC AGG GTG GAA

2 Normal cell processes recreate the plasmid double helix

Strand 1   Strand 2

    CTT TGT GGG TCC CAC CTT
    GAA ACA CCC AGG GTG GAA

Figure 46 The cell recognises the small piece of synthetic DNA and incorporates it into a plasmid.

Figure 47 The cell tolerates a change to one of the bases in the piece of DNA and incorporates the synthetic DNA into a plasmid. When the cell divides, two different plasmids are formed – one carries the normal human insulin gene, the other the gene for modified insulin.

In fact, the chance of finding the 18-base sequence anywhere else on Strand 2 is so unlikely that the cell will tolerate a mistake. If the chemist builds the small piece of DNA with the codon CTG in place of AGG, it will still stick onto the same place – even though two of the bases are wrong. The cell will then complete the double helix as usual. What the chemist has done, of course, is to replace the Ser codon (AGG) by the Asp codon (CTG).

Figure 47 shows what happens when the bacteria multiply with this new DNA inside them. A new strain of bacterium is produced which carries the gene for producing monomeric insulin. It all comes about as a result of hydrogen bonding and molecular recognition.

Back to Christopher

Monomeric insulin has been tried out on patients such as Christopher. Figure 48 shows plots of blood-insulin concentrations like those in diabetic patients who have received insulin injections. Two curves are shown: for injection of human insulin and for injection of the modified insulin just described with Asp^{B9} – aspartic acid in position B9.

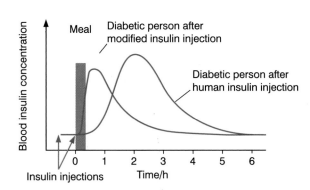

Figure 48 Insulin concentrations in diabetic patients using human insulin and modified insulin: human insulin is injected 30 min before the meal; modified insulin is injected immediately before the meal.

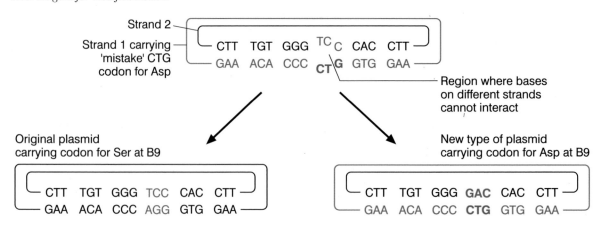

The modified insulin gets into the bloodstream more quickly than injected human insulin because it is monomeric and does not need time to break down into hexamers. Its release more closely resembles that of natural insulin in a non-diabetic person, shown in Figure 6 on p. 139.

The trials using the modified insulin have been so successful that more and more diabetic patients are using it. This means that the patients, and not the injection time, are in control. Patients can take the injection just before their meal, rather than having to guess when a meal will appear. In addition, the diet recommended for diabetics is now much more relaxed, with the emphasis on healthy eating.

It is the combination of scientists – the chemists, with their array of X-ray crystallography, computer simulations and protein engineering, and nutritionists – that have guaranteed Christopher and many others a much more comfortable lifestyle.

A breakthrough

A breakthrough came when it was observed that the presence of phenol encouraged the formation of insulin monomers. Small traces of phenol are added to insulin solutions because of the anti-bacterial properties of phenol.

X-ray crystallography of insulin crystals from such a solution show molecules of phenol incorporated between the 'dimers' (Figure 49). When the crystals dissolve in water, the phenol molecules disperse very quickly, breaking down the hexamer into dimers. Thus, as the modified insulins also incorporate phenol, there is a second reason for the insulin molecules to exist as monomers.

It is ironic that in this story of the development of such a complicated molecule as insulin, phenol plays such as important role. For phenol is a very simple molecule, C_6H_5OH, and was one of the first known antiseptics, known originally as carbolic acid.

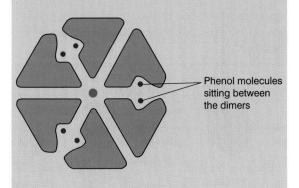

Figure 49 Phenol molecules sitting between the 'dimers' in an insulin hexamer.

Phenol molecules sitting between the dimers

EP6 *Enzymes*

There is a simple test you can do to check whether you are diabetic – although to be really certain you must go and see your GP. You can buy some reagent strips from a pharmacy and test for glucose in your urine. (When glucose builds up to a high level in the blood, it 'spills over' into the urine.)

The test strips illustrate *four* important points about enzymes. Enzymes are

1. *catalysts*
2. *highly specific*: for example, the test strips only work with glucose – they give no response with other sugars
3. *sensitive to pH*: many work best at a particular pH and become inactive if the pH becomes too acidic or too alkaline
4. *sensitive to temperature*: many enzymes work best at temperatures close to body temperature; most are destroyed above 60°C–70°C.

Testing for glucose in urine

The fresh reagent strip has a small coloured square at one end which turns to a different colour in the presence of glucose. The square is impregnated with four reagents:

- glucose oxidase, an enzyme which catalyses the reaction

  glucose + oxygen →
  gluconic acid + hydrogen peroxide

- an indicator, sometimes 2-methylphenylamine (orthotoluidine); this is present in its reduced form (XH_2) which is colourless, but it turns into a coloured form (X) when it is oxidised

- peroxidase, an enzyme which catalyses the oxidation of the indicator by hydrogen peroxide:

  hydrogen peroxide + XH_2 → water + X

- a buffer: a mixture of chemicals which keeps the reagents at a fixed pH during the test.

If the manufacturer's instruction sheet is available, you will also see that it recommends storing the test strips below 30°C but not in a refrigerator.

Figure 50 A simple urine test gives a measure of blood-glucose concentration.

You can use some glucose test strips with different sugar solutions in **Activity EP6.1**. The activity illustrates the important points about enzyme behaviour.

Active sites

Enzymes are so *specific* because they have a precise tertiary structure which exactly matches the structure of the **substrate** – the molecule which is reacting. It's another example of molecular recognition. Scientists often think of the 'lock and key' analogy, in which the enzyme is the 'lock' and the substrate is the 'key'. There may be lots of related keys, and several may fit into the lock, but only one will work.

The analogy is a good one. Figure 51 shows a space filling model of the enzyme *lysozyme* which catalyses the breakdown of the cell walls of bacteria and helps protect us from infection. There is a cleft in the enzyme surface formed by the way the protein chain folds. The shape of the cleft is tailored for the substrate molecules to fit into. Within the cleft are chemical groups – some of the side chains on the amino acid residues – which bind the substrate and possibly react with it. This region of the enzyme is called its **active site**.

Active site

Figure 51 Space filling model of the enzyme lysozyme, showing the cleft which forms the active site.

The bonds which bind the substrate to the active site have to be weak so that the binding can be readily reversed when the products need to leave the active site after the reaction. The bonds are usually hydrogen bonds or interactions between ionic groups. The binding may cause other bonds within the substrate to weaken or it may alter the shape of the substrate so that atoms are brought into contact to help them to react.

After reaction, the product leaves the enzyme which is then free to start again with another molecule of substrate. The whole process is summarised in Figure 52.

This is the simplest model of enzyme catalysis

$$E + S \rightleftharpoons ES \rightarrow EP \rightarrow E + P$$

What happens, of course, is more complex. In many cases, scientists believe that the substrate is not quite a perfect fit, and must alter its shape to fit into the active site. This means that both the substrate and the active site will be in strained arrangements, and this can help the reaction to occur.

Activity EP6.2 is a model-building exercise which illustrates the way in which substrate molecules fit onto an enzyme. It also shows how other molecules can *inhibit* the enzyme's action.

Enzymes as catalysts

In a catalysed reaction, reactants need less energy before they can turn into products than they do in an uncatalysed reaction. Their **activation enthalpy** is lower. This is illustrated in Figure 53 for the case of enzymes.

When the activation enthalpy is lower the reaction takes place more quickly.

You were introduced to the idea of activation enthalpy in **Chemical Ideas 10.1** and **10.2** and to catalysts in **Chemical Ideas 10.4** and **10.5**.

You can learn more about the effect of concentration on the rates at which chemical reactions take place in **Chemical Ideas 10.3**.

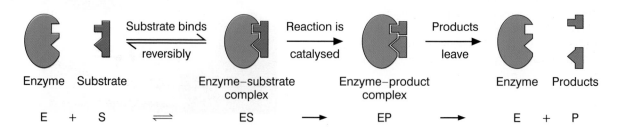

Figure 52 Illustration of the 'lock and key' model of enzyme catalysis.

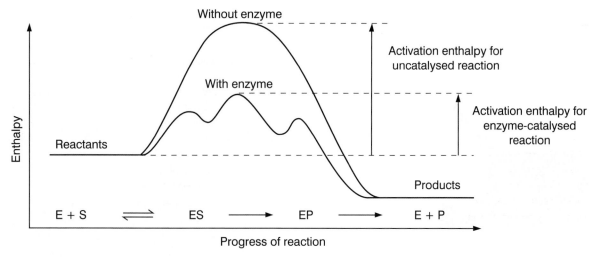

Figure 53 Lowering of the activation enthalpy barrier in an enzyme-catalysed reaction.

Enzymes are usually only present in minute traces. If the substrate concentration is high enough, all the enzyme molecules will have substrate molecules attached. If the substrate concentration is increased further, no more enzyme–substrate complexes can be formed, and the rate at which substrate molecules pass through the reaction pathway and change into products remains the same. In this situation the reaction rate does not depend on the substrate concentration. Chemists say the reaction is **zero order** with respect to substrate.

When the substrate concentration is low enough, not all enzyme active sites will have a substrate molecule bound to them. The overall reaction rate will now depend upon how frequently enzymes encounter substrates, which will depend upon how much substrate there is – twice as much substrate, twice as many encounters. Reaction rates which depend upon concentrations in this way are called **first-order reactions**.

If the enzyme's active site contains ionisable groups, the enzyme's action will be affected by a change in pH. For example, if there is a –COOH group which acts by donating an H^+ to the substrate, raising the pH will turn it into –COO$^-$ and the enzyme will not be able to work.

The enzyme will also become inactive if its shape is destroyed. Figure 54 illustrates how the active site of an enzyme can be made up from the side chains of amino acids in different parts of the protein molecule. They are held close together by the enzyme's tertiary structure.

If the tertiary structure is broken, the enzyme loses its shape and the side chains are no longer close together. The active site is destroyed and the enzyme is said to be **denatured**.

The tertiary structure is held together by weak dipole–dipole bonds and hydrogen bonds. These can easily be broken by raising the temperature, which causes them to vibrate more vigorously. Ionic interactions holding the tertiary structure together can be broken by changing the pH.

So enzymes are *sensitive* to relatively small changes in *temperature* or *pH*.

In **Activity EP6.3** you can investigate how changing the concentration of the enzyme and the substrate affects the rate of a reaction.

You can study the effect of concentration on reaction rate in **Activity EP6.4** and determine the order of the reaction.

You can investigate enzyme kinetics further in **Activity EP6.5** which uses *urease* as an example.

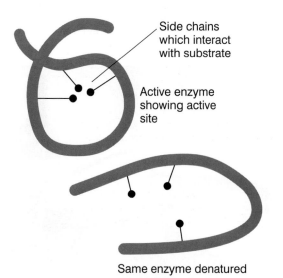

Figure 54 The shape of an enzyme is lost when it is denatured.

Figure 55 *The proteins in egg white are denatured when the egg is cooked.*

Enzymes at work

A diabetes test strip uses an enzyme to detect glucose. It is not the only medical application of enzymes. Many medical diagnostic kits are based on the use of enzymes.

However, medical applications use only small quantities of enzymes. Very much larger amounts are used in the food industry and in the manufacture of washing powders, and almost all of these are *hydrolases* – enzymes which hydrolyse fats, proteins or carbohydrates.

In the food processing industry, two of the major uses of enzymes are

- producing glucose syrup (used as a sweetener in food products) by breaking down starch with enzymes like α-amylase
- making cheese using rennet enzymes: these break down the milk protein casein and cause the separation of the curds (solid) from the whey (liquid).

Other uses include baking, brewing and fruit processing.

Figure 56 *Enzymes are important in the production of bread, cheese and wine.*

Figure 57 *Proteases are added to 'biological' detergents to remove protein stains such as grass, blood, egg and sweat.*

Many biological washing powders contain one or more enzymes to assist in the removal of stains. The enzyme is usually a *protease* to hydrolyse proteins in blood and food, but a *lipase* may also be added to break down fats. More recently, cellulases have been added too. These break down the tiny surface fibres that give older clothes a fluffy, dull look. Protein engineering is being used to make enzymes which are more stable in hot washes, or to create more active enzymes which will do their job at lower temperatures.

Enzymes are finding applications in wider areas of waste treatment. For example, an enzyme is being used to destroy cyanide ions which are left over after gold extraction or after the production of some polymers. Enzymes can also be used to help break up oil spillages. They also find their way into fashion, where they are used, for example, to give the 'stonewash' appearance in jeans.

EP7 *Summary*

This unit has introduced you to some of the things chemists have found out about proteins and their building blocks, amino acids. Amino acids (except glycine) are chiral molecules and this introduced you to optical isomerism, a second type of stereoisomerism. Studying amino acids led you to revisit earlier work on carboxylic acids, amines and amides and introduced you to a new analytical technique for structure determination, nuclear magnetic resonance spectroscopy.

Proteins play a wide variety of roles in our bodies: they are molecules of almost infinite diversity, despite all being made by the same type of condensation reaction which results in the formation of a secondary amide group.

This diversity can be explained in terms of

- the sequencing of the different structural units of the 20 naturally occurring α-amino acids (the primary structure)
- the interactions between these units which give proteins their precise three-dimensional shapes (the secondary, tertiary and quaternary structures).

Understanding these points led you into deeper consideration of molecular geometry and intermolecular forces. You also saw how a combination of these factors can lead to specific recognition of one molecule by another – a process which is essential in the building up of proteins according to the information contained in the genes on the DNA molecules stored in the nuclei of cells.

Consideration of one particular protein – the hormone insulin – provided a setting for this chemistry. It also provided an example of how chemists' understanding of proteins has reached a level enabling them to manipulate the processes which go on in a cell, in order to produce new proteins which are modified versions of those we have inherited. Central to this new field of protein engineering are the techniques of computer molecular graphics and recombinant DNA technology. Monomeric insulin is one protein produced by such technology. The dynamic equilibrium which exists between insulin hexamers and monomers led you to study the effect of concentration on chemical equilibrium.

The unit ended with a more general look at one class of proteins – the enzymes – which control the rates at which many of the chemical reactions in our bodies take place. This led to the need to develop a more detailed understanding of catalysis and how reaction rates are affected by the concentrations of chemicals.

Activity EP7 will help you to summarise what you have learned in this unit.

THE STEEL STORY

Why a unit on THE STEEL STORY?

This unit tells the story of the production of steel, with emphasis on the redox reactions involved and the huge scale of the process. You will learn about the large variety of steels possible and discover how the composition of a steel is related to the job it has to do.

The story of steel continues with the problems of corrosion. Rusting is introduced as an electrochemical process and various methods of rust prevention are considered. You will use standard electrode potentials to explain observations and make predictions about redox reactions. The importance of recycling steel is then discussed, together with some of the problems which must be overcome.

Looking at the composition of steel leads into a detailed study of the properties of iron and other d-block transition metals. These are the structural metals used by engineers to make things we need in everyday life. The metals and their compounds are also of great importance as catalysts, both in industry and in biological systems. The unique chemistry of transition metals is closely related to their electronic structure: variable oxidation state, catalytic activity, complex formation and the formation of coloured compounds. These ideas are built on in the final part of the unit.

Overview of chemical principles

In this unit you will learn more about these ideas introduced in earlier units in this course …

- extraction of metals (**From Minerals to Elements**)
- metallic bonding (**The Elements of Life**)
- electron energy levels in atoms (**The Elements of Life** and **From Minerals to Elements**)
- atomic absorption and emission spectra (**The Elements of Life**)
- redox reactions and oxidation numbers (**From Minerals to Elements**)
- chemical equilibrium (**The Atmosphere** and **Engineering Proteins**)
- catalysis (**Developing Fuels, The Atmosphere** and **Engineering Proteins**)

… as well as learning new ideas about

- the nature and production of steel
- how the properties and uses of steel are related to its composition
- redox reactions and electrochemical cells
- predicting the direction a redox reaction can take
- the properties of the d-block elements
- why compounds are coloured
- complex formation
- the effect of complexing on redox reactions.

THE STEEL STORY

SS1 *What is steel?*

There is no one material called steel: just as there are many plastics, so there are thousands of different steels. Steel is the general name given to a large family of **alloys** of iron with carbon and a variety of different elements.

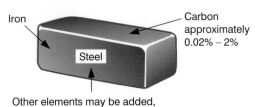

Iron

Carbon approximately 0.02% – 2%

Steel

Other elements may be added, eg manganese, chromium, nickel

Figure 1 The composition of steel.

The composition of a steel is determined by the job it has to do. Even small differences in the composition have a dramatic effect on its properties. This is particularly true in the case of carbon. Iron containing 4% carbon is extremely brittle and has a limited range of uses. With 0.1% carbon, however, it is easily drawn into wire form and is ideal for making staples or paper-clips. With 1% carbon, steel is stronger, without being too brittle, and was used to make the huge cables supporting the Humber Bridge.

Alloys

An alloy is a mixture of a metal with one or more other elements. The components are mixed together while molten and allowed to cool to form a uniform solid. The presence of other elements in a metal changes its properties and can often increase its strength. For example, brass is an alloy of copper and zinc and is stronger than either pure metal. Duralumin is used for building aircraft: it is an alloy of aluminium and copper with smaller amounts of magnesium, iron and silicon.

The presence of other elements is very important too. The steel cables on the Humber Bridge (Figure 2) contain manganese, chromium and silicon in addition to iron. Some of the metals commonly added to steels are shown in Table 1.

Metal	Symbol	Metal	Symbol
aluminium	Al	molybdenum	Mo
chromium	Cr	nickel	Ni
cobalt	Co	niobium	Nb
copper	Cu	titanium	Ti
lead	Pb	tungsten	W
manganese	Mn	vanadium	V

Table 1 Metals commonly added to steels.

Figure 2 The Humber Bridge: the cables are made of steel.

Not all elements are good news for steelmakers. Even small amounts of phosphorus, sulphur or dissolved gases (such as oxygen, nitrogen and hydrogen) can lead to poor-quality material. Brittle steel is a problem anywhere, but in an oil-rig or oil pipeline in the North Sea it could lead to catastrophe.

Changing the composition of steel is not the only method steelmakers use to adjust its properties. They can subject the final steel to varying degrees of heating and cooling (heat treatment) and work it by rolling or hammering (work treatment). All these processes modify the metal structure and affect its properties.

Steel is such a versatile material because both its composition and structure can be adjusted to tailor its properties exactly to the uses you have in mind. The possibilities for variety are almost endless.

In this unit, you will be concerned only with its composition and how this is controlled during manufacture.

Every batch of steel is destined for a particular customer who specifies the requirements for its composition and other treatments necessary, according to its eventual use. This detailed recipe is called the *specification* for that batch of steel.

ASSIGNMENT I

Write down the names of 10 items made from steel. Select items which illustrate the wide range of steels available.

In **Activity SS1.1** you can find out for yourself how much of one element, manganese, there is in a familiar item, such as a staple or a paper-clip.

You can read about why some compounds are coloured in **Chemical Ideas 6.7.** Many of the reactions in this unit are **redox reactions**. You can revise ideas about redox reactions in **Chemical Ideas 9.1**.

In **Activity SS1.2** you can carry out a redox titration.

SS2 *How is steel made?*

Nearly all new steel, whatever its final composition, is made from the same starting material. This is impure iron from a blast furnace, produced as a molten liquid. The iron has many other elements dissolved in it. The most important are carbon, silicon, manganese, phosphorus and sulphur.

You can see the composition of a typical sample of blast furnace iron in Table 2.

Element	Fe	C	Si	Mn	P	S
% by mass	94.0	4.42	0.66	0.41	0.085	0.027

Table 2 An analysis of typical blast furnace iron.

Figure 3 Iron is extracted from iron ore in a blast furnace. Here molten iron is being tapped from the base of the furnace.

Activity SS2.1 will help you to remember what happens in a blast furnace and understand why the iron produced is so impure.

Metal of this composition is very brittle. To turn it into steel, the carbon content must be lowered, most of the phosphorus and sulphur removed, and other elements added before the material is allowed to solidify.

All of this is achieved in the **Basic Oxygen Steelmaking (BOS) process,** in which batches of about 300 tonnes of high-quality steel are made in just 40 minutes. If possible, watch a video of the steelmaking process. You may be fortunate enough to visit a steelworks.

Making steel on a large scale with the right composition is a highly skilled business, involving sophisticated technology. The chemistry is spectacular!

ASSIGNMENT 2

The percentage of carbon by mass in the iron may seem rather low and insignificant (see Table 2). If we look at it in a different way its importance becomes clearer.

Use the information in Table 2 to calculate the relative amount in moles of iron and carbon in blast furnace iron. Neglecting the contributions from the other elements, work out the *percentage of moles* of the two elements, iron and carbon.

Approximately how many atoms of carbon are there in every 100 atoms of product from a blast furnace?

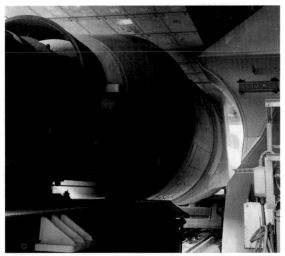

Figure 4 Molten impure iron from the blast furnace arrives at the BOS plant in special rail wagons called torpedoes. Hot metal is tipped from these into the ladle.

Removing unwanted elements

About 300 tonnes of molten iron from a blast furnace are poured into a huge container called a *ladle*.

Removing sulphur

Sulphur is the first element to be removed. This is done in a separate reduction process before the main steelmaking reactions take place. Several hundred kilograms of powdered magnesium are injected through a vertical tube, called a *lance,* into the molten iron in the ladle. In a violent exothermic reaction the sulphur is reduced to magnesium sulphide, which floats to the surface and is raked off:

$$Mg + S \rightarrow MgS$$

ASSIGNMENT 3

a Write an ionic equation to show what happens to sulphur during the reaction with magnesium and explain why this process is a reduction.

b Draw a dot–cross diagram to show the bonding in magnesium sulphide.

Removing other elements

Carbon, phosphorus and other elements are removed by direct oxidation with gaseous oxygen (the O in BOS).

$$C + \tfrac{1}{2}O_2 \rightarrow CO$$
$$Si + O_2 \rightarrow SiO_2$$
$$Mn + \tfrac{1}{2}O_2 \rightarrow MnO$$
$$4P + 5O_2 \rightarrow P_4O_{10}$$

Also, unavoidably
$$Fe + \tfrac{1}{2}O_2 \rightarrow FeO$$

The huge ladle brings the molten desulphurised iron to the steelmaking vessel or *converter*, which already contains some scrap steel (Figure 5). Now is the time for some very violent chemistry!

(a)

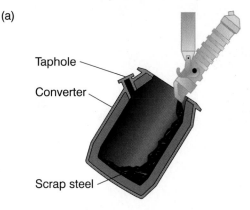

(b)

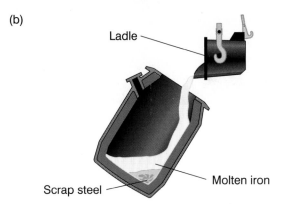

Figure 5 Adding (a) scrap steel and (b) molten iron to the converter.

Figure 6 Scrap steel being added to the converter.

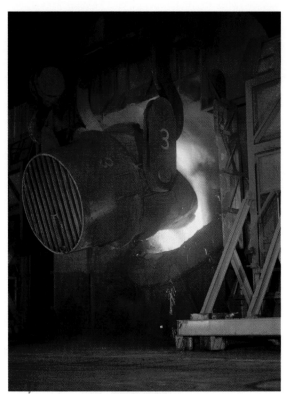

Figure 7 Molten iron being transferred from the ladle to the converter.

The converter turns into a vertical position and a water-cooled lance gradually inches its way down close to the surface of the iron. A supersonic blast of oxygen under pressure forces its way into the vessel and creates a seething foam of molten metal and gas which is blasted up the walls of the converter (Figure 8). Over the next 20 minutes or so, most of the impurities of carbon, silicon, manganese and phosphorus, as well as some of the iron, are oxidised.

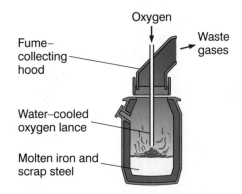

Oxygen

Fume-
collecting
hood

Waste
gases

Water–cooled
oxygen lance

Molten iron and
scrap steel

Figure 8 During the oxygen blow, most of the impurities are oxidised and a slag forms on the surface of the molten iron.

Look again at the equations on p. 167. Carbon monoxide escapes as a gas and is collected via a hood over the vessel. The other oxides remain in the converter and must be separated from the molten metal.

The oxides of phosphorus and silicon are acidic and will react with bases to form salts. So, soon after the oxygen blow has started, a mixture of calcium oxide and magnesium oxide (made by heating limestone and dolomite) is added to the converter.

These·are **basic** oxides (the B in BOS) and react with the **acidic** oxides to form a molten 'slag' which floats to the surface. The oxides of manganese and iron also collect in the slag.

The slag can be separated from the molten metal because the two have different densities and form two layers. You may have used the same principle to separate two immiscible liquids in the laboratory using a separating funnel.

ASSIGNMENT 4

a Suggest how the carbon monoxide collected may be used elsewhere on the plant.

b Apart from economic reasons, why is it not a good idea to release the carbon monoxide into the air?

c When calcium oxide reacts with the acidic oxides SiO_2 and P_4O_{10}, the products are *salts*. Write down possible names and formulae for the two salts formed.

d Explain why CaO is considered to be a basic oxide and P_4O_{10} an acidic oxide.

Keeping track

The process is closely monitored in the control room where a computer models the conditions in the converter. Two minutes before the predicted end of the oxygen blow an automatic sampling device, called a *sublance,* descends into the converter. It measures the temperature and carbon content and removes a sample of metal for analysis.

The computer uses the up-to-date information on temperature and carbon content to predict exactly how much more oxygen is needed to reach the target composition and temperature. The sample withdrawn is rushed to the analysis laboratory to determine the percentages of the elements in the steel.

Measuring the composition

Analytical chemists quickly measure the composition of the steel using *atomic emission spectroscopy*. This involves making the steel sample into an electrode for an electric arc, so that each element present emits a characteristic line spectrum. The intensities of the lines are proportional to the concentration of atoms of each element. At this stage the analysis involves only a few elements, but the method can be used to monitor up to 20 elements.

Figure 9 The progress of the blow is monitored in the control room.

You can revise earlier work on atomic emission spectroscopy in **Chemical Ideas 6.1**.

During the oxygen blow the elements are oxidised in a sequence illustrated in Figure 10. When most of the impurities have been removed, some of the iron is also oxidised. This is unavoidable but is kept to a minimum.

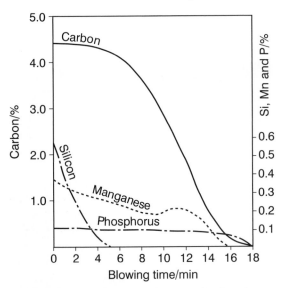

Figure 10 Removal of elements during steelmaking (note that %C and %Si, Mn and P are plotted on different scales; the rise in %Mn after 10 min is because the scrap steel used had a relatively high manganese content).

Controlling the temperature

Careful control of temperature is a vital part of making steel. At the end of the oxygen blow a temperature of about 1700 °C–1740 °C must be reached. This is the *target tapping temperature*. Higher temperatures waste expensive energy and can cause severe damage to the

converter linings. Lower temperatures can be even more costly if the metal solidifies before it is supposed to.

Remember, there is no external heating. The oxidation reactions are all highly exothermic and generate a tremendous amount of energy. Some of this energy is of course lost to the surroundings. The rest is absorbed in raising the temperature of the converter contents. Some energy is absorbed by the scrap steel (which was added to the converter at the start) as it melts and heats up, maintaining a 'heat balance'. The scrap therefore neatly serves both as a coolant and a source of recycled steel.

ASSIGNMENT 5

During the oxygen blow, the elements in the converter compete for the oxygen. The order of their removal as oxides depends on their affinity for oxygen at the high temperature involved and on the amount of each element present.

Use Figure 10 to answer these questions.

a Which element is the first to be removed?

b Which element is the last to be removed?

c Is this what you would have expected from what you know about the reactivity of these elements with oxygen at lower temperatures?

d At the high temperatures in the converter, sulphur has a similar affinity for oxygen to that of iron. Suggest why it is better to remove most of the sulphur *before* the oxygen blow in a separate process.

At the end of the blow

At the end of the blow, the converter is rotated to pour off the molten steel through a hole near the top into a ladle (Figures 11 and 12), and then tilted in the opposite direction to remove the slag (Figure 13).

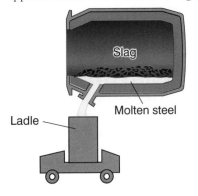

Figure 11 Tapping off the molten steel.

Lumps of aluminium are thrown into the ladle as it is being filled with the molten metal. The aluminium combines with excess oxygen that dissolved in the metal during the oxygen blow. Aluminium oxide forms and floats to the surface.

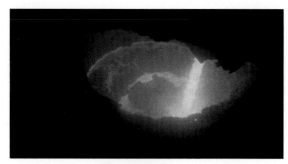

Figure 12 Tapping off molten steel from the converter.

Figure 13 Removing the slag.

Carbon, manganese and silicon, which were removed in the blow, are often added at this stage. Chromium and more aluminium are other common additions. More exotic elements such as niobium, molybdenum and tungsten may also be added. Argon is blown into the liquid steel, using a lance, to stir the mixture and make sure the composition and temperature are uniform.

Several processes are now available for refining the metal while in the ladle. The *ladle arc furnace*, for example, allows steel in the ladles to be reheated electrically so that it is possible to control the temperature precisely. The ladle arc furnace is particularly important in producing high C/low P steels which are used to make steel rods.

Throughout the addition processes, the temperature is closely monitored and samples are taken frequently for analysis. With this information the computer constantly updates its predictions.

A final 'trimming' adjusts the concentrations of elements in the steel to the required values and the molten steel is now ready for casting. This is usually a continuous process, by casting into very long strands of solid metal (Figure 14) or it may be poured into moulds in which it solidifies into ingots, rather like making ice cubes.

The effects of various elements on the properties of steel are illustrated in **Activity SS2.5.**

Activities **SS2.2** to **SS2.4** look more closely at some of the changes which happen during steelmaking.

Meeting the specification

By now the elements which are not wanted – and also some that are – have been removed from the iron. The elements which are wanted have to be put back. The computer model predicts the exact quantity of each substance needed to achieve the **specification** for the particular batch of steel.

ASSIGNMENT 6

It may seem strange to you that elements such as carbon which have just been removed in the converter are now re-added.

a Why do you think the carbon content is adjusted in this way, rather than by stopping the oxygen blow at the appropriate carbon concentration?

Steelmaking is a batch process. It is found to be most cost-effective to produce batches of around 300 tonnes. However, it is possible to make steel by a continuous process.

b Why do you think this is less economical?

Figure 14 Continuous casting of steel into narrow strands of metal.

Can steelmaking be improved?

World production of steel continues to rise even though plastics and other synthetic materials are taking over the role of steel in some areas. There is a tremendous and growing demand for even higher-quality specialist steels with uniform and consistent compositions.

The BOS process is kept constantly under review and new technology is introduced to produce batches of steel to these very precise specifications as efficiently and economically as possible. For example, recent advances in the continuous casting process have resulted in great energy savings. There are plans too for direct rolling of the hot steel from the caster – another way of saving energy.

The process is continuously monitored by computers and there are quality measurements at all stages of the process. Computers alert plant engineers when there are problems and the programs give advice on possible solutions – or perhaps correct the problem themselves.

As we become more concerned about the environment and recycle more and more steel, larger and more efficient **electric arc furnaces** are taking over a greater share of the production. Over 30% of the steel worldwide is now produced in these furnaces.

The electric arc furnace

An electric arc furnace uses old scrap steel which is melted by the heat generated when a spark is produced between carbon electrodes. Lime is added and the impurities removed as a slag.

By carefully selecting the scrap and making necessary additions, relatively small batches of steel are made to meet given specifications.

On a much smaller scale

Steel making has traditionally been a large-scale process, making batches of 300 tonnes at a time. If only small quantities of a particular steel are required, it is uneconomic to use a large-scale operation and the *mini-mill* route is becoming increasingly used.

This involves an electric arc furnace, using scrap steel, but as few as 75 tonnes can be used. Adjacent to the furnace is a casting machine to make the product.

Dawn's Story

"My name is Dawn Herridge and I took Chemistry, Physics and Maths A levels. I then went to Manchester University to study for a degree in Chemistry. Following this I moved to UMIST and completed an MSc in the 'Advanced Chemistry of Materials'. I remained at UMIST for a further three years to obtain my doctorate. My research subject involved using spectrometry to understand the chemistry of ozone depletion.

"I have now been working for Corus (formerly British Steel) for almost two years and I am currently working at the Steelmaking Plant in Scunthorpe.

Figure 15 Dawn Herridge on site at the Corus BOS plant in Scunthorpe.

"I am based in the Development Department where I look at ways of improving current practice to produce a better final product. A lot of my work is associated with tyre cord, which is used in tyres supplied to the automotive industry. This is a very demanding product as the steel is drawn down to the same diameter as a human hair!

"A major part of my work involves looking at ways to optimise the chemical composition of the steel. This can be achieved at the ladle arc furnace by the addition of a designed synthetic slag containing lime, silica and dolomet (a mineral of calcium oxide and magnesium oxide). Slag is added to remove impurities in the steel such as sulphur and phosphorus. The slag is designed to have sufficient MgO content to prevent erosion of the ladle lining. I spend a lot of time on plant, taking molten slag samples which are then analysed by a technique known as X-ray fluorescence. Taking the sample is no easy matter, for the temperature of the slag is usually between 1580 °C and 1620 °C. From the analysis I am given, I use a computer program to determine the liquid content and reactivity of the slag. After numerous trials have been carried out and the slag has been found to meet all our requirements, we introduce the process into the plant on a routine basis.

"Besides this work, I receive phone calls and e-mails asking me to investigate problems that have arisen on plant. Something new happens every day!

"Training is important for both management and personal development. I have already attended management, technical and personal development courses. These have included an 'Outward Bound' course in the Peak District, where we undertook hiking, climbing, abseiling and potholing to improve our leadership and team-working skills. Great fun!"

SS3 *Rusting*

A return to nature

Many metals, including iron, occur in the Earth's crust as oxides. This is because the change from a metal to its oxide is an energetically favourable process – in other words, the oxide is more stable than the metal. Indeed, to reverse the process and extract the metal from its ore requires a great deal of energy. No wonder then, that given the opportunity iron tends to re-form its oxide – ie it **rusts**. Rusting is a common name for the **corrosion** of iron.

Figure 16 shows the redox cycle involved when a metal is extracted from its ore and then corrodes.

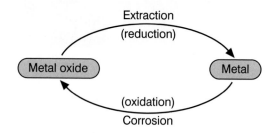

Figure 16 Redox cycle involved in the extraction of metal from oxide ores and subsequent corrosion of the metal.

Rusting is a redox reaction. To understand this section, you will need to find out about electrode potentials and electrochemical cells. You can do this by reading **Chemical Ideas 9.2**. In **Chemical Ideas 9.3** you can see how electrode potentials can be used to predict the direction a redox reaction can take.

Activities SS3.1–SS3.3 will help you understand these ideas.

The rusting of steel is a familiar problem to many car owners; the iron the simply returns to nature.

Cars rust because the steel they are made from reacts with oxygen and water in the atmosphere. When iron or steel rusts a hydrated form of iron(III) oxide with variable composition ($Fe_2O_3.xH_2O$) is produced. This oxide is permeable to air and water and does not form a protective layer on the metal surface. So the metal continues to corrode under the layers of rust.

Iron and steel will rust whenever they are in contact with moist air, but the rate of rusting is greatly influenced by other factors, such as impurities in the iron, the presence of acids or other electrolytes in the solution in contact with the iron, and the availability of dissolved oxygen in this solution.

What happens during rusting?

Rusting is an electrochemical process. Electrochemical cells are set up in the metal surface, where different areas act as sites of oxidation and reduction.

Two half-reactions involved in rusting are

$$Fe^{2+}(aq) + 2e^- \rightarrow Fe(s) \qquad E^\ominus = -0.44\,V$$
$$\tfrac{1}{2}O_2(g) + H_2O(l) + 2e^- \rightarrow 2OH^-(aq) \qquad E^\ominus = +0.40\,V$$

The reduction of oxygen to hydroxide ions occurs at the more positive potential, and so electrons flow to this half-cell from the iron half-cell in which iron is oxidised to iron(II) ions.

Figure 17 shows what happens when a drop of water is left in contact with iron or steel.

The concentration of dissolved oxygen in the water drop determines which regions of the metal surface are sites of reduction and which regions are sites of oxidation.

At the edges of the drop, where the concentration of dissolved oxygen in the water is higher, oxygen is reduced to hydroxide ions.

$$\tfrac{1}{2}O_2(g) + H_2O(l) + 2e^- \rightarrow 2OH^-(aq)$$

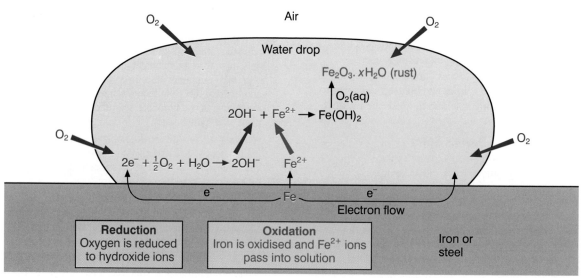

Figure 17 Rusting is an electrochemical process.

The electrons needed to reduce the oxygen come from the oxidation of iron at the centre of the water drop, where the concentration of dissolved oxygen is low. The $Fe^{2+}(aq)$ ions pass into solution:

$$Fe(s) \rightarrow Fe^{2+}(aq) + 2e^-$$

The electrons released flow in the metal surface to the edges of the drop.

This explains why corrosion is always greatest at the centre of a water drop or under a layer of paint: these are the regions where the oxygen supply is limited. Here 'pits' are formed where the iron has dissolved away.

Rust forms in a series of secondary processes within the solution, as Fe^{2+} and OH^- ions diffuse away from the metal surface. It does not form as a protective layer in contact with the iron surface.

$$Fe^{2+}(aq) + 2OH^-(aq) \rightarrow Fe(OH)_2(s)$$
$$Fe(OH)_2(s) \xrightarrow{O_2(aq)} Fe_2O_3.xH_2O(s)$$

Some ionic impurities, such as sodium chloride from salt spray near the sea, promote rusting by increasing the conductivity of water.

Other ionic compounds can interfere with the electrochemical reactions involved and actually inhibit rusting. This might happen if the positive ions form an insoluble hydroxide with the $OH^-(aq)$ ions produced in the oxygen half-reaction, or if the negative ions form an insoluble iron(II) compound. Sodium chloride is clear on both these fronts.

The pH of the solution is also important. Rusting is accelerated under acidic conditions but inhibited under alkaline conditions.

In **Activity SS3.4** you can investigate some of the chemistry involved in rusting and in one method of rust prevention.

Keeping nature at bay

The simplest way of protecting steel against rust is to provide a barrier between the metal and the atmosphere. The barrier may be oil, grease or a coat of paint.

*Figure 18
The 'Angel of the North' sculpture adjacent to the A1 road near Gateshead is made from weather-resistant high-strength COR-TEN steel. The steel contains 0.25%–0.40% copper and weathers to a rich brown colour.*

Figure 19 Early advertising signs were saved from the damaging effects of weather and pollution by a corrosion-resistant coating of vitreous enamel.

A barrier made from an organic polymer is increasingly used. The steel is coated with a plastic film – a colourful and flexible solution to the rusting problem. A quick look around at home will provide a wealth of examples – sink drainers, refrigerator and dishwasher shelves, milk bottle carriers and bicycle baskets.

Sometimes iron is covered with a thin layer of another metal. Now, most car manufacturers make their car bodies from **galvanised** steel. This has a protective coating of zinc. When the galvanised surface is undamaged the zinc layer is protected from corrosion by a firmly adhering layer of zinc oxide.

Even if the coating is scratched the protection is still maintained because the zinc corrodes in preference to the iron. The zinc is being used as a **sacrificial metal**.

Great improvements in protecting cars from rusting have been made over the last few years (Figure 20). Indeed, some manufacturers give a 7-year warranty against rusting. Even so, if steel is used in car bodies, the rusting problem is only postponed, not eliminated.

Figure 20 The whole car body is dipped into an aqueous dispersion of tiny particles of polymer. The coating is then heat treated to form a film on the steel which will protect it from rusting for about 10 years.

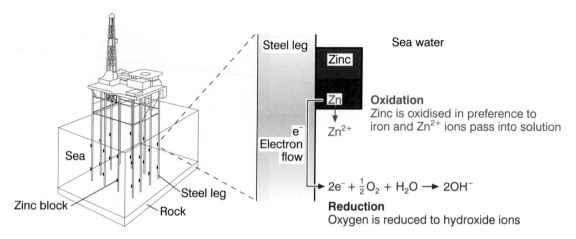

Oxidation
Zinc is oxidised in preference to iron and Zn^{2+} ions pass into solution

$$2e^- + \tfrac{1}{2}O_2 + H_2O \rightarrow 2OH^-$$

Reduction
Oxygen is reduced to hydroxide ions

Figure 21 In a North Sea oil-rig sacrificial protection of the steel supports is achieved by using zinc blocks: zinc is oxidised in preference to iron and so protects the steel legs from corrosion.

Figure 22 Constructing an oil rig. Zinc blocks are attached to the steel posts for sacrificial protection.

One of the earliest examples of using a sacrificial metal was suggested by Humphry Davy in 1824 to protect the metal sheathing on sailing ships from corrosion.

ASSIGNMENT 7

Use the standard electrode potentials given below to explain why zinc and magnesium, but not tin, can be used as sacrificial metals to protect steel.

Half-reaction	$E^\ominus/V$
$Mg^{2+}(aq) + 2e^- \rightarrow Mg(s)$	−2.36
$Zn^{2+}(aq) + 2e^- \rightarrow Zn(s)$	−0.76
$Fe^{2+}(aq) + 2e^- \rightarrow Fe(s)$	−0.44
$Sn^{2+}(aq) + 2e^- \rightarrow Sn(s)$	−0.14

In the sea, conditions are far from standard, but the *order* of electrode potentials for these reactions is not changed.

Today blocks of zinc are used to protect North Sea oil-rigs by transferring corrosion from a valuable complex steel structure to a readily replaceable metal lump (see Figures 21 and 22).

You can see the reason for using zinc by comparing the standard electrode potentials of the iron and zinc half-cells. Any metal with a more negative $E^\ominus$ value than iron could be used as a sacrificial metal.

Tin cannot be used as a sacrificial metal to protect steel, but as long ago as 1812, iron coated by tin was used to make containers, *tin cans*, to preserve cooked food.

A prize-winning invention

Napoleon remarked at the beginning of the 19th century that an army marches on its stomach. His armies were widely separated on campaigns in Russia and Spain and were severely limited by long supply lines from France and the lack of fresh provisions.

Napoleon offered a prize to anyone who could suggest a solution to his food supply problem. In 1812 he awarded 12 000 Francs to a French confectioner, Nicolas Apert, for inventing a method of preserving cooked food by sealing it in an airtight glass jar while it was still hot.

Meat, vegetables and fruit could be kept palatable for long periods of time. Later that year an Englishman, Peter Durand, adapted Apert's method by using a tin-plated iron canister instead of a glass bottle – with obvious advantages. So the tin canister or 'tin can' was born.

In 1824 Captain Sir Edward Parry set out on his third voyage in search of the 'North West Passage' from the Atlantic to the Pacific Ocean, taking with him a good supply of canned food. His aim was to find an Arctic route to India and the Far East. One of his ships became ice-bound and he was forced to abandon it together with a large quantity of stores.

Some of the cans were recovered by Captain Ross in a similar expedition 4 years later (Figure 23). In 1939 one of these cans was opened. After more than a century, the roast veal and gravy inside were still wholesome.

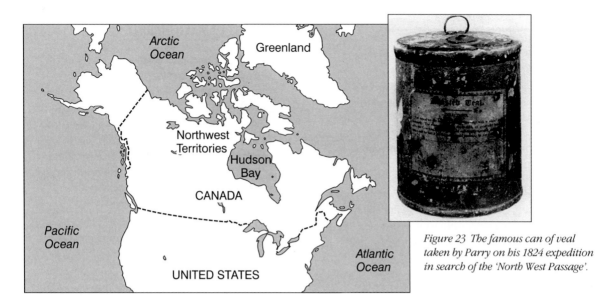

Figure 23 *The famous can of veal taken by Parry on his 1824 expedition in search of the 'North West Passage'.*

ASSIGNMENT 8

a When preserving food, why is it important that the jar or can is airtight and sealed while the food is still hot?

b Peter Durand used iron plates dipped into molten tin and soldered together to make a container. What was the purpose of the tin?

c What would happen if a tin can were scratched on the outside so that the iron showed through? Refer to the standard electrode potentials in Assignment 7.

d Suggest reasons why it is not a good idea to use zinc instead of tin to coat the can.

Stainless steel – the perfect solution?

Many of the steel items you use at home, particularly those which come into regular contact with water – such as the kitchen sink, cutlery and the drum of your washing machine – are made of stainless steel which needs no further protection.

Stainless steel was developed in 1913 by a Sheffield chemist called Harry Brearley. He was investigating the rapid wear of rifle barrels and decided to try a steel containing a high level of chromium to see if this would prolong their life.

Routine analysis of steel at that time involved dissolving it in acid, and here Brearley met an unexpected difficulty. His high-chromium steel would not dissolve. He also noticed that samples of it left lying around the laboratory stayed shiny.

Brearley immediately realised that he had found a steel which would make excellent cutlery. It would not need to be dried carefully after washing or need frequent polishing.

He did encounter some prejudice about the idea. One of the foremost cutlers in Sheffield thought the idea 'contrary to nature', while another is said to have remarked that 'rustlessness is not so great a virtue in cutlery, which of necessity must be cleaned after each using'! We now take it for granted that our knives and forks stay shiny and are not attacked by the acids in food.

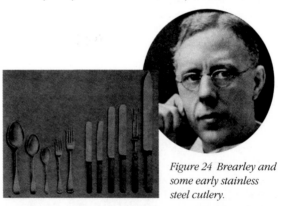

Figure 24 *Brearley and some early stainless steel cutlery.*

Stainless steel does not rust because it forms a surface layer of chromium(III) oxide (Cr_2O_3). Unlike rust, this oxide is not hydrated and adheres closely to the metal surface. The oxide layer is invisible to the naked eye, being only a few nanometres thick, and allows the natural brightness of the metal to show through. Even so, it is impervious to air and water and so protects the metal beneath it. Furthermore, if you scratch the surface film it quickly reforms and restores the protection.

Is stainless steel then the perfect solution to the rusting problem? Unfortunately things are not that simple. Stainless steel is expensive and we have to take this into consideration when deciding which steel to use.

Activity SS3.5 will help you to make sure you have understood the ideas in this section.

SS4 *Recycling steel*

Why recycle?

About 40% of the world's steel production is from recycled steel. This means that over 200 million tonnes of iron are recovered each year, with an energy saving equivalent to 160 million tonnes of of coal or 100 million tonnes of oil – about 40% of the UK's annual energy consumption.

You saw earlier in the unit that recycled scrap is an integral part of the BOS process. Scrap steel makes up about 18% of every cast of 'new' metal. In the electric arc process for making steel, *only* scrap is used.

Much of the scrap used in steelmaking comes from the steelworks itself – waste from previous batches, miscasts, etc – or from industries which make the steel products. The composition of this type of scrap is well known. In this respect, the steel industry consumes its own waste.

Scrap from discarded products, such as cars and washing machines, however, must be carefully graded and selected. Steelmakers need to have a good idea of the content of the scrap metal to avoid adding unwanted elements to the steel. Some of the elements present improve the properties of the steel. For example, many mild steels now contain low concentrations of transition metals such as nickel and chromium from the added scrap. Other elements such as tin and copper can cause problems if incorporated into the 'new' steel.

Figure 25 An electric arc furnace recycles scrap steel.

ASSIGNMENT 9

a Explain why steel manufacturers prefer to use scrap from their own steel works or from manufacturers who make products from their steel?

b Decommissioning nuclear power stations raises the issue of what to do with the steel used in the construction, about 5000 tonnes per site. Much of this steel is mildly radioactive because of the presence of an isotope of cobalt, ^{60}Co. This is formed from the stable ^{59}Co present in the original steel by *neutron capture*.

$$^{59}\text{Co} + {}^{1}\text{n} \rightarrow {}^{60}\text{Co} + \gamma$$

^{60}Co is a gamma (γ) emitter with a half-life of 5.3 years.

If this radioactive steel is diluted with ordinary scrap steel in the ratio of 1:25, it would bring the activity down well below permitted levels. It is estimated that using this diluted steel to make everyday objects would contribute an additional 1 microsievert (μSv) per year to the average dose received by an individual. The average dose of radiation received by an individual in the UK is 2600 μSv per year (2200 μSv from natural sources and 400 μSv from artificial sources, mostly medical).

Write a brief report (200 words) summarising the factors which need to be considered when deciding how to dispose of waste steel from nuclear reactors.

Recycling used 'tin' cans

Recycling used 'tin' cans involves removing the tin coating from the steel. This has been done for a very long time using the waste metal from the tin plating works. Only since the 1980s have attempts been made to extract used 'tin' cans from household waste and recycle them on a large scale.

Figure 26 In the UK, we use 13 billion steel cans a year – almost enough for everyone to use a can a day. The cans weigh 40% less than those 30 years ago and the tin coating is less than 6×10^{-3} mm thick, an enormous saving in mineral resources, energy and waste. About 25% of the cans are made from recycled steel.

Shredding the cans and removing unwanted paper and residual food is a vital part of the preparation before detinning. One easy way of cleaning the cans is to burn off the unwanted material: unfortunately, during burning, the tin diffuses into the steel and makes it less useful. Mechanical shredding devices now shred and clean the 'tin' cans and the steel fragments are picked out magnetically. About 98% of the unwanted material can be removed in this way.

The cleaned and shredded tin cans are treated with a hot solution of sodium hydroxide in the presence of an oxidising agent. The tin dissolves as a compound of tin(IV), as shown in the half-equation below:

$$Sn(s) + 6OH^-(aq) \rightarrow [Sn(OH)_6]^{2-}(aq) + 4e^-$$
$$\text{stannate(IV) ion}$$

The steel left behind is rinsed and pressed into bales for transport to a steel plant. The tin can be recovered by electrolysis.

SS5 *A closer look at the elements in steel*

There may be many different elements present in a steel in addition to iron. A few, such as carbon and silicon, are non-metals but most are metals. Look back at Table 1 on p. 165, which shows the metals commonly added to steels.

If you look at the positions of these elements in the Periodic Table shown in Figure 27, you will see a clear pattern emerging. Many of the elements present in steel, including iron, are **d-block elements**.

Understanding the chemistry of d-block elements

To understand how steel behaves when exposed to weathering and what can be done to prevent corrosion,

or to understand how fruit juices can affect the inside of a food can, you need to know more about the chemistry of d-block elements.

These elements are sometimes called **transition metals** because they show a transition in properties between the reactive s-block metals and the less reactive metals on the left of the p-block. Their chemistry is very characteristic and is a direct result of their electronic structure.

***

Chemical Ideas 11.5 tells you about the properties of d-block elements.

You can remind yourself about energy levels in atoms and how electrons are arranged in these energy levels by reading **Chemical Ideas 2.4**.

***

Typical chemical properties of transition metals include

- the formation of compounds in a variety of oxidation states
- catalytic activity of the elements and their compounds
- a strong tendency to form complexes
- the formation of coloured compounds.

The elements on the edges of the d-block, such as scandium and zinc, do not show many of these properties and are not usually classed as transition metals.

Variable oxidation states

Metals like sodium or magnesium have just one oxidation state in all of their compounds, but transition metals form compounds in a range of oxidation states, many with beautiful and characteristic colours (Figure 28).

***

In **Activity SS5.1** you can explore a number of the colourful oxidation states of vanadium and use electrode potentials to make predictions about changes between them.

***

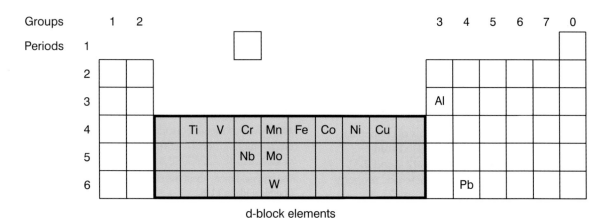

Figure 27 *The position in the Periodic Table of the metals commonly present in steel.*

Figure 28 *The oxidation states of vanadium: from left to right, the beakers contain solutions of vanadium in oxidation states +5, +4, +3 and +2.*

Catalysis

Another property characteristic of transition metals and their ions is that many can act as catalysts.

For example, iron is the catalyst in the manufacture of ammonia in the Haber process. Nickel is used in the hydrogenation of vegetable oils to make margarine. A platinum–rhodium mixture (sometimes with other transition metals) is the catalyst in the catalytic converters in car exhaust systems.

Catalysts also play a vital role in biological systems, enabling complex reactions to occur quickly in dilute aqueous solution at moderate temperature and pH – the conditions in a living cell. Many transition metal ions are required by living things in minute but definite quantities. Cobalt, copper, manganese, molybdenum and vanadium are all ultra-trace elements essential for the catalytic activity of various enzymes.

ASSIGNMENT 10

Make a list of some of the catalysts you have met already together with the reactions they catalyse.

How many are transition metals or compounds of transition metals?

You can also read about catalysis in **Chemical Ideas 10.4** and **10.5**.

In **Activity SS5.2** you will investigate the catalytic activity of cobalt(II) ions.

Complex chemistry

You may have met the blue complexes of copper(II) ions with water and ammonia ligands in **From Minerals to Elements (Activities M2.2** and **2.3).** Complex formation is a further characteristic property of transition metals.

Typically, a central metal ion is surrounded by six electron-donating ligands, although complexes with two and four ligands are also common.

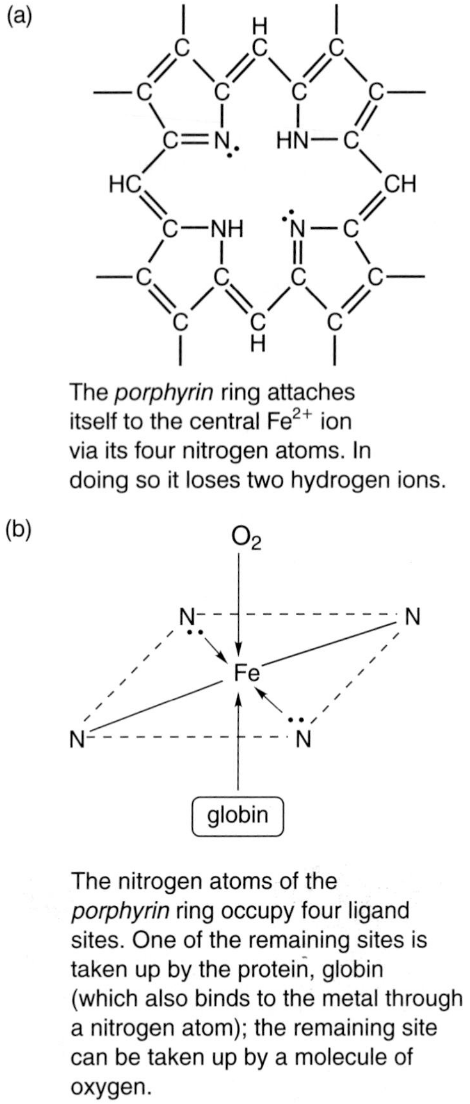

(a)

The *porphyrin* ring attaches itself to the central Fe^{2+} ion via its four nitrogen atoms. In doing so it loses two hydrogen ions.

(b)

The nitrogen atoms of the *porphyrin* ring occupy four ligand sites. One of the remaining sites is taken up by the protein, globin (which also binds to the metal through a nitrogen atom); the remaining site can be taken up by a molecule of oxygen.

Figure 29 *(a) The porphyrin ring system; (b) haemoglobin bound to an oxygen molecule to form oxyhaemoglobin.*

Iron forms the red complex, haemoglobin, responsible for transporting oxygen in the blood (**The Elements of Life** storyline, Section **EL2**). The structure of this complex is shown in Figure 29.

The oxygen molecule is relatively loosely attached to the iron. It is carried around the body in the bloodstream and released from the complex when needed.

The complexing of metal ions in solution affects electrode potentials of metal–metal ion half-cells and can affect the way metals corrode. Figure 30 shows how the environment inside a fruit can may cause problems.

You can read more about the chemistry of complexes in **Chemical Ideas 11.6.**

Chemical Ideas 9.4 looks at how complexing affects electrode potentials.

Peaches in Fruit Juice

A can of fruit contains carboxylic acids such as citric acid. The acids are partially dissociated into ions in solution and their anions – such as citrate ions – can form complexes with Sn^{2+} ions. The complexing removes Sn^{2+}(aq) ions from the solution and upsets the equilibrium between Sn(s) and Sn^{2+}(aq) ions. Under these conditions, the relative values of the electrode potentials of the tin and iron half-cells are reversed and the tin is now the more readily oxidised of the metals. So, in the presence of peaches and other fruit, the tin corrodes in preference to the iron.

The Sn^{2+} ions which become part of the food contents are not toxic – nor do they have an adverse effect on the flavour. In fact, they are responsible for some of the tangy taste we expect of canned fruit!

There may, however, be problems if we leave the tin on the shelf too long. As the thin layer of tin is sacrificially oxidised it will eventually be stripped off revealing the steel beneath. The acidic juices will then begin to react with the iron. Peaches are perhaps better eaten sooner rather than later!

Figure 30 The chemical environment inside a can of peaches reverses the relative values of the electrode potentials of the tin and iron half-cells.

In **Activity SS5.3** you will meet some complexes of d-block elements.

Coloured compounds

The colour of a transition metal compound depends on the oxidation state of the metal ion, the nature of the ligands surrounding it, and the spatial arrangement of these ligands. A wide variety of colours is observed (Table 3).

Gemstone	Colour	Ion present
blue sapphire	blue	V^{3+} or Co^{2+}
emerald	green	Cr^{3+}
topaz	yellow	Fe^{3+}
turquoise	blue-green	Cu^{2+}
amethyst	purple	Mn^{3+}
ruby	red	Cr^{3+}

Table 3 The colours of many gemstones are due to the presence of traces of d-block metal ions.

Figure 31 The Rose Window at York Minster: glass is coloured green with either chromium(III) or iron(II); cobalt(II) gives blue glass and copper(II) gives a blue-green colour.

d-Block metals can be expensive

The most abundant d-block element in the Earth's crust is iron which is relatively cheap. Some of the other d-block elements, however, can be expensive. The cost of the steel rises if these are used in steelmaking.

For example, stainless steel contains a minimum of 12% chromium (by mass) and usually nickel as well to make it corrosion resistant. A typical stainless steel might contain 74% iron, 18% chromium and 8% nickel. It is as much as five or six times more expensive than ordinary steel and so is used selectively.

The price of a metal depends on many other factors besides its abundance in the Earth's crust. These include the cost of mining the ore, the ease of extraction of the metal from the ore, the demand for the metal, transport costs and political factors in the countries involved.

Some elements are said to be *strategically critical* because one or two countries have a monopoly over their supply. For example, the Republic of South Africa holds more than 70% of the world's known reserves of chromium, with Zimbabwe having over half of the remainder, whereas most of the world's deposits of molybdenum are found in the US and Canada.

SS6 *Summary*

In this unit you have followed the story of steel production from the iron ore fed into the blast furnace to a finished steel product. This is not one story but many. Each starts with iron from the blast furnace but leads to one of a multitude of different steels, each tailored to the job it has to do. Steel production has to be a batch process and must be closely monitored. Recent advances have been aimed at producing higher quality steels to exact specifications, quickly and reliably – both on a huge scale and in small batches.

Many of the reactions involved in the BOS process are redox reactions and this allowed you to revise earlier work in this area. You also saw how atomic emission spectroscopy is used to check the composition of the steel during the process.

The story of steel next led to the problems of corrosion and the inevitable return to nature. Here you found out about the mechanism of rusting as an electrochemical process and what steps can be taken to slow it down. This led to a more detailed study of electrode potentials and electrochemical cells. You learned how electrode potentials can be used to predict the direction of a redox reaction.

Next, you considered the importance of recycling steel and some of the problems which must be overcome, for example, when recycling 'tin' cans.

Looking at the composition of steel led you into a more detailed study of iron and other d-block metals. These metals form compounds in a variety of oxidation states and you were able to use electrode potentials to predict the relative stability of these states. You learned about ligands and complex formation and the effect complex formation can have on electrode potentials.

Activity SS6 will help you to summarise what you have learned in this unit.

ASPECTS OF AGRICULTURE

Why a unit on ASPECTS OF AGRICULTURE?

Growing crops for food has long been a major human activity. The rapidly increasing world population means that the need to provide enough food without destroying our environment is one of the biggest challenges facing us.

The biological, chemical and physical processes occurring in soil are highly complex. They involve a number of interrelated reactions. This unit looks at the chemical nature of soil, and at some of the processes which occur as plants grow and decay. It then looks at ways in which a knowledge of these processes can be used to optimise crop yields and ensure our food supply.

Finally, chemical methods of pest control are studied. In particular, the unit covers the development of pesticides which do not persist in the environment, and also explores the applications of some herbicides and fungicides.

In considering these aspects of agriculture you will apply knowledge and understanding of chemical principles developed earlier in the course. These include ideas about intermolecular forces, rates of reaction, redox chemistry, chemical equilibrium and the behaviour of functional groups in organic molecules. You will also learn about the structures of alumino-silicate minerals, ion-exchange processes, and the redox chemistry of nitrogen, the first element in Group 5. The industrial manufacture of ammonia is covered, and there is a general overview of bonding, structure and properties.

Overview of chemical principles

In this unit you will learn more about …

ideas introduced in earlier units in this course

- intermolecular forces (**The Polymer Revolution**)
- the relationship between properties of substances and their structure and bonding (**From Minerals to Elements** and **The Polymer Revolution**)
- the Periodic Table and periodicity (**The Elements of Life**, **From Minerals to Elements** and **The Steel Story**)
- redox reactions (**From Minerals to Elements** and **The Steel Story**)
- rates of reaction (**The Atmosphere** and **Engineering Proteins**)
- the behaviour of functional groups in organic molecules (various units)
- chemical equilibrium (**The Atmosphere** and **Engineering Proteins**)

… as well as learning new ideas about

- structures and properties of silicates and clays
- ions in solution and ion-exchange processes
- the redox chemistry of nitrogen
- chemical equilibria involving gases and partition equilibria
- the selection of optimum conditions for the industrial manufacture of ammonia.

ASPECTS OF AGRICULTURE

AA1 *What do we want from agriculture?*

The impact of agriculture is enormous and has changed the face of the Earth. It is hard to take in the *scale* of agriculture today, as we try to provide food for an ever increasing number of people.

In this unit you will look at the production of crops for food, but remember that agriculture also involves the cultivation of plants such as cotton, grown for fibre, and sugar cane, grown to make alcohol for fuels, and forests producing timber.

As they grow, plants take carbon dioxide from the air into their leaves, and water and nutrients from the soil via their roots. When we harvest crops we disturb the natural nutrient cycles by removing large quantities of plants before the natural processes of decay take place.

Our planet cannot produce an *unlimited* supply of food for its human population. The problem of rapid population growth needs to be tackled, but the challenge to agriculture remains. The ultimate goal is to feed everyone adequately, without harming the environment. This involves producing enough food, of the right kind, in the right place, and at the right time.

How can we increase food production *without* encroaching on the world's remaining forests and wildernesses? It can be done by making the most efficient use of *existing* agricultural land – particularly by improving crop varieties and planting techniques, and making sensible use of added plant nutrients (in manures and fertilisers) and pesticides.

Mistakes have been made. Growing one crop again and again on the same soil can destroy its fertility. Agricultural technology appropriate in one place cannot just be exported to other regions with different soil types and growing conditions.

To avoid repeating past mistakes we need to use knowledge and understanding, gained both from scientific research and from the experience of farmers, to develop *sustainable* systems of agriculture – that is, agriculture that can go on indefinitely without degrading the environment.

Figure 2 The ultimate goal is to feed everyone adequately, without harming the environment.

Figure 1 Harvesting wheat in Montana, US.

What do plants need for growth?

The nutrient elements essential for the growth of plants are shown in Table 1.

Elements used in relatively large amounts		Elements used in relatively small amounts
carbon	nitrogen	iron
hydrogen	phosphorus	manganese
oxygen	potassium	boron
	calcium	molybdenum
	magnesium	copper
	sulphur	zinc
		cobalt
		chlorine
mostly from carbon dioxide and water		from soil

Table 1 *Essential nutrient elements and their sources; other minor nutrients (sodium, fluorine, iodine, silicon, strontium and barium) are not needed by all plants.*

Table 1 and Figure 3 show the importance of soil as the source of nutrients for plants. So what *is* soil, and how does it support plant life?

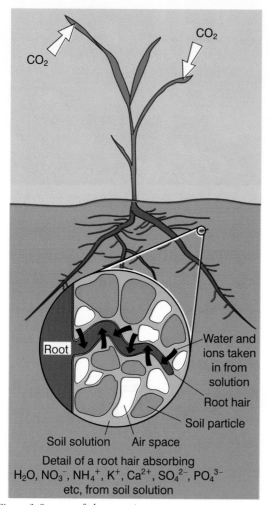

Figure 3 *Sources of plant nutrients.*

Detail of a root hair absorbing H_2O, NO_3^-, NH_4^+, K^+, Ca^{2+}, SO_4^{2-}, PO_4^{3-} etc, from soil solution

AA2 *The world at your feet*

The Earth's crust is just 1.5% of the volume of the planet (Figure 4). Table 2 gives the composition of the Earth's crust by mass. In this unit we are concerned with the thin layer, typically 1 m–2 m thick, on the top of the crust. This is the soil.

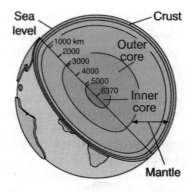

Figure 4 *The layers of the Earth.*

Road cuttings often reveal to us the underlying layers of a soil (Figure 5). Plants take nutrients and water from the soil which also supports them as they grow.

Element	% by mass
oxygen	47
silicon	28
aluminium	8
iron	5
calcium, sodium, potassium, magnesium	2–4 each

Table 2 *Composition of the Earth's crust by mass.*

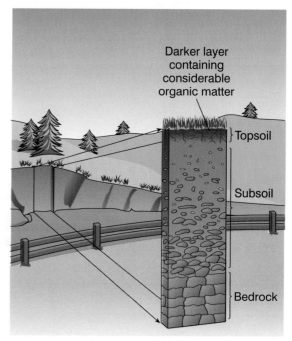

Figure 5 *Layers of soil exposed in a road cutting.*

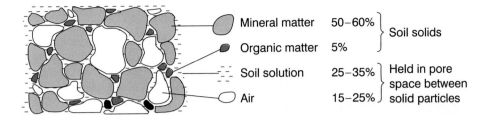

Figure 6 The four major components of soils, showing the approximate proportions by volume for a typical moist soil.

What is soil?

Soil is a mixture of weathered rock fragments and organic matter. Together they make a porous fabric which can hold both air and water. The approximate proportions by volume for a typical moist soil are shown in Figure 6.

The quantities of air and water in a soil are variable and the ratio is important in determining how well the soil supports plant growth. The minerals play a vital part in making nutrient ions available to plants.

Soil minerals are produced by *weathering* of the rocks which make up the Earth's crust. The fragments of rock and minerals vary enormously in size, from coarse stones and gravel to fine clay particles.

The coarse fragments are relatively inactive chemically. It is the very fine clay fraction which is most active because clay minerals can bind positive ions to their surfaces. In this way, essential ions are held in the soil.

Weathering

Weathering involves the action of wind, rain, frost and sunlight – and nothing on the surface of the Earth's crust escapes it. Weathering breaks up the rocks. It changes their physical and sometimes their chemical composition. It carries away soluble materials and some of the solid fragments as well. Biological processes also contribute to weathering. For example, roots can produce acids which speed up the chemical decomposition.

But weathering is creative as well as destructive: it makes a *soil* from the uppermost layers of weathered rock. Figure 7 summarises the main pathways of weathering in the moderately acidic conditions of humid temperate climates. In different climates the rates of the reactions alter, altering the composition of the soils.

ASSIGNMENT I

In different climates soil compositions can be very different. For example, many tropical soils are high in iron oxide and aluminium oxide, which come from the chemical decomposition of rocks, as shown in Figure 7.

Use your knowledge of the factors affecting reaction rates to explain why the iron oxide and aluminium oxide concentrations are higher in tropical soils than in temperate soils.

Most chemical reactions take place more quickly at higher temperatures. You can remind yourself about the effect of temperature on the rate of a reaction by reading **Chemical Ideas 10.2**.

You can read about rate equations and rate constants in **Chemical Ideas 10.3**.

In **Activity AA2.1** you can use the 'iodine clock' method to investigate the effect of temperature on the rate of a reaction.

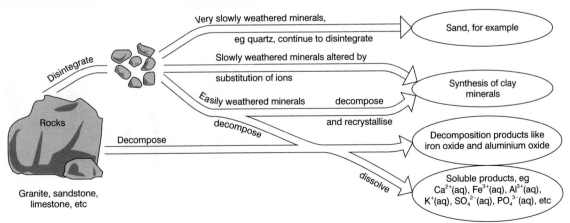

Figure 7 Processes of weathering in moderately acidic conditions common in humid temperate climates (granite and sandstone are mixtures of silicate minerals such as quartz, feldspars and micas).

Figure 8 A view of the Scottish coastline at Sheigra, Sutherland, showing sand and fertile soil produced by the weathering of rocks.

You will need to learn something about the structure of the clay minerals to understand how soils function. But first, something about the rocks from which clay minerals are formed: the **silicates.**

You can revise earlier ideas on molecular and network covalent structures in **Chemical Ideas 5.2**.

You can find out more about the relationship between the properties of a substance and its structure and bonding in **Chemical Ideas 5.6**.

Silicates

An isolated silicate ion SiO_4^{4-} is shown in Figure 9. Each silicon atom is covalently bonded to four oxygens at the corners of a tetrahedron. The silicon atom is at the centre of the tetrahedron and has an oxidation state of $+4$.

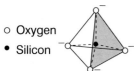

○ Oxygen

● Silicon

Looking from above, this can be represented as

Figure 9 The silicate(IV) ion SiO_4^{4-}.

Many minerals in the Earth's crust are silicates, in which SiO_4 tetrahedra are linked by sharing oxygen atoms between adjacent tetrahedra. A simple example is shown in Figure 10. Note that the *unshared* oxygen atoms carry a single negative charge.

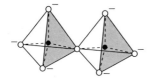

Figure 10 The $Si_2O_7^{6-}$ ion.

ASSIGNMENT 2

Draw dot and cross diagrams for an isolated SiO_4^{4-} ion and for the ion, $Si_2O_7^{6-}$, made by linking two silicate tetrahedra together. Show clearly the charges on individual oxygens.

Pure quartz is a continuous three-dimensional network of silicon and oxygen atoms, with each oxygen atom shared between two adjacent tetrahedra. (Because all the oxygen atoms are shared, none of them carry a negative charge.) The number of oxygens per silicon is

$$\tfrac{1}{2}+\tfrac{1}{2}+\tfrac{1}{2}+\tfrac{1}{2} = 2$$

and quartz has the overall composition SiO_2. It is very slowly weathered to sand. This type of weathering is a purely physical process.

(a)

(b)

Figure 11 (a) Crystals of quartz, SiO_2; (b) sand is made up of small grains of SiO_2 (the pale brown colour is due to iron impurities).

The silicate tetrahedra can join up in a number of other ways. We will represent a silicate tetrahedron as

looking from above. Remember that the corners represent oxygen atoms. The tetrahedra can join up to form **chains** by sharing two corner oxygens.

Figure 12 A single chain.

They can also form double-stranded chains in which two or three oxygens are shared (Figure 13). Asbestos is an example of this kind of arrangement (Figure 14).

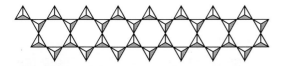

Figure 13 A double-stranded chain.

Figure 14 Fibres of asbestos are made up of double-stranded silicate chains.

Sheets of silicate tetrahedra can be built up by joining more strands together (Figure 15). These occur in micas and clays (Figure 16).

In sheets like this all the unshared, negatively charged oxygens point in one direction (upwards, out of the plane of the paper in Figure 15). In each silicate tetrahedron one oxygen is wholly owned by a silicon, and the other three are shared. The number of oxygens per silicon is

$$\tfrac{1}{2} + \tfrac{1}{2} + \tfrac{1}{2} + 1 = \tfrac{5}{2}$$

In whole numbers, the ratio Si:O is 2:5. The net charge on each Si_2O_5 unit is -2 (each oxygen has an oxidation state of -2 and silicon has an oxidation state of $+4$). The overall structure of the sheet can be represented as $(Si_2O_5^{2-})_n$.

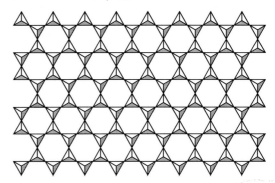

Figure 15 A silicate sheet.

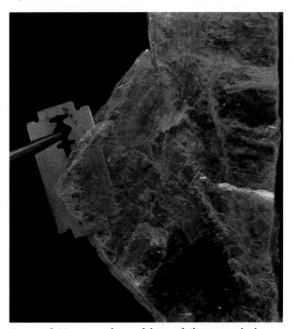

Figure 16 Mica is made up of sheets of silicate tetrahedra.

What balances the negative charge? Many positive ions (cations) such as K^+, Ca^{2+}, Mg^{2+} and Al^{3+} are held to the silicate sheets to balance the negative charge. Some of these cations fit into the hollows formed by the rings of tetrahedra.

During weathering, aluminium(III) can replace some of the silicon(IV) atoms at the centre of the tetrahedra in the silicate sheets to give a variety of different minerals. When this happens each aluminate tetrahedron has an extra negative charge because the aluminium is in oxidation state $+3$ compared with silicon's $+4$. Extra cations are then needed on the surface of the sheet to balance the extra negative charges. The physical properties of the silicate minerals are very dependent on how many Si(IV) atoms are replaced, and therefore how many extra cations are needed.

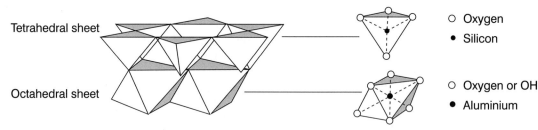

Figure 17 The sheet structure of clay minerals.

Activity AA2.2 allows you to build models of silicate and clay structures to help you to understand the structures represented in the diagrams in this section.

Clay minerals

Clay minerals contain two different kinds of **sheets:**

- **tetrahedral sheets** based on silicate tetrahedra with varying amounts of Al(III) substituted for Si(IV)
- **octahedral sheets** based mainly on Al^{3+} ions surrounded by six oxide or hydroxide ions in an octahedral arrangement. Some Mg^{2+} ions substitute for Al^{3+} ions in the sheet. The octahedra link together in the sheet by sharing oxygens (see Figure 17).

These sheets form into **layers** in different ways. Clays are classified into types based on the arrangements of the sheets.

Kaolinite is a common example of a **1:1 type** clay. Each layer in the kaolinite is made up of one tetrahedral and one octahedral sheet as shown in Figure 18.

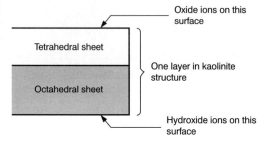

Figure 18 A layer in kaolinite.

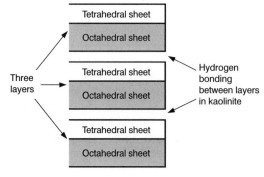

Figure 19 Hydrogen bonding between the layers in kaolinite.

The layers are held closely together by hydrogen bonding between the hydroxide ions on the surface of the octahedral sheets and the oxide ions on the surface of the tetrahedral sheet in the next layer (Figure 19). Water and cations cannot enter *between the layers* of a crystal, so kaolinite does not expand very much on wetting. (Kaolinite does expand to a small extent as water is taken up into the spaces between the crystals.)

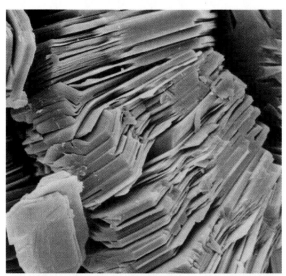

Figure 20 Scanning electron microscope picture of kaolinite showing the crystalline layer structure.

Figure 21 Modelling clay, containing kaolinite, absorbs comparatively small amounts of water when wet – the water gets between the small plate-like crystals which slide easily over one another, so wet clay is pliable and slippery; however, water cannot penetrate between the layers within a crystal.

In **2:1 type** clays an octahedral sheet is sandwiched between two tetrahedral sheets (see Figure 22). The *smectite* group of minerals, of which **montmorillonite** and **vermiculite** are examples, have this structure and are common in soils.

There is little attraction between the oxygens at the bottom of one layer and those at the top of the next layer. This means that water and cations can easily enter the interlayer space in 2:1 type minerals.

When water enters the interlayer space in a clay mineral such as montmorillonite, it forces the layers apart and exposes a large internal surface. This internal surface is much greater than the external surface area of the crystals (see Table 3).

In montmorillonite, Mg^{2+} ions substitute for some of the Al^{3+} ions in the octahedral sheet, and Al^{3+} ions substitute for some of the silicon in the tetrahedral sheet. The individual layers of this substance therefore have a high negative charge. In vermiculite, even more aluminium has substituted for silicon so that the negative charge is even higher than in montmorillonite.

A swarm of cations is attracted to both the external and internal surfaces to balance the negative charge (see Figure 23).

These cations are hydrated: each one is surrounded by layers of water molecules. This water held round the clay crystals gives wet clay its characteristic sticky feel.

The cations held to the surfaces of clay minerals are very important for agriculture because they provide a source of nutrient ions for use by plant roots. There is continuous movement of ions between the surfaces of clays and the soil solution in a process called **ion exchange**.

You can investigate the physical properties of some silicate and clay minerals like quartz, mica and vermiculite in **Activity AA2.3.**

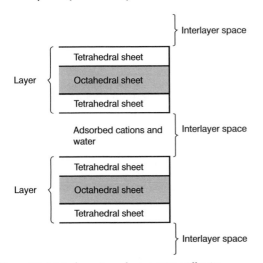

Figure 22 *A 2:1 clay mineral, eg montmorillonite.*

	Montmorillonite (2:1 expanding mineral)	Kaolinite (1:1 non-expanding mineral)
total surface area per unit mass/m² g⁻¹	700–800	up to 15

Table 3 *The total surface areas (internal + external surface areas) of an expanding and a non-expanding mineral.*

Figure 24 *Clay soils containing montmorillonite or vermiculite are heavy to plough and dig when wet; when dry, they shrink and crack.*

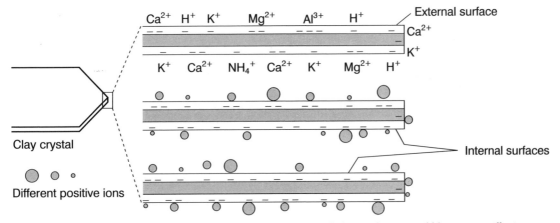

Figure 23 *Exchangeable cations at the inner and outer surfaces of a crystal of a 2:1 clay mineral like montmorillonite.*

Soil organic matter

The organic matter in soil is made up of plant debris, animal remains and excreta, and the products formed by decomposition of all these things. Soil organic matter forms a store from which the next generation of plants will get their nutrients. A variety of organisms living in the soil make use of the debris deposited. During the processes of decomposition, elements in organic compounds are converted into inorganic ions such as ammonium, nitrate(V), sulphate and phosphate ions. This process is called **mineralisation.**

Some of the carbon in decomposing organic matter is released into the atmosphere as carbon dioxide. A range of new organic molecules is synthesised to make **humus.** There are many macromolecules in humus with relative molecular masses up to 500 000. Major components of humus are esters of carboxylic acids, carboxylic acid derivatives of benzene and phenolic compounds. The following two groups are therefore common:

carboxylic acid group   *phenol group*

Both these groups can lose H^+ ions. If they do then negatively charged groups are left on the molecules in humus. These are able to form ionic bonds with metal cations so that humus can hold a variety of nutrient ions in a similar way to clays.

How are cations retained by soils?

Many of the cations held to the inner and outer surfaces of clay minerals can be replaced by different ions from the soil solution. The process is called **cation exchange**. For example NH_4^+ ions can exchange with Ca^{2+} ions. The balance of charge must be maintained, so two singly charged ions will replace one doubly charged ion.

$$\text{clay–}Ca^{2+}(s) + 2NH_4^+(aq) \rightleftharpoons$$
$$\text{clay–}2NH_4^+(s) + Ca^{2+}(aq)$$

The exchange is rapid and reversible. The direction of the exchange reaction depends on the concentrations of the ions involved and on how strongly they are held by the clay.

The ability of a clay mineral to exchange ions is measured by its **cation exchange capacity**. This is the amount in moles of exchangeable positive charge (mol_c) held by 1 kg of the clay mineral. (One mole of singly charged ions is equivalent to half a mole of doubly charged ions or to one-third of a mole of triply charged ions.)

ASSIGNMENT 3

In a laboratory experiment, a solution of ammonium nitrate(V) is allowed to seep slowly through a sample of a clay soil.

a Describe what happens to the ammonium ions and the nitrate(V) ions.

b Ammonium nitrate(V) is commonly added to soils as a nitrogen fertiliser. Use your answer to part **a** to explain why heavy rain can leach out nitrate(V) ions from clay soils more quickly than ammonium ions.

Cation exchange capacities vary widely, depending on the surface area of the mineral, and the number of charges on its inner and outer surfaces.

Substance	Cation exchange capacity/$mol_c\,kg^{-1}$
kaolinite	0.1
montmorillonite	1.0
vermiculite	1.5
humus	0.3–1.5

Table 4 Cation exchange capacities of some clay minerals and humus.

Soils vary in the quantities of clay and organic matter which they contain. Typical soil cation exchange capacities are in the range $0.02\,mol_c\,kg^{-1}$–$0.6\,mol_c\,kg^{-1}$.

The ions held by the clay or humus are in equilibrium with the free ions in the soil solution. Plant roots withdraw nutrients from the soil solution. They are replaced from this pool of exchangeable cations.

You can find out more about ion exchange by reading **Chemical Ideas 7.5**.

The size of ions in solution is discussed in **Chemical Ideas 3.2**.

In **Activity AA2.4** you can investigate the process of ion exchange.

Controlling soil acidity

Under natural conditions H^+ ions from rain, and from plant roots and microbe activity, displace Ca^{2+} and other ions from soil solids. This has two effects: the soil becomes more acidic, and its store of nutrients in the form of exchangeable cations is reduced.

Exchangeable cations are lost for two reasons. Firstly, cations held to the surfaces of the clay layers are replaced by H^+ ions. Secondly, under acidic conditions (at low pH), weathering of clay minerals takes place more quickly (see Figure 7 on p. 185).

Aluminium ions are released into the soil and some aluminium oxide is formed. The surface of the aluminium oxide binds H^+ ions, so that it becomes positively charged. It repels cations and holds anions.

The release of aluminium ions from clays at low pH causes another problem. High aluminium concentrations in the soil solution are toxic to crops.

Table 5 shows soil pH values below which plant growth is restricted.

Crop	pH below which growth is restricted
beans	6.0
oats	5.3
potatoes	4.9
wheat	5.5
lettuces	6.1
cabbages	5.4
apples	5.0
blackcurrants	6.0

Table 5 Soil pH values below which plant growth is restricted.

Basic carbonates, such as ground limestone or chalk ($CaCO_3$), or bases, such as lime $Ca(OH)_2$, can be added to the surface to make a soil less acidic. The amount needed to reach a desired pH depends on the soil's capacity to resist the neutralising action of the base. This is called its **buffering capacity.** A clay soil at pH 5 will need more lime to bring its pH to 7 than the same mass of sandy soil. The clay soil has a higher buffering capacity.

As lime is added to a clay soil, the pH of the soil changes very little at first, and then slowly rises. The soil acts as a **buffer** and is able to resist changes in pH to some extent. It can do this because H^+ ions bound onto the soil solids replace some of the H^+ ions in the soil solution as soon as they are removed.

Figure 25 shows the effect of adding alkali to some clay minerals.

Plots like Figure 25 for different soils allow the **lime requirement** of each soil to be calculated.

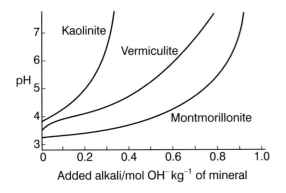

Figure 25 The effect of adding alkali to some clay minerals.

ASSIGNMENT 4

Look at Figure 25.

a Which of the three minerals has the lowest buffering capacity?

b Suggest a reason why the mineral you have chosen in part **a** should have a lower buffering capacity than the other two.

In **Activity AA2.5** you can measure the pH of a soil, and estimate its buffering capacity and lime requirement.

Figure 26 Farmers add lime to reduce soil acidity.

AA3 *Keeping soil fertile*

How can we grow crops again and again on the same soil without decreasing its fertility?

The fertility of a soil depends on many complex interactions between the biological, chemical and physical processes occurring. It can be damaged by destroying the arrangement of particles and pore spaces in the soil, and by removing too many nutrients.

Nutrient cycling

Apart from carbon dioxide from the air, plants get all their nutrients from the soil (see Section **AA1**, Figure 3).

These nutrients are drawn from an **inorganic store** in the soil, and an **organic store** partly on top of and partly in the soil. Elements are cycled continuously between living systems, the organic store and the inorganic store. The general routes are shown in Figure 27.

The organic store is replenished by organic manures, animal excretions, and by death and decay of living organisms. Microorganisms act on organic matter and convert it into humus. They also convert it into inorganic ions (mineralisation), producing ammonium, nitrate(V), phosphate and sulphate ions (see Section **AA2**).

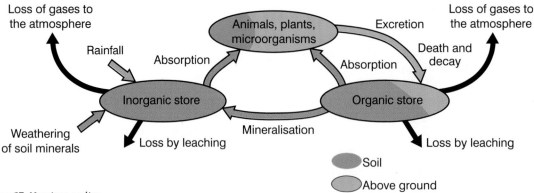

Figure 27 Nutrient cycling.

Weathering of soil and rock minerals releases more ions into the inorganic store.

Nutrients can be lost by being leached out of the top layers of soil by rainwater, and nitrogen can be lost by conversion into gases, such as NH_3, N_2 and N_2O, which disperse into the atmosphere. Figure 28 shows typical values for the total quantities of some elements in different soils.

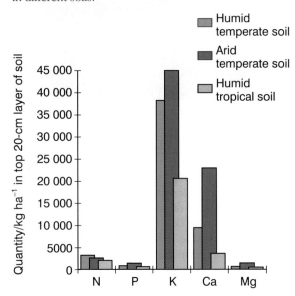

Figure 28 Quantities of some elements present in different soils.

the nitrogen content of soil in the form of nitrate(V) and ammonium ions extracted from a known mass of soil. Extraction methods are also used to determine quantities of potassium and phosphate ions. The extracted quantities are used to classify soils as low, medium and high in nutrients present in forms which plants can use. Figures are given for three major nutrients in Table 6.

Soil range	Nutrient extractable in forms which plants can use/kg ha^{-1} in top 20-cm layer of soil		
	Nitrogen	Potassium	Phosphorus
low	40	0–150	0–23
medium	90	150–600	23–63
high	140	over 600	over 63

Table 6 Nutrients in soils in forms which plants can use.

ASSIGNMENT 5

Refer to Table 6 and Figure 28. In a humid temperate soil rated high in nutrients, approximately what percentage of

a the total potassium content

b the total nitrogen content

is in a form which plants can use?

Hectares

Land areas are conveniently measured in **hectares** (ha).

$1 \text{ ha} = 1 \times 10^4 \text{ m}^2 (= 2.47 \text{ acres})$

A football pitch is about 0.5 hectares. Fields range from about 6 hectares to about 60 hectares.

Only a small fraction of each element in the soil is actually available to plants in the form of ions which can be absorbed through their roots. You can measure

Figure 29 A German chemist, Justus von Liebig, used this barrel to illustrate to his students that a deficiency of any single plant nutrient is enough to limit growth – in the diagram, potassium is the nutrient limiting the yield.

Figure 30 A leaf from (a) a healthy plant and (b) a plant with potassium deficiency.

Healthy crop growth depends on the ability of the soil to supply nutrients. This makes the rates at which nutrients are interconverted in the cycles very important.

The rate of supply of ions to plant roots has to be rapid enough during the peak growing period in May and June to meet the demands of the crops. Crop yields are reduced if there is a shortage of even one nutrient (Figure 29.) Figure 30 shows one of the effects of a shortage of potassium ions.

In the next section, you will look at one nutrient cycle in more detail.

You can measure the nitrogen content of a soil in **Activity AA3.1.**

The nitrogen cycle

Almost all the nitrogen in soil is present in complex organic compounds and so is not readily available to plants. Various processes convert gaseous nitrogen and organic nitrogen compounds into the soluble ammonium and nitrate(V) ions which plants *can* use. The main processes in the nitrogen cycle are listed below. As you read through, refer to the diagram of the nitrogen cycle in Figure 32 on p. 194.

Chemical Ideas 11.3 gives a summary of the chemistry of nitrogen and other elements in Group 5.

It may help you to revise earlier work on redox reactions and oxidation states in **Chemical Ideas 9.1**.

Gains of nitrogen to the soil

Biological fixation Some kinds of bacteria in soil, and in root nodules in legumes, such as peas and beans, can convert nitrogen gas to ammonium ions. A reducing agent is needed to provide the electrons. The half-equation for the reduction is

$$N_2(g) + 8H^+(aq) + 6e^- \rightarrow 2NH_4^+(aq)$$

Other additions to soil nitrogen Lightning, burning of hydrocarbon fuels and natural fires all produce nitrogen oxides which are released into the atmosphere and are then deposited on the soil.

In Europe about $20\,kg\,N\,ha^{-1}$–$40\,kg\,N\,ha^{-1}$ are deposited from the air onto the soil each year. Some of this is in the form of nitrogen oxides and some as ammonium ions which have come from ammonia emitted by animal excreta.

Figure 31 Lightning generates enough energy to convert nitrogen and oxygen in the air to nitrogen monoxide, $N_2 + O_2 \rightarrow 2NO$.

Transformations in the soil

Mineralisation Soil bacteria and other microorganisms break down organic nitrogen compounds into simpler molecules and ions. Any nitrogen not needed by the organisms themselves is released into the soil as ammonium ions:

$$\text{organic N} \xrightarrow{\text{several steps}} NH_4^+(aq)$$

The NH_4^+ ions are held by clay minerals as exchangeable cations, but are readily converted to NO_3^- by other microorganisms.

A well-drained soil is best for mineralisation, and the reaction is much faster at the higher temperatures of tropical soils. Radioactive labelling with ^{15}N shows that only 1%–3% of soil nitrogen is mineralised each year.

ASSIGNMENT 6

Overall, mineralisation has a first-order rate equation

$$\text{rate of mineralisation} = k[N]$$

where k is the rate constant at a particular temperature and $[N]$ is the quantity of organic nitrogen per hectare in the top 20 cm of soil.

The rate constant k varies from 0.01 yr^{-1} to 0.06 yr^{-1}. For a time interval of 1 year, the quantity of organic nitrogen mineralised in the top 20 cm of soil equals $k[N] \text{ kg ha}^{-1}$.

a Use the above equation to calculate the quantity of nitrogen mineralised in 1 year (in the top 20 cm of 1 ha) in the three soils A, B and C. The soils have different organic nitrogen contents and different temperatures. (Remember, the value of k depends on the temperature.)

Soil	Soil organic nitrogen/kg ha^{-1}	Rate constant k/yr^{-1}
A	1000	0.01
B	2000	0.03
C	2000	0.06

b Refer to Figure 28 on p. 192 for information about the nitrogen content of different types of soil. Think about the factors that affect the rate of mineralisation, and explain why a humid tropical soil has the highest rate of mineralisation.

Nitrification Ammonium ions can be oxidised by certain aerobic bacteria in the soil. The bacteria carry out the reactions as a means of obtaining respiratory energy. The overall process is called nitrification because the end product is the nitrate(V) ion, formed via the nitrate(III) (nitrite, NO_2^-) ion. Nitrification occurs in several stages. The formation of nitrate(III) can be represented by:

$$NH_4^+(aq) + 1\tfrac{1}{2}O_2(g) \rightarrow$$
$$NO_2^-(aq) + 2H^+(aq) + H_2O(l)$$

The bacteria which do this are called *Nitrosomonas*. The optimum pH is between 7 and 9, and the reaction stops in dry conditions.

The nitrate(III) ion, NO_2^-, produced is rapidly oxidised further. This can be represented by:

$$NO_2^-(aq) + \tfrac{1}{2}O_2(g) \rightarrow NO_3^-(aq)$$

The bacteria converting nitrate(III) to nitrate(V) are called *Nitrobacter*. They can tolerate dry conditions and higher acidity than *Nitrosomonas*.

Some nitrification occurs in soils at temperatures down to $0\,°C$, but it ceases in waterlogged soils where the oxygen content is too low.

Losses of nitrogen from the soil

Denitrification Where oxygen content is low, anaerobic bacteria reduce nitrate(V) ions in the sequence:

$$NO_3^-(aq) \rightarrow NO_2^-(aq) \rightarrow NO(g) \rightarrow N_2O(g) \rightarrow N_2(g)$$

In flooded soils, like those used in rice cultivation, losses by denitrification can be high.

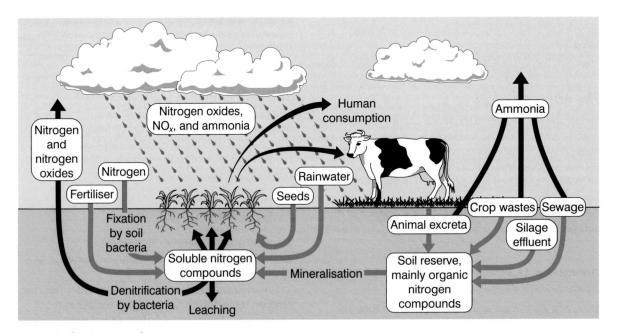

Figure 32 The nitrogen cycle.

Leaching Nitrogen is lost by leaching mainly as the nitrate(V) ion NO_3^- which is not held by clays or humus in temperate soils. The quantities lost depend on the soil structure and the amount of rainfall, as well as the nitrate(V) concentration in the soil.

Loss of ammonia gas Ammonium ions are converted into ammonia under alkaline conditions. The ammonia then evaporates into the atmosphere.

Uptake by plants In natural systems the quantity of nitrogen removed per year is relatively small. For example, a coniferous forest takes up 25 kg–78 kg nitrogen per hectare each year. However, crops cultivated to give high yields take up much more nitrogen, 100 kg–500 kg nitrogen per hectare each year. Soil nitrogen cannot be mineralised to ammonium ions and nitrate(V) ions fast enough to meet this demand, even when the reserves of organic nitrogen are high.

In order to maintain or increase crop yields, nitrogen must be added to the soil in a form that plants can readily use.

In **Activity AA3.2** you can use the information in this section to investigate the nitrogen cycle in more detail.

Adding nutrients

Nutrients are added to soil in two ways:

- as manure from livestock
- as inorganic compounds, produced by industrial processes

Long-term experiments have been conducted at the Rothamsted Experimental Station in Hertfordshire since 1843, investigating the effects of adding nutrients to the soil (Figure 33). They have shown that yields can be increased, and high productivity maintained, by adding nutrients to the soil and by controlling weeds pests and diseases.

Figure 33 Famous fields: researchers at Rothamsted Experimental Station study plots set up more than a century ago.

After a century, the application of farmyard manure to plots of land has more than doubled the organic nitrogen content of soil, increasing its nitrogen reserve. In addition, farmyard manure has encouraged flourishing populations of small animals like earthworms, and soil microorganisms. These assist in converting organic nitrogen into forms usable by plants.

The water content of farmyard manure is high (around 75%–80%). The water content of animal slurry is even higher. So large quantities of manure or slurry have to be used to supply adequate nutrients (see Table 7). Because of the bulk, long-distance transport of manure is uneconomical. Drying processes have been developed, but processing costs are high.

	Manure produced/ $t\ yr^{-1}$	Nutrient content/$kg\ t^{-1}$		
		N	P	K
1 dairy cow	23	4.7	0.6	4.4
10 pigs	21	6.3	1.5	2.9

Table 7 The nutrient content of some farmyard manures.

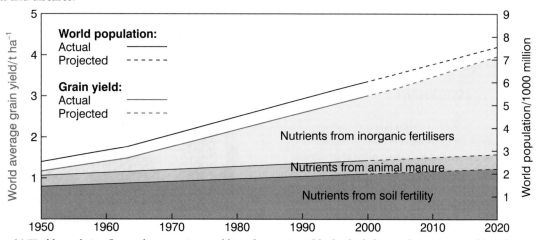

Figure 34 World population figures shown against world trends in grain yields; the shaded areas show estimates of the relative contributions of the different sources of plant nutrients.

ASSIGNMENT 7

A farmer plans to add 144 kg nitrogen per hectare to a small field of wheat. The area of the field is 7 ha.

a Calculate the total mass of nitrogen needed.

b Calculate the mass of cow manure needed to supply the nitrogen for the field.

While manure is a resource to be used wherever possible it only returns some of the nutrients removed by livestock back to the soil. It cannot *completely replace* nutrients removed in crops and exported from a farm. Another source of nutrients is needed in order to maintain the productivity of the soil.

Long-term experiments as at Rothamsted show that high levels of productivity can also be sustained using inorganic fertilisers.

You can see the increasing importance of inorganic fertilisers over the last 50 years in Figure 34. During this time average grain yields have almost tripled. Today, inorganic fertilisers support about half the world's cereal production. Without them we cannot hope to feed the world.

The quantities of inorganic fertilisers being used in the developing world is increasing particularly rapidly. For example, in China, fertiliser applications have risen from 1% of plant nutrients in 1950 to over 60% in 2000.

How are inorganic fertilisers produced?

The range of fertilisers produced includes
- ammonium nitrate(V)
- ammonium sulphate
- ammonium phosphate
- triple superphosphate (a form of calcium phosphate)
- urea ($CO(NH_2)_2$)
- potassium chloride.

The compounds are sold individually, or mixed to produce a range of products with different N:P:K ratios to meet farmers' needs.

ASSIGNMENT 8

The major costs involved in distributing and applying nitrogen fertilisers are to do with the mass of material which must be transported for a given mass of nitrogen.

a Calculate the percentages by mass of nitrogen in the three nitrogen fertilisers ammonium nitrate(V), ammonium sulphate and urea. (Make sure that you write the correct formula for each one.)

b List the fertilisers in order of increasing transport costs.

Figure 35 The percentages of N, P (as P_2O_5) and K (as K_2O) are clearly shown on this bag of fertiliser.

Fritz Haber's discovery

At the heart of fertiliser production is the synthesis of ammonia from nitrogen and hydrogen. The apparatus shown in Figure 37 was used by the German chemist Fritz Haber to develop the method of synthesis, and with it he managed to produce 100 g of ammonia by 1909.

The process was scaled up by Carl Bosch, a chemical engineer employed by BASF (Badische Anilin und Soda Fabrik) near Mannheim in Germany. In 1913 the first industrial plant went into production with a capacity of 30 tonnes of ammonia per day. Modern plants use the same basic design principles, but with capacities of about 1500 tonnes per day.

Figure 36 Extracting ammonia from coal tar around 1860. Haber's process was a great improvement.

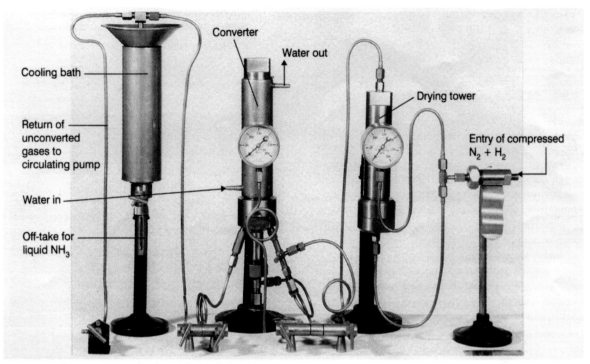

Figure 37 Haber's apparatus for the synthesis of ammonia.

About 6500 experiments were carried out between 1910 and 1912 to discover an effective catalyst. Today, finely divided iron is used. Small amounts of potassium, aluminium, silicon and magnesium oxides are added to improve its activity.

Both Haber and Bosch were awarded Nobel Prizes in Chemistry: Haber in 1918 for his academic work, and Bosch in 1931 for inventing and developing the high-pressure technology.

Figure 38 The patent taken out for the manufacture of ammonia by BASF in 1908.

The modern Haber process plant

In modern factories the hydrogen needed to make ammonia is usually made by reacting water with natural gas. Nitrogen from the air is purified and mixed with the hydrogen. The heated gases pass over the catalyst where they react to form ammonia:

$$N_2(g) + 3H_2(g) \rightleftharpoons 2NH_3(g)$$

The reaction is reversible. Ammonia produced in the forward reaction can decompose to nitrogen and hydrogen in the reverse reaction. If a mixture of nitrogen and hydrogen is placed in contact with the catalyst and left for a sufficient time then an equilibrium mixture containing all three gases will be obtained.

To investigate how the highest yields can be obtained, nitrogen and hydrogen in the volume ratio 1:3 were mixed and held at different temperatures and pressures. The equilibrium yield of ammonia was recorded. Some results are shown in Table 8.

The equilibrium position depends on the temperature and pressure chosen for the reaction. As you can see from the data in Table 8, the percentage of ammonia in the mixture increases with increasing pressure and falling temperature. Maximum conversion to ammonia is obtained at high pressures and low temperatures.

However, high pressures are very expensive to achieve, and at low temperatures ammonia is produced only very slowly. The reaction conditions chosen are those which give the most economic production of ammonia.

Pressure/atm	NH₃ present at equilibrium/%					
	100 °C	200 °C	300 °C	400 °C	500 °C	700 °C
10	–	50.7	14.7	3.9	1.2	0.2
25	91.7	63.6	27.4	8.7	2.9	–
50	94.5	74.0	39.5	15.3	5.6	1.1
100	96.7	81.7	52.5	25.2	10.6	2.2
200	98.4	89.0	66.7	38.8	18.3	–
400	99.4	94.6	79.7	55.4	31.9	–
1000	–	98.3	92.6	79.8	57.5	12.9

Table 8 Volume percentage of NH₃ in equilibrium mixtures in the reaction $N_2(g) + 3H_2(g) \rightleftharpoons 2NH_3(g)$.

It is better to get moderate yields of ammonia rapidly than to wait a long time for a higher yield. Ammonia is separated from unreacted nitrogen and hydrogen before equilibrium is reached. The unreacted gases are recycled over the catalyst.

Most reactors now operate at pressures between 25 atm and 150 atm, and at temperatures between 400 °C and 500 °C. Energy consumption in modern factories is about 35 MJ for every kilogram of nitrogen converted to ammonia.

For reactions involving gases, it is more convenient to write equilibrium constants in terms of pressures (K_p), rather than concentrations (K_c). You can read about gaseous equilibria in **Chemical Ideas 7.3**.

You can read about the effect of catalysts on the rate of a reaction in **Chemical Ideas 10.4** and **10.5**.

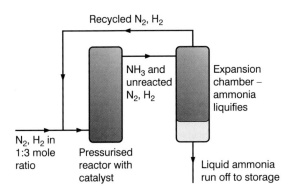

Figure 39 A flow diagram for the Haber process.

Figure 40 The Norsk Hydro NPK fertiliser factory at Glomfjord in Norway.

ASSIGNMENT 9

a The forward reaction in the Haber process is exothermic ($\Delta H^\ominus = -92 \text{ kJ mol}^{-1}$). Explain how the experimental results in Table 8 are in agreement with Le Chatelier's principle.

b Pick out the conditions of temperature and pressure from those listed in Table 8 which would give
   i the highest yield of ammonia
   ii the fastest rate of conversion to ammonia.

c What practical reasons can you think of for not using very high pressures?

d The boiling points of N₂, H₂ and NH₃ are −196 °C, −253 °C and −33 °C respectively. Explain how the ammonia is separated from unreacted nitrogen and hydrogen.

Saving money and protecting the environment

Fertilisers cost money. Farmers do not want to apply fertiliser nitrogen just to have it leached out of the soil as nitrate(V) ions. To avoid wastage, they need to match the addition of fertiliser to the needs of the crop they are growing, and also apply it when the crop is most likely to take up the nitrogen. Any excess leached out of the soil could eventually get into drinking water. Concern over nitrate(V) levels in drinking water has led to an EC limit of 50 mg dm⁻³, which is considered to be well inside the safety margin.

To reduce the risk of nitrate(V) leaching, winter crops can be grown. Figure 41 shows the soil nitrate(V) levels during a year for a crop sown in the autumn and harvested in August.

Nitrate(V) levels are low in January and February. In spring the rate of mineralisation increases and fertiliser is added. Higher soil nitrate(V) levels result, but the rapidly growing crop takes up nitrogen, and nitrate(V) levels decrease again. After harvesting there is an increase in soil nitrate(V).

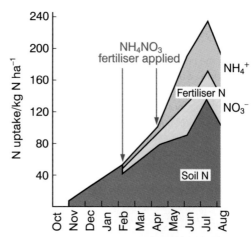

Figure 42 *Uptake of nitrogen by winter wheat measured at Jealott's Hill Research Station; the shaded areas show the relative contribution of nitrogen from the soil and from fertilisers.*

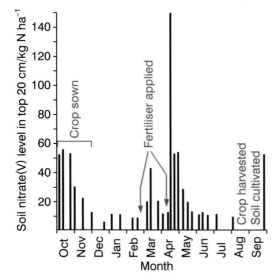

Figure 41 *Soil nitrate(V) levels throughout the year measured at Jealott's Hill Research Station.*

Ploughing introduces air into the soil and encourages microbial activity. Warm moist soil in autumn leads to rapid mineralisation of organic matter. Soil nitrate(V) levels rise. As temperatures fall and the soil becomes waterlogged, denitrification causes a decrease in soil nitrate(V) levels. Leaching reduces the levels even further from December onwards.

Figure 42 shows the uptake of nitrogen by winter wheat. Over $40 \, kg \, ha^{-1}$ of soil nitrogen have been absorbed by late February. Ammonium nitrate(V) fertiliser is then applied in two stages to match the needs of the crop. The fertiliser used in this study was labelled with the radioactive ^{15}N isotope so that nitrogen entering the crop could be identified as either soil nitrogen or fertiliser nitrogen. Such experiments show that when the correct amount of fertiliser is applied to cereals at the correct time, less than 2% remains in the soil as nitrate(V) by autumn to be at risk of loss by leaching. This means that nearly all the nitrate(V) in the soil at this time of the year comes from the mineralisation of soil organic matter.

Work at Rothamsted has shown that, in plots where no crops have been grown and no fertiliser applied, $45 \, kg \, ha^{-1}$ nitrate(V) are leached from the soil each year. This comes from the large reserve of organic nitrogen in soil, converted to nitrate(V) ions by microbes.

ASSIGNMENT 10

In answering these questions you may need to refer to the data in Figures 41 and 42, as well as the nitrogen cycle in Figure 32 (p. 194).

a What effect does ploughing have on the rate of mineralisation of soil organic matter to inorganic ions? Why does it have this effect?

b In uncropped land when do you think the greatest loss of nitrate(V) will occur? What two processes account for this loss? What are the environmental consequences?

c Suggest why denitrification tends to be the most important factor in determining soil nitrate(V) levels in early winter.

d Organic farmers grow crops without the use of any fertilisers or pesticides. However, they must be very careful about when they apply farmyard manure. If applied in the autumn, it can lead to far greater nitrate(V) loss by leaching than inorganic nitrogen fertilisers. Explain why farmyard manure increases the rate of nitrate(V) production.

e 'Catch' crops can be grown between main crops. What do you think they are catching? When should catch crops be ploughed into the soil?

Inorganic fertilisers are designed containing different proportions of nutrients. This makes it easier for farmers to apply just the quantities needed by the crops. If used correctly, inorganic fertilisers can supplement the nutrient levels in the soil, so that high yields of crops can be maintained without damaging the environment.

Activity AA3.3 will help you to think about revising for your end of course exams and to draw up your revision timetable.

AA4 *Competition for food*

Increasing the crop yield in the field by improving the concentration of nutrients in the soil is but one of the concerns of an agricultural chemist. Another is to protect the crop, both before harvest and in storage.

Other organisms compete with us for the food we grow. Worldwide, for example, 50% of rice and 40% of maize are lost each year to disease, pests and weeds. Even the potato crop is savaged, with losses of about 40%.

Figure 43 Some common pests: (a) 'rust' disease; (b) lupin aphids; (c) a cotton bollworm caterpillar; (d) flowering weeds in a crop of wheat.

Control of diseases and pests is now easier by a variety of scientific advances, such as selective breeding of plant species more resistant to attack and chemical control using pesticides. Nevertheless, the problem is compounded by the variety of species that can attack crops. For example, there are at least two types of bacteria, five types of fungi and four types of viruses that are responsible for the diseases of rice. By using new plant species and chemcial control, a developed country such as Australia can contain losses to these to about 20%, but a developing country such as Bangladesh may still suffer losses up to 70%.

Pesticides (insecticides, herbicides and fungicides) kill insects which eat our crops, weeds which compete with the crops for soil nutrients, and moulds which rot plants and seeds. Many disease-carrying organisms such as mosquitoes are also controlled by pesticides.

DDT

DDT is an **organochlorine** insecticide (the initials stand for dichloro-diphenyl trichloroethane). It was discovered in 1939 in the laboratories of the Swiss chemical company, Geigy. It has prevented millions of deaths from diseases such as typhus (which killed 2 500 000 Russians during the First World War) and malaria (one of the most fatal and debilitating human diseases).

DDT is highly toxic to insects but has very low toxicity to mammals. It is made in a one-stage synthesis from cheap raw materials, which keeps its cost down.

Excellent insecticide though it is, DDT has its disadvantages. It is chemically very stable, and so it accumulates in the environment where it can persist for many years and becomes concentrated up food chains. Its use has been banned in many countries, but, because it is cheaper than newer insecticides, it is still widely used in developing countries.

Some insects have developed a resistance to DDT by increasing their production of enzymes which catalyse the removal of HCl from DDT to give a non-toxic product, DDE.

DDT

removal of HCl by enzymes in mammals and resistant insects

DDE

When the double bond forms in DDE the molecule becomes planar. The change in shape is enough to make DDE biologically inactive because it changes the way in which the molecule interacts with the receptor site in insects.

While there are undoubted benefits from using pesticides, there can also be problems. Many pesticides can be damaging to human health and to the natural environment if used incorrectly. Pesticides may leach into our water supplies. Organisms other than the target ones can be killed. If these are predators which would eat the pests we wish to destroy, then the pests might actually benefit.

Some older pesticides can remain in the soil, and then build up through food chains, affecting predators such as birds and contaminating human food supplies. Fat-soluble molecules such as DDT can accumulate in the fatty tissues of animals and become concentrated up food chains (Table 9).

	Concentration of DDT/mg kg^{-1}
sea water	3×10^{-6}
fat of plankton	4×10^{-2}
fat of minnows	0.5
fat of needlefish	2
fat of cormorants	25

Table 9 DDT concentrations up a food chain (data from Long Island, US).

Therefore the challenge to the modern agricultural chemist is an enormous one – to find substances which are *specific* to the target organism, which kill at low dosages so that only small amounts need be applied to fields, and which do not persist in the environment or travel into the water supply. Great advances have been made in the last 20 years.

Pests can build up resistance to chemicals, so chemists need to keep finding new products to overcome this.

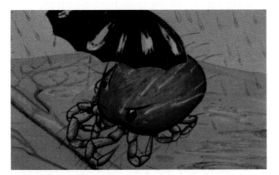

Figure 44 Pests can develop resistance to chemicals.

Activity AA4.1 gives you an opportunity to think about and discuss some of the issues involved in using pesticides.

The search for a new pesticide

The research and development involved in producing a new product is lengthy and expensive, requiring the collaboration of a great many scientists. Chemists, biologists, toxicologists, chemical engineers and process engineers are all involved.

A large company may invest sums of the order of £100 million each year in research and development, from which only one or two new products may result.

When an interesting compound is discovered which is active against pests, chemists will usually try to improve its activity by making systematic changes to the structure, continually testing and working out the 'best' substitutions in various parts of the molecule.

The compounds are tested on target pests and compared with existing products for potency and for the range of pests affected. The compounds which come out best in laboratory tests may be tried out to see if they will work in real field situations. Then hundreds of field trials are conducted on substances chosen for development.

The successful compound will be judged on a range of factors which will include

- ease of manufacture
- specificity
- persistence in soil
- cost of the final product
- marketability
- leaching losses into drainage water
- toxicity to humans
- comparison with known compounds
- who owns the patents surrounding the invention.

Patents are important because the company needs to be able to make a profit in return for its investment in research and development.

The pyrethroid story

For many centuries the dried flower heads of a chrysanthemum, *Chrysanthemum cinerariaefolium*, have been used to ward off insects, particularly mosquitoes. The structures and stereochemistry of the natural insecticides present in the flower heads were worked out between 1920 and 1955. One of them is *pyrethrin 1*.

pyrethrin 1

Pyrethrins have some of the qualities of the ideal insecticide. They are powerful against insects but are harmless to mammals under all normal circumstances. However, natural pyrethrins are unstable in light. A photochemical oxidation reaction occurs which means that they are of limited use in agriculture since they break down so quickly.

Figure 45 The pyrethrum flower.

ASSIGNMENT 11

a A carbon atom with four different groups attached to it is described as a chiral centre. (If the carbon atom is part of a ring, and the structure of the ring is different on each side of the carbon atom, the ring counts as two different groups.) Copy the structure of pyrethrin 1 (p. 201) and identify the chiral centre(s) by marking them with an asterisk (*).

b List the functional groups present in pyrethrin 1.

Michael Elliott and his co-workers spent many years working at Rothamsted on the synthesis of **pyrethroids**, compounds related to natural pyrethrins. They were looking for compounds like the natural compounds which would be active against insects, but more stable in light and air.

They made an important breakthrough in 1977, with the synthesis of *permethrin*, the first pyrethroid sufficiently stable to be used widely in agriculture. Permethrin is a mixture of stereoisomers. One of these, biopermethrin, is shown below.

biopermethrin

A later discovery, *cypermethrin*, is a more active insecticide, so smaller quantities can be applied to achieve the same effect. One isomer is shown below.

biocypermethrin

ASSIGNMENT 12

a Compare biopermethrin with pyrethrin 1. The changes Elliott made decreased the sensitivity of the molecule to photochemical reactions. List all the changes you can identify.

b Make a model of the cyclopropane ring in biopermethrin so that you can identify the geometric isomers. The structure shown is a *trans*-permethrin. Draw the structure of the corresponding *cis*-permethrin, and mark on it any chiral centres. (You can read about stereoisomerism in **Chemical Ideas 3.5** and **3.6**.)

How do pyrethroids work?

Pyrethroids are able to penetrate insects very rapidly to reach their site of action. These are sites in the membranes of nerve cells. Cell membranes are made up of a double layer of molecules, aligned with their long non-polar hydrocarbon chains inside the membrane and their ionisable groups on the surfaces of the membrane. You can see the structure of a cell membrane in Figure 46.

Proteins embedded in the nerve cell membrane act as channels for the passage of ions in and out of the cells. The channels open and close selectively, so that a difference in the concentration of Na^+ and K^+ ions builds up across nerve cell membranes. This causes an electrical potential difference across the membrane and is vital to the functioning of nerve cells. If the system of transporting ions across the membrane is disrupted the nervous system cannot work properly. You will find out more about the functioning of nerve cells in **Medicines by Design**.

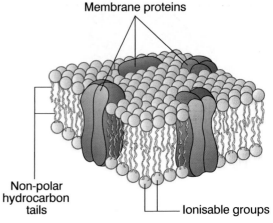

Figure 46 The structure of a cell membrane.

Pyrethroids work because they penetrate the cell membrane and block open the sodium channels. This leads to massive disruption of the nervous system of the insects.

A key factor in the activity of pyrethroids is that they are much more soluble in fats than in water. This means that they can pass readily from the aqueous solution used for spraying into the fatty tissues of the insect.

This difference in solubility can be measured as a **partition coefficient**. Partition coefficients are equilibrium constants which measure the ratio of concentrations of a substance dissolved in two immiscible solvents in contact with one another at the same temperature:

$$\text{partition coefficient, } K = \frac{\text{concentration in solvent A}}{\text{concentration in solvent B}}$$

For studies of biologically active substances, the most common measurement is the partition coefficient, K_{ow}, between octan-1-ol (solvent A) and water (solvent B)

$$\text{partition coefficient, } K_{ow} = \frac{\text{concentration in octan-1-ol}}{\text{concentration in water}}$$

If K_{ow} is high the compound will move out of aqueous solution into fatty tissue.

Values of K_{ow} for pyrethroids are very high so $\lg K_{ow}$ is usually quoted. For example, if $K_{ow} = 1 \times 10^5$, then $\lg K_{ow} = 5$; if $K_{ow} = 1 \times 10^7$, $\lg K_{ow} = 7$.

Pyrethroids with $\lg K_{ow}$ less than 5 are not insecticidal. In active compounds, $\lg K_{ow}$ is usually between 6 and 7, or even higher.

In practice, when K_{ow} has been measured for one compound, tables of data can be used to calculate the effect on K_{ow} of changing substituents, so that molecules with the desired properties can be selected.

The partition coefficient also shows us the extent to which pesticides will be taken up by organic matter in the soil. This is the major method of binding. (The herbicide, *paraquat*, which you will meet later in this section, binds to clays and is an exception.)

Concentrations of pyrethroids needed for insecticidal activity can be very low. For example, deltamethrin is active at concentrations as low as $1 \times 10^{-12}\,\text{mol dm}^{-3}$ in cell membranes.

deltamethrin

Chemical Ideas 7.4 will tell you more about partition equilibria.

In **Activity AA4.2** you can analyse data for pyrethroids and other pesticides to investigate the link between values of partition coefficients and the concentration of pesticides in living organisms.

In **Activity AA4.3** you can discuss what structural features are important in making pyrethroids active against insects.

What happens to pyrethroids in the environment?

In mammals, pyrethroids are rapidly broken down into polar products, either by oxidation or by hydrolysis of the ester group. These polar products are not attracted to the fatty membranes, but remain in aqueous solution and are excreted before they can reach sensitive sites in the body.

Commercially available pyrethroids are effective in the field at applications of $200\,\text{g ha}^{-1}$ or less. With the most active compounds, only $20\,\text{g ha}^{-1}$ may be needed. This is much less than the quantities required when using other insecticides such as organochlorines, organophosphates and carbamates, for which applications of $1\,\text{kg ha}^{-1}$ are common.

Synthetic pyrethroids persist on crops for 7–30 days (Figure 47). Any pyrethroid residues reaching the soil are attracted into the soil organic matter. Once there they are rapidly hydrolysed and oxidised by routes like those in mammals. The products are inactive polar compounds, so residues of active, non-polar compounds do not build up in the environment.

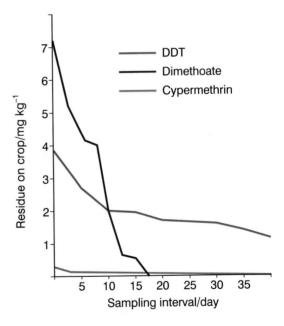

Figure 47 Decay of pesticides – the graph shows the large reduction in pesticide residues on crops when newer pesticides such as cypermethrin replace older ones such as DDT and dimethoate; less pyrethroid is needed, and it decays faster to inactive products.

ASSIGNMENT 13

This assignment requires you to make use of chemical ideas met in earlier parts of the course. You may wish to refer to **Chemical Ideas 13.5** for the reactions of esters, **Chemical Ideas 7.6** for thin-layer chromatography and **Chemical Ideas 10.3** for information about half-lives

Biopermethrin is safe to mammals because enzymes called esterases catalyse the hydrolysis of the ester linkage.

a   Look at the structure of biopermethrin on p. 202, and draw the structural formulae of the hydrolysis products. Explain why these are more soluble than biopermethrin in water.

b   The course of the hydrolysis can be followed by thin-layer chromatography. The R_f values in the eluting solvent used in one experiment are shown below:

	Biopermethrin	Alcohol derivative
R_f value	0.6	0.15

The acid derivative did not move much above the base line.

Describe in outline the procedure you would use to follow the progress of the reaction. Sketch how the chromatograms would look (i) when the hydrolysis reaction was incomplete and (ii) when it had reached completion.

c   The hydrolysis of biopermethrin in the soil is a first-order reaction. Calculate the half-life of biopermethrin if there is 2% of the insecticide left in the soil 2 months after application.

Figure 48 Measuring the effects of an agrochemical.

Herbicides

Herbicides can increase crop yields by destroying weeds. There are two main types of herbicide: **total herbicides** and **selective herbicides**. Total herbicides destroy all green plant material and are used in fields before a crop is planted.

One example of a total herbicide is *paraquat*. Pure paraquat is highly toxic, but in the concentrations applied to kill weeds it is relatively harmless to humans and is rapidly inactivated on contact with soil.

The structure of the paraquat ion is

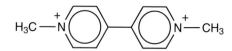

paraquat ion

As soon as the positive paraquat ions contact the soil they are removed by adsorption onto the soil solids. This is a particularly useful property of paraquat because it means that it is inactivated as soon as it reaches the soil. It only kills those plants whose *leaves* it touches.

Adsorption and absorption

Paraquat is deactivated when it comes into contact with soil because it is adsorbed onto the soil solids. A substance is **adsorbed** when it is bound to the *surface* of another substance.

Figure 49 Adsorption.

Be careful not to confuse this with **absorption**. In absorption, the absorbed substance diffuses *into the bulk* of another substance. In coloured plastics, for example, dye molecules are absorbed into the bulk of the plastic.

Figure 50 Absorption.

Each type of soil is capable of holding irreversibly a particular amount of paraquat. This amount is called the **strong adsorption capacity** of the soil. Some values are given in Table 10. If the paraquat concentration rises above the strong adsorption capacity of the soil, paraquat can be displaced into soil water and can damage growing plants.

Table 10 Strong adsorption capacities of some soils.

Source	Soil type	Composition			Strong adsorption capacity/ mol kg^{-1}
		% sand	% silt	% clay	
Newark	clay	34	24	42	0.137
Oakham	clay loam	33	37	29	0.047
Jealott's Hill	sandy loam	49	33	18	0.041
Sutton Coldfield	sandy loam	78	15	8	0.0078
Bagshot	loamy sand	85	11	4	0.0019

Table 10 Strong adsorption capacities of some soils.

Paraquat can be used as an alternative to ploughing as a means of destroying weeds before a crop is sown, but farmers also need to destroy weeds in growing crops without harming the crops themselves. They therefore want a range of selective herbicides.

Another very effective 'kill-all' herbicide is *glyphosate*, which is sold under several names, such as 'Roundup' and 'TUMBLEWEED'.

It is sprayed onto the outside of leaves and spreads through the plant to the roots. It is rapidly degraded in the soil, so it only affects the plants that it is sprayed on.

Unfortunately, it too is non-selective so it will destroy the wanted plants along with the weeds. Scientists are attempting to produce plants, such as maize, which are resistant to specific herbicides (see **Engineering Proteins** storyline, Section **EP3**).

Some herbicides have been developed which can kill broad-leaved plants without damaging grasses. Others will kill grasses but not broad-leaved crops such as soya bean and sugar beet. Some can even control grassy weeds in grass-like crops such as rice, maize and other cereals. Their selectivity depends on differences in metabolism between the plants.

ASSIGNMENT 14

Refer to Table 10. (A loam is a rich soil consisting mainly of clay, sand and humus.)

a How is the strong adsorption capacity of a soil related to soil type?

b You learned about soil solids in Section **AA2**. Use your knowledge to explain the correlation between soil type and strong adsorption capacity.

c What physical method would you use to confirm your suggestions in your answer to part **b**?

d What are the implications of the figures in Table 10 for farmers and gardeners in Bagshot?

Fungicides

Various fungi can reduce the yields of cereal crops, such as wheat and corn, by as much as 20%, by attacking the roots, the stem, the leaves and the ears. Even the germination of the seeds can be affected.

About a hundred years ago, Bordeaux mixture (a mixture of copper(II) sulphate and calcium hydroxide in water) was introduced as a fungicide and some gardeners still spray potatoes with it. This and other later fungicides simply coat the leaves as a protection against fungal spores which settle there. They do not protect any new growth produced after spraying.

In the late 1960s, DuPont introduced the first **systemic fungicides**. These are absorbed by the plant, through the leaves, moving through the plant, thereby protecting even new growth against fungal spores.

Over the last few years, a new systemic fungicide called *Azoxystrobin* has been developed by Zeneca Agrochemicals based on a chemical found in a natural fungus, *Strobilurus tenacellus* (Figure 51).

strobilurin A

The natural product is *strobilurin A*, which inhibits respiration in fungi.

Azoxystrobin

By synthesising compounds similar to this, Azoxystrobin was discovered. It has properties superior to the natural product, is not harmful to the environment and has a very low toxicity to mammals.

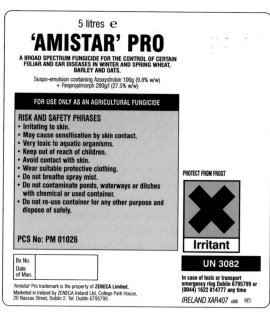

Figure 53 *This fungicide contains Azoxystrobin. The label clearly outlines the hazards when using it.*

Figure 51 Strobilurus tenacellus.

Figure 52 The fungus Septonia nodorum *growing on wheat. Above, untreated; below, treated with Azoxystrobin.*

This is a classic example of the way in which both new agrochemicals and new pharmaceuticals can be found. Something in nature is known to have special properties. Chemists then determine what the substance is that confers these properties and next find a means to synthesise it to test its effectiveness when in a pure form. They may then develop compounds which have an even greater effect, based on the structure of the natural product. At this stage, extensive trials are made and methods of transferring the synthesis from laboratory to manufacturing scale are explored. No wonder it is so expensive to develop an agrochemical or a medicine.

AA5 *Summary*

This unit has introduced you to some of the issues involved in developing and maintaining a system of agriculture which is able to feed the world's rapidly increasing population. There is a constant dilemma on how to do this without damaging the environment.

Between 1960 and 2000, the world's food supply has increased threefold, without increasing the area used for farming. However, between 2000 and 2035, the population is set to increase from 4 billion to over 8 billion. Further, it is hoped and expected that the quality of life for the poorest will increase significantly. How will it be possible to increase the agricultural output even more? Is the environment bound to suffer?

To understand these issues, you first needed to learn about the composition and structure of soil, in particular the nature of silicate and clay minerals, and to understand the relationship between the properties of a substance and its structure and bonding.

Consideration of the weathering processes which produce soil provided a setting in which to revisit ideas about the effect of temperature on chemical reactions.

Clay soils can bind positive ions – such as metal ions, NH_4^+ ions and H^+ ions – to the surface of the negatively charged silicate sheets. These exchangeable cations form a pool of nutrients from which plants can draw. This led to a study of ion-exchange reactions and the behaviour of ions in general. Understanding the processes which occur as plants grow and decay is vital to understanding how nutrients are cycled. Central to this is the nitrogen cycle and the redox chemistry of nitrogen.

The use of inorganic fertilisers is vital in maintaining soil fertility and high crop yields. The Haber process for the manufacture of ammonia from nitrogen and hydrogen is the key step in the production of nitrogen fertilisers. An understanding of the effects of temperature and pressure, both on the position of the chemical equilibrium and on the rate of the reaction, is necessary to select the optimum conditions for this process. The use of a catalyst is also important.

In the last section of the unit, you studied some methods of pest control. A study of the action of pyrethroids, modern insecticides which do not persist in the environment, provided a setting in which to learn about partition equilibria.

The secret to increasing the production of food, and keeping the food in good condition, lies in keeping the soil in good condition and in the judicious use of compounds which act as fertilisers and pesticides, with no deleterious effect on the environment.

Activity AA5 will help you to summarise what you have learned in this unit.

COLOUR BY DESIGN

Why a unit on COLOUR BY DESIGN?

Coloured compounds are everywhere. The ability to make synthetic dyes and pigments to colour an enormous variety of things is one of the great achievements of modern chemistry.

The first part of the unit describes work carried out in the Scientific Department of the National Gallery, London, on the conservation of two paintings: *The Incredulity of S. Thomas* painted by Cima da Conegliano in 1504 and *A Wheatfield, with Cypresses* painted by Vincent van Gogh in 1888.

The chemists at the National Gallery use a variety of analytical techniques to investigate pigments and paint media. This provides an ideal setting in which to learn about ultraviolet and visible spectroscopy, atomic emission spectroscopy and gas–liquid chromatography. Analysis of the drying oils used in the 16th-century Italian altarpiece requires an understanding of the structure of oils and fats. This links in well with your previous work on acids, alcohols and esters.

The second part of the unit traces the development of synthetic dyes for cloth. In this part of the unit, you will extend your knowledge of organic chemistry further and learn about the special structure of the benzene ring and the types of reactions arenes undergo.

Finally, the unit provides an opportunity to draw together some ideas you have met earlier to build up a simple theory of colour.

Overview of chemical principles

In this unit you will learn more about …

ideas introduced in earlier units in this course

- why compounds are coloured (**The Steel Story**)
- the interaction of radiation with matter (**The Atmosphere**, **What's in a Medicine?** and **The Steel Story**)
- spectroscopy (**What's in a Medicine?**)
- atomic emission spectra (**The Elements of Life** and **The Steel Story**)
- aromatic compounds (**Developing Fuels** and several other storylines)
- esters (**What's in a Medicine?** and **Designer Polymers**)
- reaction mechanisms (**The Atmosphere** and **The Polymer Revolution**)
- intermolecular forces (**The Polymer Revolution**)
- the relationship between structure and bonding, and properties (**Aspects of Agriculture** and several other storylines)

… as well as learning new ideas about

- ultraviolet and visible spectroscopy
- gas–liquid chromatography
- oils and fats
- the structure of benzene and the reactions of arenes
- the chemistry of dyes
- theories of colour.

COLOUR BY DESIGN

CD1 *Ways of making colour*

The natural world is full of colour. Some colours, like the blue of the sky or the colours in a rainbow, are produced by the scattering or refraction of light. But in most cases, colour is due to the presence of *coloured compounds* and arises from the way these compounds interact with light.

Figure 1 Cave paintings in Lascaux in France thought to date back 17 000 years.

From the earliest times people have used the natural substances around them to colour themselves and their possessions. We know that some Neanderthal tribes roaming Europe 180 000 years ago prepared their dead for burial by coating them with *Red Ochre* (iron(III) oxide). For tens of thousands of years, humans made colouring agents from minerals they found in rocks, so the colours produced were mostly dull and earthy. These mineral **pigments** were mixed with oil or mud to form a paste which would stick to surfaces.

Pigments and dyes

The way a coloured substance is used determines whether it is called a **dye** or a **pigment**. You can dye your hair or your clothes, but when you paint a picture you are using pigments.

The main thing to remember is that *dyes are soluble substances whereas pigments are insoluble.*

Pigments can be spread in a surface layer (as in a paint or a printing ink) or mixed into the bulk of a material (as when making a coloured plastic bowl).

Dyes are always incorporated into the bulk of a material and they attach themselves to the molecules of the substance they colour. This attachment can be the result of hydrogen bonding or of weaker intermolecular forces, such as instantaneous dipole–induced dipole forces or other dipole–dipole forces. Sometimes stronger ionic or covalent chemical bonds are involved.

You will find that there is another general distinction. Most dyestuffs are organic compounds. Pigments can be organic or inorganic.

This was fine until people learnt to weave and make fabrics. When the paste-like pigments were applied to fabrics, the cloth became stiff and the colouring material soon fell out. Pigments were no good for colouring cloth. Cloth could only be coloured by soaking it in a solution of a **dye**.

You can find out why some compounds are coloured by reading **Chemical Ideas 6.7**.

Figure 2 Modern dyes and pigments make life very colourful.

Many of the early dyes came from crushed berries or plant juices. The early Britons used a blue dye which they extracted from the woad plant (*Isatis tinctoria*). The main coloured component of woad is *indigo*. The same dye is used today to colour blue denim jeans.

Some dyes came from animals. Mexican dyers around 1000 BC discovered the red dye *cochineal*. They extracted it from small insects which live on the *Opuntia* cactus. Only female insects produce the dye, and they had to be collected by hand: about 150 000 insects were needed to make 1 kg of dye! It was the Spaniards who brought cochineal to Europe in 1518 AD. It was used until 1954 to produce the bright red jackets of the Brigade of Guards. You may have eaten it as a food colouring.

indigo

cochineal

Figure 3 A late 19th-century dyehouse.

Figure 4 A gas holder built by the Imperial Gas Company at Bethnal Green in London. 'Coal gas' was made by heating coal in the absence of air. Coke, coal tar and a liquid rich in ammonia were also produced (see Figure 36 on p. 196).

The coal gas industry provided the raw materials for the production of synthetic dyes – and later for the production of pharmaceuticals, plastics, perfumes and explosives. Coal tar was then replaced by oil as the source of organic chemicals.

Until about 150 years ago, dyes and pigments were expensive and colours were mainly for the wealthy. Ordinary people wore clothes dyed with cheap vegetable dyes – the colours were often drab and quickly faded.

The great breakthrough came when chemists learnt how to make coloured substances in the laboratory. The starting materials for the new **synthetic** dyes were cheap. They came from **coal tar**, an unwanted by-product of the new coal gas industry. In the second half of the 19th century, as towns and cities were lit up by gas lamps, chemical companies producing dyes from coal tar flourished – and Europe exploded in a riot of colour!

Not only did chemists learn to copy natural colours, but they also used the compounds they obtained from coal tar to make a whole range of entirely new coloured compounds.

Modern colour chemists now have a vast range of coloured substances available to them. They need to understand not only why compounds are coloured, and which structures lead to particular colours, but also how to bind coloured substances to different types of

fibres and surfaces. Once the chemistry is understood, it becomes possible to *design* a coloured molecule for a particular purpose: colour by design.

Many of the advances in colour chemistry have been made as a result of intensive research and painstaking 'trial and error'. Every so often, though, an important step forward seems to happen by chance.

In **Activity CD1** you can produce coloured compounds by a variety of different processes.

ASSIGNMENT I

a Look carefully at the structure of cochineal on p. 210. Make a list of the different functional groups in the molecule, giving the name and formula of each.

b What structural feature present in both indigo and cochineal do you think could be responsible for their colours?

CD2 *The Monastral Blue story*

A chance observation

Monastral Blue is one of the best blue pigments ever made. It is widely used to colour plastics, printing inks, paints and enamels because it gives a very pure blue. The paint on most blue cars, for example, contains Monastral Blue. Its discovery arose from a chance observation at a Scottish dyeworks. Fortunately, the people who made the observation took the trouble to investigate further.

Figure 5 Monastral Blue is the pigment in the blue paint on this classic steam locomotive.

In 1928 in Grangemouth, Scotland, the firm Scottish Dyes (which later became part of ICI) was making phthalimide, a compound needed for dye manufacture.

In this process, a white substance, phthalic anhydride, is melted in a glass-lined iron vessel, and ammonia is passed into it:

ammonia(g) + phthalic anhydride(l) → phthalimide(l)

The product, too, is white, but on this occasion they found that some batches contained traces of a blue substance. The blue compound seemed to have formed where the reaction mixture came into contact with a part of the vessel where the lining was damaged.

Later, when the blue substance was analysed, all samples were found to contain 12.6% iron. The structure of the blue substance, although complex, was interesting for a number of reasons – not least because it was closely related to that of some naturally occurring substances called **porphyrins**.

At first it was thought that the new blue compound, named **iron phthalocyanine**, would be of academic interest only, but further investigation suggested that it may be a useful blue pigment.

Porphyrins everywhere

You have already met an important example of a coloured substance which contains a porphyrin ring system. *Haemoglobin* is responsible for the red colour of blood and is involved in transporting oxygen around the body (see **The Elements of Life**, Section **EL2** and **The Steel Story**, Section **SS5**). *Chlorophyll a* gives leaves their green colour and is responsible for harvesting light and initiating photosynthesis. It too is a porphyrin pigment, but with magnesium as the central metal.

ASSIGNMENT 2

Read the account of the discovery of iron phthalocyanine in the previous column.

a Why did the blue pigment form?

b What is the significance of the fact that *all* samples of the blue pigment contained 12.6% iron?

The structure of the blue compound

The blue compound is made up of large flat molecules. Each molecule contains a 16-membered ring of alternating carbon and nitrogen atoms, with an iron atom at the centre. Look carefully at Figure 6 which shows the structure of iron phthalocyanine and pick out the 16-membered ring surrounding the iron atom.

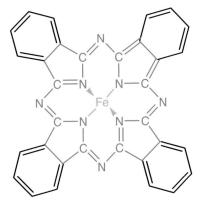

Figure 6 Structure of iron phthalocyanine (the porphyrin ring system is shown in blue).

Don't be put off by the size of the molecule. You won't be expected to remember its structure, but you should be able to recognise some of its important features.

Four of the nitrogen atoms act as **ligands** to form a planar complex with the iron atom. If you have studied **The Steel Story**, you will remember that complexes of d-block metals are often highly coloured.

The 16-membered ring in iron phthalocyanine contains alternating single and double bonds. This type of arrangement is called a **conjugated system of bonds**. For each double bond, one pair of electrons is not confined to linking two particular atoms like those in a single bond, but is spread out or **delocalised** over the whole conjugated system. The lone pairs of electrons on two of the nitrogen atoms are also involved, so the delocalisation extends over the whole 16-membered ring – and over the four benzene rings too. The result is a huge delocalised structure.

You met a conjugated bond system when you made a conducting polymer in **Activity PR6, The Polymer Revolution.**

The more that electrons are delocalised like this, the lower the energy of the molecule. So, iron phthalocyanine is a very stable pigment.

Molecules with extended conjugated systems also tend to be coloured. You will find out more about this later in the unit.

It's even better with copper

Molecules with similar structures have similar properties, so it is always worth making a series of related compounds in the hope that you come across one or two with even better properties than the ones you already have.

Once the structure of the blue iron compound was known, a team of chemists led by R. P. Linstead at Imperial College, London, set to work to investigate other phthalocyanines. They repeated and modified the original process, using other d-block metals instead of iron as the central atom for the molecule.

Within a few years of the original discovery, they had shown that a whole range of similar pigments could be made. Copper phthalocyanine was particularly promising. Writing about this in 1934, Linstead said:

'… it is even more stable than the other compounds of this series, and in this respect must be classed among the most remarkable of organic compounds. It resists the action of molten potash (potassium hydroxide) and of boiling hydrochloric acid. … It is exceptionally resistant to heat, and at about 580 °C it may be sublimed … in an atmosphere of nitrogen or carbon dioxide. … The vapour is deep pure blue and the crystals, which have the usual purple lustre, may be obtained up to 1 cm in length.'

By 1939, copper phthalocyanine was on the market as a pigment with the trade name **Monastral Blue** (Colour Index Pigment Blue 15: CI 74160).

The commercial importance of Monastral Blue and other phthalocyanine pigments is due to a combination of factors:

The Colour Index

Most pigments and dyes are known by the name originally given to them by the manufacturer, rather than by a systematic name. So it's quite common for these compounds to have several names.

To avoid confusion, The Society of Dyers and Colourists in Bradford have drawn up a *Colour Index* in which coloured compounds are classified according to their application and colour, and given a *generic name*, e.g. CI Pigment Blue 15. All commercial products having the same structure are classified under a generic name. Where the structure is known, the compound is also given a Constitution Number (eg CI 74160).

- they have very beautiful bright blue to green shades and high colour strengths (which means that only a relatively small amount of pigment is needed to give a good colour)
- they are very stable pigments and have excellent fastness to light – they don't fade.

CD3 *Chrome Yellow*

Artists are always on the look out for new and better pigments from which to make their paints. By the middle of the 18th century a variety of yellow pigments were being used. *Orpiment* (arsenic(III) sulphide) and *Yellow Ochre* (hydrated iron(III) oxide) had been known since earliest times. *Massicot* (lead(II) oxide) was introduced in the 16th century.

But there was still a need for a pigment capable of giving a bright lemon-yellow colour.

The French chemist, Louis Vauquelin, discovered chromium in 1797 whilst investigating a mineral called *Siberian red lead spar* from the Beresof gold mine in Siberia. The colours of samples of the mineral varied

Figure 7 Louis Vauquelin (1763–1829) discovered metallic chromium, Chrome Yellow and other artists' pigments.

from orange-yellow to orange-red. We now know that the mineral was a form of lead chromate(VI) called *crocoite*, but at the time its composition was unknown.

Colourful compounds

A contemporary of Vauquelin described how the new metal, chromium, got its name (*chroma* is the Greek word for colour):

'[The new element] ... had the property of changing all its saline or earthy combinations to a red or orange colour. This property, and that of producing variegated and beautiful colours when combined with metals, induced him to give it the name of "chrome".'

By 1809, a new source of crocoite had been found in the Var region of France and the mineral became readily available in Europe. Within a few years, the enormous potential value of lead chromate(VI) as a yellow or orange pigment became clear. Painters particularly valued the dark lemon-yellow colour which lasted better than the pigments previously available.

In 1809, Vauquelin carried out a series of investigations to determine the best conditions for making the lead chromate(VI) pigment in the laboratory. He obtained it by **ionic precipitation**, using solutions of a lead salt and a chromate(VI), eg

lead nitrate(V)(aq) + sodium chromate(VI)(aq) →
sodium nitrate(V)(aq) + lead chromate(VI)(s)

Chemical Ideas 5.1 will remind you about ionic precipitation.

ASSIGNMENT 3

a For the reaction between lead nitrate(V) and sodium chromate(VI), write
 i a full balanced equation
 ii an ionic equation which shows only the ions involved in the precipitation reaction.
 Include state symbols in both your equations.

b Give equations for other possible ways of making lead chromate(VI) by precipitation.

In **Activity CD1** you had an opportunity to repeat Vauquelin's work of 1809 and investigate factors which affect the colour of the Chrome Yellow produced.

Chrome Yellow was the first of a range of metal chromates containing lead, barium, strontium or zinc which can be used to produce pigments with different shades of yellow. Chrome Yellow is still widely used today in paints and printing inks.

What is a paint?

A paint is usually made up of two main parts:
- the colouring matter (the pigment)
- a liquid which carries the pigment, allows it to spread and helps to bind it to the surface.

In modern paints, the liquid is made up of a polymer or resin (known as the *binder*) and a solvent which can evaporate. In spray paints, the volatile organic solvent evaporates quickly once the paint has been applied.

In watercolours, the finely ground pigment and a binder are suspended in water. The paint dries as the water evaporates and the pigment and binder are absorbed into the paper.

In oil paints, the drying process is slow. Oil paints usually contain some solvent, such as white spirit, to give the paint a workable consistency. The oil is the binder (sometimes called the *binding medium*) and does not evaporate. Instead, it slowly hardens as it reacts with air to produce a flexible film. The oil protects the pigment and helps to bind it to the surface being covered.

Figure 8 Yellow road markings contain Chrome Yellow pigment.

Pigments can be poisonous

Many of the artists' pigments used in the past were hazardous in one way or another, and the artists were almost certainly unaware of the risks they were taking.

Some pigments, such as *Red* and *Yellow Ochre* (both forms of Fe_2O_3) and many organic pigments are harmless – others like the green and yellow arsenic pigments that used to be used can be deadly. We now know that lead, cadmium and mercury compounds are highly toxic, and certain soluble chromates have been shown to be **carcinogenic** (they can induce cancers) and are now rarely used in paints. So like any chemist, handle pigments with care and follow the safety instructions.

Modern paints tend to be much less toxic. Lead compounds, for example, are no longer included in household paints. In many cases, inorganic pigments are being replaced by less toxic organic ones.

ASSIGNMENT 4

Modern cosmetics are carefully formulated to be beneficial to the skin – but this was not always the case. The pigments used in ancient Egyptian, Greek and Roman make-up read like a list from a poison cupboard!

- Face make-up (*White Lead*, $2PbCO_3.Pb(OH)_2$)
- Rouge (*Red Phosphorus*)
- Lipstick (*Cinnabar*, HgS)
- Eye-shadow (*Orpiment*, As_2S_3)
- Mascara (*Stibnite*, Sb_2S_3)

a What is the oxidation state of Pb in White Lead?

b Suggest two solutions which could be mixed to give a precipitate of $PbCO_3$.

c What are the modern names for Orpiment and Stibnite?

d Look up the hazard warning data for some of the substances above in a chemicals catalogue.

e Try to find out what pigments are used to colour modern lipsticks.

A famous wheatfield

During 1888, the Dutch artist van Gogh painted three pictures of *A Wheatfield, with Cypresses*. One of these (Figure 9) is in the National Gallery in London. If you are in Trafalgar Square, you can easily go in and see it.

During the 1980s this picture was cleaned and remounted. This gave scientists at the National Gallery a welcome opportunity to examine the materials and techniques which van Gogh used.

The pigments on the surface of paintings are coloured because they absorb some of the white light falling on them. **Activity CD3** will help you to understand this.

They photographed the picture in daylight and in both ultraviolet and infrared light. These photographs give information about areas in the painting where particular pigments have been used. For example, *Zinc White* (zinc oxide, ZnO) fluoresces under ultraviolet light. *Emerald Green* (a pigment containing copper and arsenic) absorbs in the infrared region, so areas painted with this pigment appear dark in infrared light.

Figure 9 A Wheatfield, with Cypresses *by Vincent van Gogh.*

Vincent Willem van Gogh (1853–1890)

Vincent van Gogh was the son of a Dutch clergyman. When he was 16, he started work in an art gallery, previously owned by his uncle. After a few years, he was sent to their London branch. While in London, he fell in love with his landlady's daughter, but the relationship was a one-sided and unhappy affair.

He abandoned the art business and eventually returned to Holland with the intention of entering the Church. But he did not complete his preparation for the ministry. Instead, he went to Belgium as a lay preacher, living there mostly as a tramp.

He was 28 and back in his parents' home town when he finally decided to become an artist. During the following 10 years he worked at his paintings with what one writer has described as 'a single-minded frenzy'. He produced no fewer than 800 paintings and as many drawings. Sadly, he ended his life in turmoil. In 1888 a piece of his ear was cut off during a quarrel with a fellow-artist, Gaugin. Shortly after that, he was admitted, at his own request, to a mental hospital. It was here that he painted *A Wheatfield, with Cypresses*. In July 1890 he shot himself.

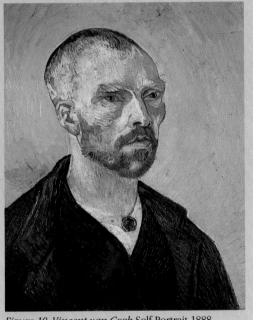

Figure 10 Vincent van Gogh Self-Portrait 1888.

These investigations, combined with historical records, showed that van Gogh used Chrome Yellow, mixed with Zinc White and other pigments, to create the different shades of yellow in the wheatfield.

Table 1 shows an extract from the data obtained by the scientists at the National Gallery.

Paint sample	Pigment
darkest yellow of wheatfield	Chrome Yellow
mid-yellow of wheatfield	Chrome Yellow + Zinc White
lightest yellow of wheatfield	Zinc White + Chrome Yellow
dull yellow of wheatfield	Chrome Yellow + Zinc White + small amounts of Emerald Green
pale green bushes	Zinc White + Chrome Yellow + Viridian (a green pigment)

Table 1 Pigment mixtures used in A Wheatfield, with Cypresses *by Vincent van Gogh.*

(a) Pure lead chromate(VI)

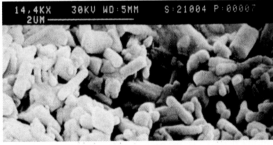

(b) Dark yellow wheatfield

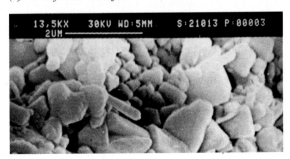

(c) Light yellow wheatfield

Figure 11 Scanning electron microscope pictures of pure lead chromate(VI) and pigment samples from the wheatfield (the scale at the top of each picture shows 2 μm, ie 2×10^{-6} m).

The scientists took very small samples of different colours of paint from the edges of the painting and analysed them to find out which elements were present. One way they did this was to look at the lines in the **atomic emission spectrum** from each sample of paint. You will learn more about this technique later in the unit. They also used a scanning electron microscope to look at the tiny individual crystals in the paint (see Figure 11).

a In what way do you think the electron microscope pictures in Figure 11 support the conclusions of the scientists shown in Table 1 concerning the pigments used in the dark yellow and light yellow areas of the wheatfield.

b Van Gogh used Cobalt Blue and Zinc White for the distant mountains. Both these pigments reflect infrared light strongly. He used Emerald Green to produce the dark green of the cypresses. In infrared light, the trees appear dark and the mountains cannot be distinguished from the sky. Explain why this happens.

CD4 *Chemistry in the art gallery*

The chemists involved in identifying the pigments used by van Gogh are *analytical chemists*. Their job is to find out which compounds are present in different substances. Some of the problems involved in restoring a painting can be very challenging. Solving them can involve a great deal of patience and ingenuity, a knowledge of chemistry and the application of some sophisticated techniques.

The Incredulity of S. Thomas

In 1504 an Italian artist, Cima da Conegliano, put the finishing touches to his altarpiece *The Incredulity of S. Thomas*. The painting now hangs in the National Gallery in London. It is considered to be a masterpiece and is highly valued because of its rarity.

By the standards of the time, Cima used an exceptionally wide range of pigments in creating the altarpiece. Some of the substances he used will be familiar to you. For example, to produce certain shades of green, he used *malachite* ($CuCO_3.Cu(OH)_2$, a basic copper carbonate). For reddish brown, he used *haematite* (an ore containing Fe_2O_3, iron(III) oxide).

Many of the pigments, such as the blue *Natural Ultramarine* made from the semi-precious stone *lapis lazuli*, would have been quite expensive. In fact, there are several areas of blue paint on the altarpiece. These vary widely in colour from the rich turquoise blue of the ceiling to the pale blue drapery worn by the apostle to the left of Jesus Christ.

How do we know which blue pigments Cima used to achieve these colours nearly 500 years ago? One way to find out is to shine light on different areas of the painting and examine the wavelengths present in the reflected light. Different pigments absorb different wavelengths

from the incident light, so the resulting **reflectance spectrum** is characteristic of a particular pigment.

If you are ever fortunate enough to stand in front of Cima's painting in the National Gallery and look at its vast array of bright colours, you will find it difficult to believe how badly disfigured and fragile the painting was when restoration was started in 1969 (Figure 12). Layers of dirt and old varnish masked or dulled the colours. The paint was badly blistered and was flaking off in many places.

Figure 12 The Incredulity of S. Thomas *by Cima da Conegliano in 1969 before restoration.*

Over the next 15 years a great deal of laborious restoration took place. Much of this was made possible by scientific analysis – not only of the pigments involved but also of the binding medium used to make the paint and of the foundation coating beneath the paint layer.

But first we will look at the story of how the painting got into such an awful state.

You can read about ultraviolet and visible spectroscopy in **Chemical Ideas 6.8**.

In **Activity CD4.1** you can find out how chemists can help in deciding which pigments Cima used to produce the different areas of blue in the altarpiece.

Figure 13 The painting in 1986 after cleaning and restoration.

The history of the painting

The start of the trouble (1504–1870)

Cima's painting *The Incredulity of S. Thomas* was originally hung above an altar in a church in the Italian town of Portogruaro, about 50 miles northwest of Venice.

For some reason the paint began to flake and blister. In 1745, what may have been the first in a series of restorations was attempted. But the condition of the painting continued to deteriorate. Eventually it was sent to the Academy of Fine Arts in Venice where one of the foremost experts of the time, Professor Guiseppe Baldassini, restored it as best he could.

Between 1822 and 1830 the painting was stored in a ground-floor room of the academy while arguments were taking place over Professor Baldassini's bill. This led to the most disastrous episode in the troubled history of the painting: a sudden and unusually high tide flooded the room and the picture was submerged in salt-water for several hours.

More flaking and blistering occurred and there were several further attempts at restoration. Despite these problems, the painting caught the eye of Sir Charles Eastlake in 1861 while he was on a picture-buying expedition for the National Gallery. After some haggling over the price and a lengthy legal dispute about the ownership of the painting, it was bought for the National Gallery in April 1870 for £1800.

1870: to the National Gallery, London

At this point the gallery authorities made the emphatic recommendation that 'no restoration work should again be attempted'. But paint continued to flake off and there were repeated treatments over the next 80 years to reattach this flaking.

During the Second World War, the whole of the National Collection of art was housed in large artificial caves in a quarry near Bangor in North Wales. The conditions of stable humidity and temperature in the caves were far superior to those in the gallery, and the problems of warping wooden panels and blistering paint were much reduced.

It was when the collection was returned to London in 1945 that serious problems arose. The effect was particularly bad in the very cold winter of 1947, when the heat had to be turned up high to maintain a tolerable temperature for visitors to the gallery. The low humidity caused many paintings to dry out, resulting in cracking and flaking. Cima's poor altarpiece flaked more than almost any other painting.

When an oil painting is produced on a wooden panel, the wood is first coated with a stable inorganic substance bound together with an animal glue. In Cima's painting, *gypsum* (hydrated calcium sulphate, $CaSO_4.2H_2O$) was used for this purpose. This coating layer is called the *gesso* (the Italian word for gypsum).

In 1969, investigators found that the repeated flaking was due to the gesso layer becoming detached from the wooden support, and not to the paint coming away from the gesso. Worse still, it was found that the wooden panel had dry rot and woodworm!

The challenge

In the light of all this, experts at the National Gallery decided to give the picture the most drastic form of treatment used in restoration. The delicate and priceless layers of paint and calcium sulphate were to be transferred from the wooden panel to a new support.

You might get some idea of the scale of the problem if you consider that the picture measures approximately $3 m \times 2 m$ and that the thickness of the brittle paint and gesso together is only a fraction of a millimetre.

The first step was to cover the painting with layers of special tissue using an organic resin as an adhesive. It was then placed face-down on a temporary support and the original 5-cm-thick wooden panel was removed from the back. No high-tech here: it was a slow laborious process achieved by hand using small gouges, chisels and finally scalpels! What remained was a priceless film of paint and calcium sulphate.

Figure 14 The painting after cleaning and transfer to a new support – but before repairing the damaged paint layer.

Next, a heat-activated adhesive was used to secure the back of the painting to a new support. This consisted of a fibreglass plate secured to an aluminium honeycomb. Now the layers of tissue on the front of the painting could be carefully peeled away. After 500 years the painting was finally attached to something from which it was unlikely to come unstuck – at least not for a few more centuries.

However, after all the years of hard work the painting was still unsuitable for exhibition (see Figure 14). The last stage was to repair the damaged paint layer.

So far, the scientific department at the National Gallery had played a relatively small role. In the final restoration, the *retouching* of the paint layer, the work of analytical chemists was crucial.

When retouching the painting, the restorers used modern pigments and modern *binding media* to make up the paints they used. They had to choose these carefully to make sure of two things:

- that any restorations could easily be removed without damaging the original work
- that the colours used in retouching were a good match to the original colours.

The first of these principles applies to the restoration of any work of art in any country. It ensures that a future generation can remove the work of the restorer if they decide that it is inappropriate or if scientific advances make better restorations possible.

Restorations can only be removed with ease if the modern binding medium dissolves in organic solvents more readily than the one used by Cima 500 years ago. So it is important to know exactly which binding medium Cima used.

The binding medium

The medium used to bind the particles of pigment together to form a paint must be viscous enough to prevent the paint from running as it is applied. But it must not be *too* sticky otherwise the artist's freedom would be restricted during painting.

Once the paint has been applied, the medium must then dry and become hard in order to produce a durable painted surface.

What medium did Cima use?

Throughout the history of painting, artists have experimented with different types of media. Egg yolk (known as *egg tempera* when used as a binding medium) was widely used in European paintings in the Middle Ages. Egg yolk is an emulsion of globules of fat and protein in water. It dries and hardens quickly as the water evaporates. You will be able to confirm this if you have ever had to wash up plates coated in dried egg.

Emulsions

An emulsion is a mixture of two liquids which do not dissolve in one another, such as oil and water.

For example, egg yolk is an oil-in-water emulsion. The oil droplets are so small that they do not settle out, and the mixture appears uniform and opaque.

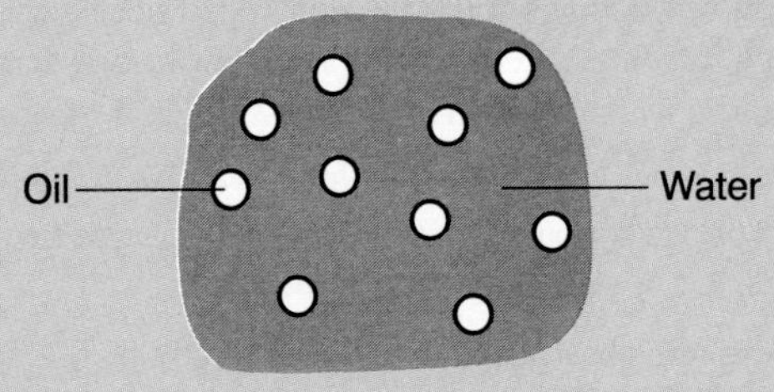

Figure 15 An oil-in-water emulsion.

By the time Cima was working on *The Incredulity of S. Thomas*, paints were more often prepared using oils as the medium. Oils dry more slowly than egg yolk and give artists more freedom. Natural oils, such as linseed or walnut oil, which dry and harden to develop a protective coating are suitable. They are called *drying oils*.

The drying oils which Cima might have used all contain significant amounts of **triesters** based on two carboxylic acids, *palmitic acid* and *stearic acid*. Different oils contain the palmitate and stearate esters in different ratios. So, if the palmitate and stearate ratio can be measured, we can tell with some confidence which oil Cima used to bind his pigments.

This ratio can be found using an analytical technique called **gas–liquid chromatography (g.l.c.)**. Figure 16 shows the gas–liquid chromatogram obtained from a small sample of paint from Cima's painting.

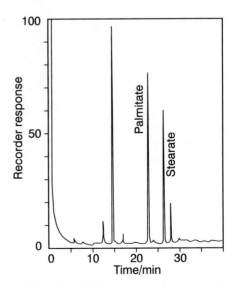

Figure 16 Gas–liquid chromatogram from a sample of paint from Cima's painting.

The area under each peak in the chromatogram is proportional to the amount of each substance present. So you can work out the ratio of palmitate to stearate esters in the paint, and then compare this ratio with those obtained from known oils.

The paints used in the restoration were made using a synthetic polymer rather than oil as the medium. The polymer is soluble in hydrocarbon solvents. All the recent retouchings could be removed, if necessary, with a solvent-soaked swab in just a few hours, leaving behind only what remains of Cima's original work.

It will help to find out about the chemistry of oils and fats. You can do this by reading **Chemical Ideas 13.6**.

You can read about gas–liquid chromatography in **Chemical Ideas 7.6**.

In **Activity CD4.2** you can investigate what factors affect the drying potential of an oil.

You can work out which oil Cima used by doing **Activity CD4.3**.

Figure 17 At home again! Cima's great painting hanging in the Sainsbury Wing of the National Gallery.

What were the pigments?

You have already met some of the methods chemists can use to identify a pigment. One of the methods mentioned earlier is a form of atomic emission spectroscopy called **laser microspectral analysis (LMA)**. It proved particularly useful when analysing paint from Cima's altarpiece. It allows chemists to identify the *elements* present in a sample of paint. From this they can usually say which pigments are present. (If you have studied **The Steel Story**, you may remember that atomic emission spectroscopy is used to monitor the composition of steel during manufacture.)

You can remind yourself about atomic emission spectroscopy by reading **Chemical Ideas 6.1**.

Laser microspectral analysis

In this technique, a pulse of laser light is focused onto a small sample of paint. Laser light is a high-energy beam of light of a single wavelength. The energy of the pulse is high enough to vaporise the high boiling point metal compounds present in the pigments. A small plume of vapour rises from the paint sample into a region between two electrodes.

The atoms and ions in the vapour are then excited to higher electronic energy levels by an electrical discharge between the two electrodes. Each chemical element present gives rise to a characteristic **emission spectrum**.

The method is very sensitive and can be carried out on very small quantities of material. The sample of paint taken from a valuable picture is usually of the order of 1×10^{-5} g. Of this, a minor pigment component might comprise only 1×10^{-7} g $- 1 \times 10^{-6}$ g.

Figure 18 shows the apparatus used at the National Gallery. The apparatus consists of four main parts:

A – a laser head to generate high-energy pulses of laser light (694 nm)

B – a microscope with a system of lenses to focus this light onto the sample

C – a pair of carbon electrodes and a power supply to generate a high-energy discharge

D – a u.v./visible spectrometer to separate the radiation emitted by the excited atoms and ions into its component frequencies and produce a record of the emission spectrum.

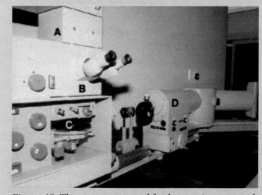

Figure 18 The apparatus used for laser microspectral analysis of pigments.

ASSIGNMENT 6

The LMA emission spectrum obtained from a paint sample consists of a complex arrangement of lines corresponding to radiation of different frequencies.

a Explain why excited atoms only emit certain frequencies of radiation.

b Why are these frequencies different for atoms of different elements?

c How do you think chemists work out which elements are present in the paint sample from the lines in the LMA spectrum?

LMA is just one of the analytical techniques used to identify the elements present in a pigment. A newer and more versatile technique, called *energy dispersive X-ray fluorescence* (EDX) is now commonly used within the scanning electron microscope.

Scientific evidence is always used alongside information from other sources. Art historians, for example, can often provide additional information by studying contemporary manuscripts and by using what is known about other paintings from the same period.

Once the composition of the original pigment is known, a modern substitute can be chosen to give a good match for use in restoration.

In **Activity CD4.4** you can combine evidence from scientific and historical sources to identify the orange-yellow pigment Cima used for the robe of S. Peter.

Now you can try your hand at finding matches for two of the blue pigments in Cima's painting, in **Activity CD4.5**.

CD5 *At the start of the rainbow*

During the 19th century, many new pigments became available as chemists investigated more and more compounds and learned to imitate the properties of natural substances.

The 19th century was also a time of great innovation in the production of dyes for cloth. Today, many of our more obviously high-tech industries are based on electronics; then, it was the colourist who enjoyed the regard we now have for software engineers and system analysts.

The story of the development of dyes is closely linked to the development of organic chemistry. Colourists learned both the methods of dyeing and the principles of chemistry as they moved about Europe selling their skills and knowledge.

To understand this section you will need to know about the special nature of benzene and the type of reactions benzene undergoes. You can do this by reading **Chemical Ideas 12.3** and **12.4**.

Three young entrepreneurs

Perkin

Figure 19 William Perkin (1838–1907).

William Perkin was only 18 years old when he invented a method of synthesising a dyestuff, later known as **Mauve**. He was a student at the Royal College of Chemistry in London and in those days the study of plant extracts dominated organic chemistry. August Hofmann, Director of the College, encouraged Perkin to try to make *quinine*, a natural product used to treat malaria. Perkin tried to do this during the Easter holiday of 1856 in his laboratory at home.

His starting material was a complex organic amino compound obtained from coal tar. Little was known of the structure of quinine, so Perkin's chance of success was minimal, and sure enough the experiment failed. But Perkin was not deterred. He repeated the process with a simpler starting material, **aniline (phenylamine)**, also obtained from coal tar. The reaction produced a purple solution – but again nothing related to quinine.

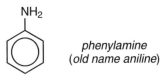

phenylamine (old name aniline)

This type of reaction was already known to analytical chemists who often used reactions involving colour changes to test for compounds. Perkin realised that the purple substance might have other uses. Purple was a popular colour of fashion: the new colour was brilliant and held fast to a piece of cloth. Why not test it as a dyestuff?

The tests were promising and Perkin filed a patent for his discovery. It was the first *synthetic* dye. With the help of his father and brother, he built a factory to manufacture the dye. The difficulties were immense. For one thing, he had to devise a way of producing aniline on a large scale, but very little was known about the behaviour of reactions when scaled up. Figure 20 shows Perkin's reaction scheme and the type of equipment he used.

The venture was an overwhelming success. The colour was in great demand from the world of fashion and both Queen Victoria and the Empress Eugénie in France wore dresses dyed with the new dye. The dye was originally called *Aniline Purple*, but it was later named Mauve after the French for the mallow flower.

Perkin's discovery led to increased interest across Europe in aniline as a starting material. Other synthetic *aniline dyes* in a variety of colours soon followed.

You can investigate some typical reactions of arenes in **Activity CD5**.

Levinstein

Figure 21 Ivan Levinstein (1845–1916).

Perkin was not the only teenager to realise the potential of his college experiments. A German student, Ivan Levinstein, worked on the new aniline dyes at the Gewerbe Akademie in Berlin. In 1864, with the support of his family, he opened a factory there to make *Aniline Green*. The following year, aged 19, he moved to England and settled near Manchester.

Here he eventually started manufacturing *Aniline Red* in a row of cottages at Blackley near Manchester (and thus in the centre of the textile industry so important in the great Industrial Revolution at this time) and gradually built up the biggest British company in the field. Since 1926, it has been incorporated, first into ICI, and is now known as Avecia. Figure 22 shows an advert for dyes made by Ivan Levinstein's company in 1871.

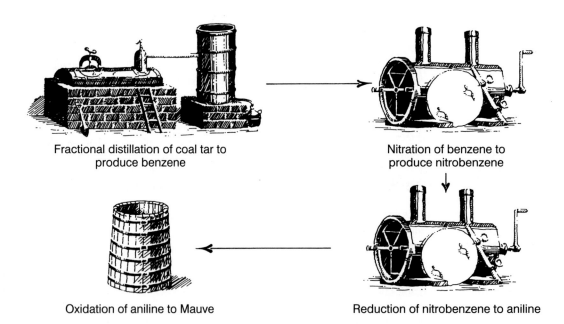

Fractional distillation of coal tar to produce benzene

Nitration of benzene to produce nitrobenzene

Oxidation of aniline to Mauve

Reduction of nitrobenzene to aniline

Figure 20 The manufacture of Mauve showing the type of equipment Perkin used.

Figure 22 An advert for dyes made by Ivan Levinstein's company in 1871. Manufacturers like Levinstein often started in their own homes using household appliances – to this day, the main building in any modern dyeworks is called the dyehouse.

Caro

Green chemistry

Aldehyde Green was the first pure green dye. Previously a mixture of yellow and blue was used to get a good green colour, but this mixture looked dull in artificial light. The Empress Eugénie dazzled everyone when she went to the opera wearing the dress. The dye provided the German company Hoechst with a product that enabled it to become one of the largest chemical companies in the world.

Figure 23 Heinrich Caro (1834–1910).

Figure 24 Empress Eugénie, wife of Napoleon III, wearing a dress dyed with 'Aldehyde Green' (a synthetic 'coal tar' dye) in the 1860s.

Another young German entrepreneur, Heinrich Caro, saw the economic potential of dye manufacture and made his way without family support. Caro was also a student at the Gewerbe Akademie before spending 7 years in Manchester as a colourist and plant manager in a dyeworks. He returned to Germany in 1866 and 2 years later joined the newly formed Badische Anilin und Soda Fabrik (BASF).

In Germany during the late 1860s, work centred on a systematic study of organic chemistry, as well as on possible commercial developments. Caro's skills in chemistry and chemical engineering enabled him to work out a key step in the synthesis of the red colour of the natural madder dye (Figure 25 overleaf). Like Perkin and Levinstein, he was able to transform laboratory experiments into a successful commercial process.

But was it chemistry?

It is important to bear in mind the state of chemical theory at the time these young entrepreneurs were making their discoveries. Ideas that we now take for granted about atoms, atomic masses and periodicity were just beginning to emerge.

For example, the idea that carbon can form four bonds and is capable of forming long chains of carbon atoms emerged in 1858. Kekulé first reported his ideas about a ring structure for benzene in 1865. It was not until much later (in the 1880s) that chemists began to write the structural formulae of dyes in their scientific papers as a matter of routine.

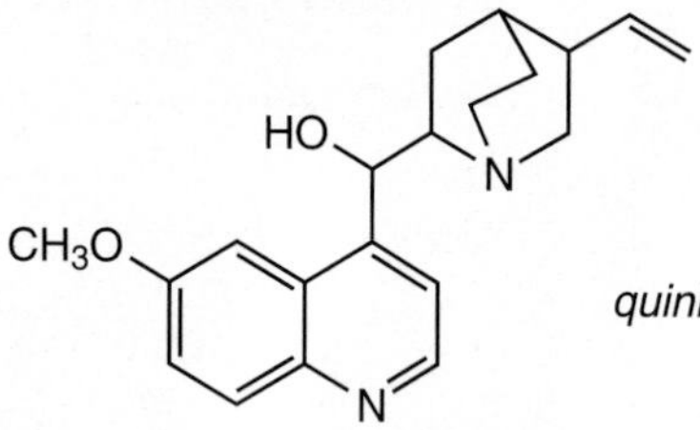

quinine

When Hofmann challenged Perkin to make quinine, it was thought that the formula of quinine was $C_{20}H_{22}O_2N_2$ (although this was actually two hydrogens short). He therefore proposed the following synthesis, using sodium dichromate(VI) as the oxidising agent:

$$2C_{10}H_{13}N + 4O \rightarrow C_{20}H_{22}O_2N_2 + 2H_2O$$

This approach to tackling a synthetic problem now seems ridiculous, but at the time it seemed sound chemical theory! When the structure of Mauve was unravelled in the late 1880s, it turned out to be a mixture of compounds. The most important one, *mauveine*, is a complex organic salt with the structure

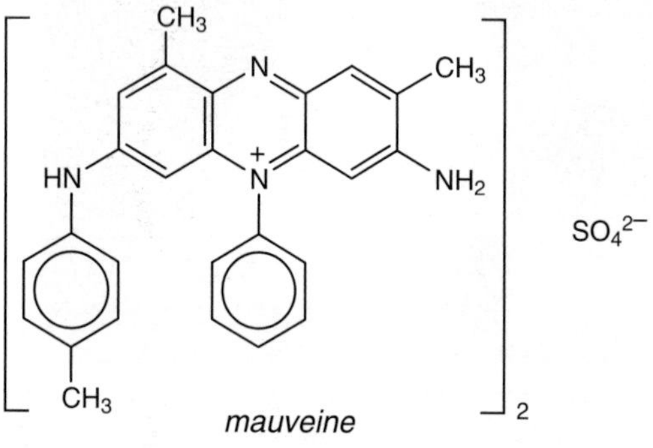

mauveine

SO_4^{2-}

So it was quite amazing that Perkin should stumble upon this molecule by chance – particularly as it was shown later that the formation of Mauve depended on the presence of impurities in the starting aniline. This also explained why Mauve was made up of more than one purple compound.

Alizarin

Until Perkin made his discovery, dyes had come mainly from plants, and vast areas around the world were

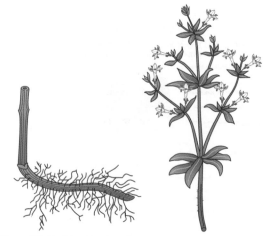

Figure 25 Madder plant and root.

given over to cultivation for the dyestuffs industry. The indigo plant was used to make blue dyes; red dyes came from the madder root. The colouring matter from the root of the madder plant (Figure 25) is called **Alizarin**. We now know that it has the structure shown below:

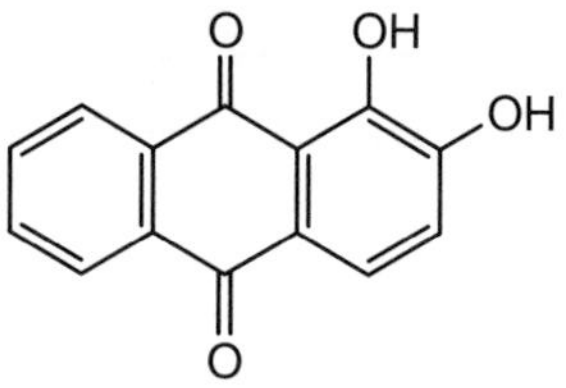

Alizarin

Alizarin only sticks fast to cloth which has been impregnated with a metal compound such as aluminium sulphate. The process is called **mordanting**. The colour of the dyed cloth depends on the metal chosen. When Alizarin is used with an aluminium mordant, the cloth is dyed red; when tin(II) is used the cloth is dyed pink; and iron(II) gives a brown colour.

The mordanting takes place under alkaline conditions so that the metal hydroxide is precipitated in the fibres. The metal ions firmly attach themselves to the cloth and then bind the dye molecules by forming chelate rings, as shown in Figure 26.

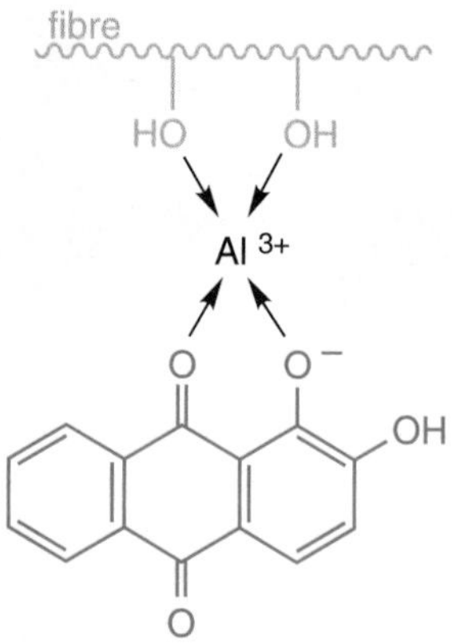

Figure 26 Chelate of Alizarin with the metal ion Al^{3+} (the two remaining ligand sites above and below the Al^{3+} ion could be taken up by OH^- ions).

ASSIGNMENT 7

a Write an *ionic equation* with state symbols for the precipitation of aluminium hydroxide by the reaction of an aluminium compound with alkali.

b What *type of reaction* takes place when mordanted cloth is dyed with Alizarin? What is the role of Alizarin in this reaction?

c Explain how the use of a mordant holds the dye in the fibres.

Alizarin had been used as a dye for thousands of years but no-one had any idea of its chemical nature until Carl Graebe and Carl Liebermann, two talented students of Adolf Bayer at the Gewerbe Akademie in Berlin, decided to find out. Bayer had recently developed a method for converting aryl compounds into their parent hydrocarbons by heating them with zinc dust. For example

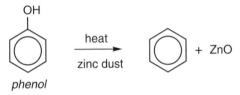

phenol

In January 1868, Graebe and Liebermann used this technique to show that Alizarin is derived from **anthracene**, a minor component of coal tar.

ASSIGNMENT 8

Graebe and Liebermann knew the molecular formula of anthracene. Its properties suggested that it was some kind of condensed aromatic system. Below are the two structures (A and B) that they suggested for anthracene in 1869:

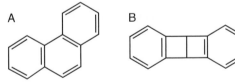

Compare structures A and B with C, the modern structural formula of anthracene:

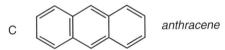

C   *anthracene*

a What is the molecular formula of anthracene?

b What is the relationship between the three structures A, B and C?

c Use your knowledge of bonding in organic compounds to explain why structure B is the least likely possibility.

Graebe and Liebermann then set out on a quest to find a method for synthesising Alizarin. They eventually devised the route shown in Figure 28 overleaf (Route I). Their method worked in the laboratory and they patented it in 1868, but the yield was very poor. Also, a route based on bromine was too expensive at that time for a larger scale process.

The last step of their route, in which the dibromo compound is heated with solid KOH, only added to the confusion of those later trying to work out the structure of Alizarin, because an isomerisation reaction takes place at the same time!

Once Graebe and Liebermann had announced their success, the race was on to find a cheaper commercial route to synthetic Alizarin – a route which was expected to bring rich returns to the first people to patent the process. Both Perkin and Caro joined in the competition.

The race turned out to be a tie. The competing chemists independently hit on the same solution in May 1869! It's the reaction scheme shown in Route II in Figure 28, but remember that the structures were not known at the time.

Perkin filed his patent in London on 26 June 1869. But an almost identical patent had been deposited at the Patent Office on the previous day on behalf of Caro, Graebe and Liebermann working for BASF. Chemists at the Hoechst company in Germany also solved the problem at the same time, but they didn't patent their process in London.

Figure 27 The 19th-century Turkey Red dyeing and printing process used Alizarin as the red dye: Turkey Red fabric was famous for its bright colour and distinctive patterns.

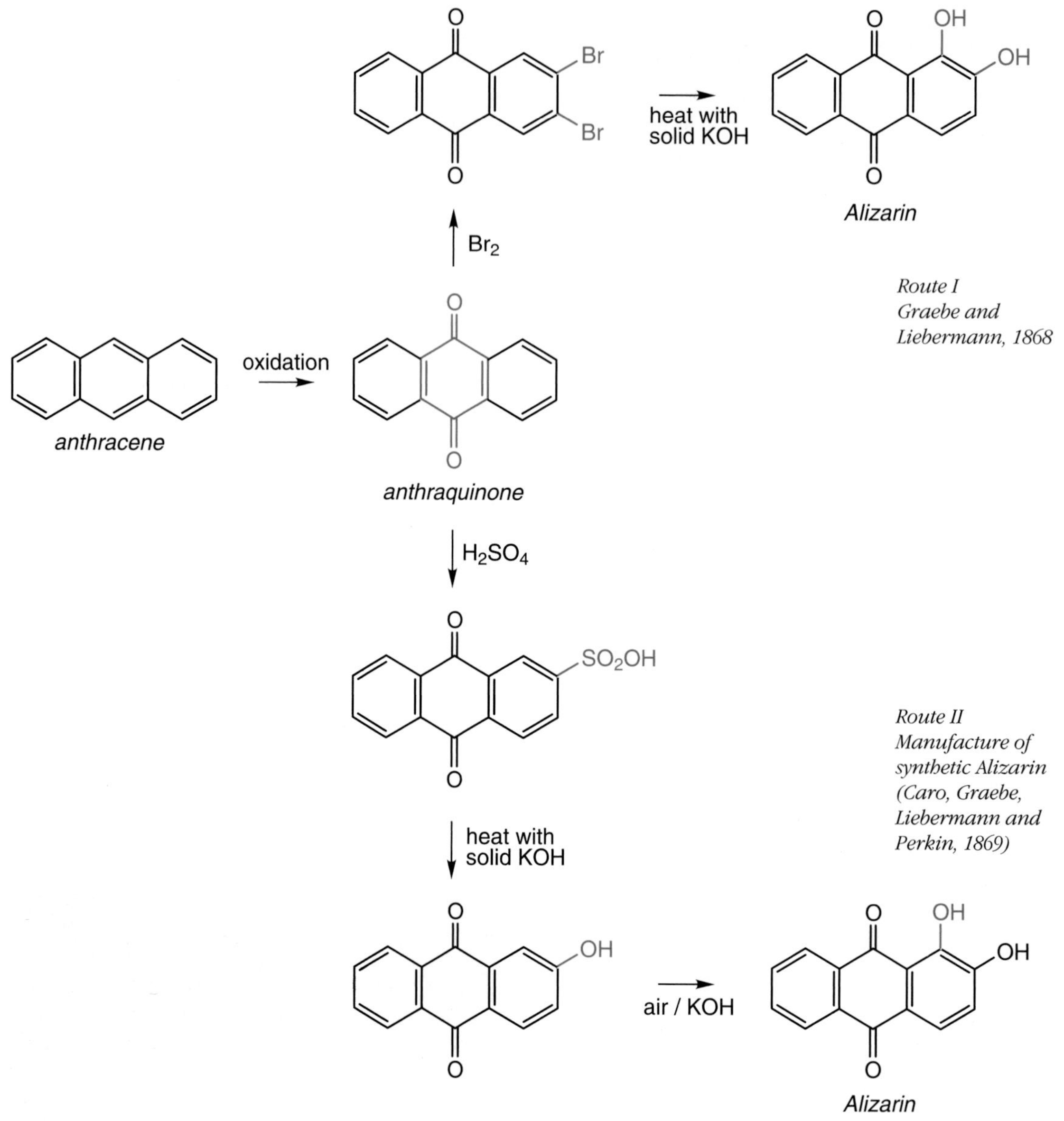

Figure 28 The conversion of anthracene to Alizarin. (The red colour shows the changes for each of the steps.)

The situation was resolved amicably later in the year, when Perkin and BASF agreed to share the market in Alizarin. Perkin retained the UK trade while BASF was to dominate Europe and the USA.

The discovery of a route to synthetic Alizarin had a devastating effect on the madder industry. Hundreds of thousands of acres in Southern Europe and eastwards towards Asia were devoted to growing madder. In 1868 a crop of 70 000 tonnes of madder root was processed to produce around 750 tonnes of Alizarin. By 1873, just 5 years later, the madder fields had disappeared – Perkin's company alone produced 430 tonnes of Alizarin that year.

ASSIGNMENT 9

Compare the two reaction schemes (Routes I and II) for the synthesis of Alizarin in Figure 28.

a What is the essential difference between the two routes?

b Why was the second route a commercial success when the first was not? Would the same apply today?

c What *type of reaction* is involved in the reaction of anthraquinone in each route?

d Use a set of molecular models or molecular modelling software to build up the structure of anthracene. Now modify your model step by step through the stages shown in Figure 28 to arrive at the structure of Alizarin.

CD6 *Chemists design colours*

Figure 29 shows a cartoon of Kekulé's ideas about the structure of the benzene ring, in which monkeys appear instead of carbon atoms. This appeared in a hoax pamphlet put out by two of Kekulé's young disciples. One of them was Otto Witt, a Swiss-trained chemist, who in 1875 was working for a dye-making company in Brentford near London.

Witt was working on a theory that related colour to structure. He wanted to know *why* certain structures led to coloured substances, and how small changes to the structure led to changes in colour.

To do this he was investigating the **diazo reaction**, known since 1858, which led to strongly coloured products which were insoluble in water.

To understand this section you will need to know about the structure and preparation of azo compounds. You can find out about these in **Chemical Ideas 13.10**.

Figure 29 Witt's cartoon lampooning Kekulé's ideas about the structure of the benzene ring: monkeys form single bonds by holding hands and double bonds by linking tails as well.

The first azo dyes

The first azo dyes were made by coupling a **diazonium salt** obtained from phenylamine with one of a variety of **coupling agents**. Figure 30 shows one example of the reaction involved, in which the coupling agent is another molecule of phenylamine.

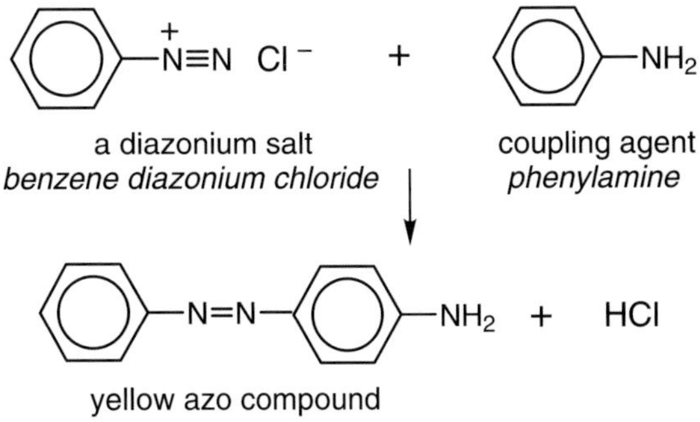

a diazonium salt
benzene diazonium chloride   coupling agent
phenylamine

yellow azo compound

Figure 30 Azo coupling reaction.

Witt knew about this reaction. He also knew that the corresponding reaction using triaminobenzene as the coupling agent in place of phenylamine gave a brown azo compound.

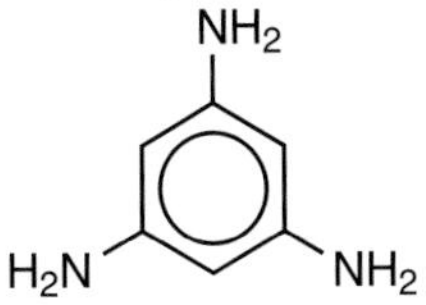

triaminobenzene

Witt thought that the colour of an azo compound was related to its structure. He predicted that the missing azo compound in the series, the one with *two* amine groups on the benzene ring, should be intermediate in colour. Sure enough, when he made this compound he found it was coloured orange, midway between yellow and brown. It proved to be a successful dye for cotton and was marketed as *Chrysoidine*. It was the first commercially useful azo dye.

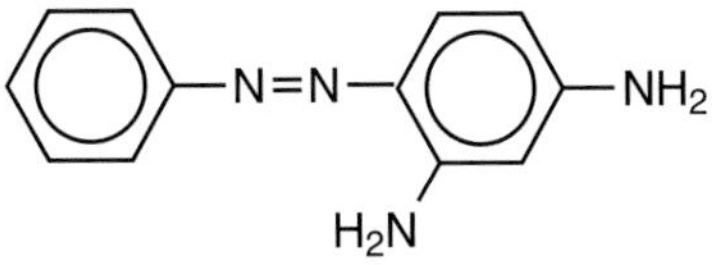

Chrysoidine

Ivan Levinstein manufactured a range of azo dyes at Blackley, but many other British manufacturers retired around this time. William Perkin sold his factory in 1874 having made his fortune. Meanwhile, the German companies Agfa, BASF, Bayer and Hoechst flourished. They set up research teams with links to the universities to invent new colours. By 1913, Germany was exporting around 135 000 tonnes of dyestuffs per year, while Britain was exporting only 5 000 tonnes.

The German companies were also using azo chemistry to produce pigments. Even today, azo pigments dominate the yellow, orange and red parts of the spectrum. Azo pigments are used to colour paints, plastics and printing inks. They provide the yellow and magenta standards used in the *three-* and *four-colour printing processes*.

Figure 31 Many azo dyes are still widely used today.

ASSIGNMENT 10

a Write an equation for the azo coupling reaction Witt used to make Chrysoidine.

b In the early years, azo dyes could only be manufactured satisfactorily in winter. How do you account for this?

c Chrysoidine is a basic dye. Which group in the molecule is responsible for its basic properties?

How does structure affect colour?

Witt's work on azo dyes helped him to put forward a theory of colour in dye molecules which still influences our thinking today. A dye molecule is built up from a group of atoms called a **chromophore**, which is largely responsible for its colour.

We now know that chromophores contain unsaturated groups such as $C=O$ and $-N=N-$, which are often part of an extended delocalised electron system involving arene ring systems.

For Chrysoidine, the chromophore is the delocalised system shown in the box below:

chromophore $\boxed{\langle\rangle-N=N-\langle\rangle-}NH_2$

H_2N functional group

Attached to the chromophore in Chrysoidine are two $-NH_2$ **functional groups** with lone pairs of electrons, which interact with the chromophore to produce the orange colour.

Other functional groups may be added which can

- modify or enhance the colour of the dye
- make the dye more soluble in water
- attach the dye molecule to the fibres of the cloth.

All azo dyes contain the $X-N=N-Y$ arrangement. Chemists worked to make as many different XY combinations as possible to find new dyes with good colours, which were fast to fabrics and commercially viable. As they did this they began to understand the effect of chromophores and functional groups on the colour and properties of the dyes they produced.

A vast range of azo dyes is now available, produced by coupling one of 50 diazonium salts with one of 52 coupling agents to give a whole rainbow of colours – although these are mostly yellow, orange or red, with relatively few blues and greens (see Figure 31).

In **Chemical Ideas 6.9** you can read more about chromophores and modern theories of colour.

In **Activity CD6** you can make a range of azo dyes using different diazonium salts and coupling agents.

CD7 *Colour for cotton*

The search for fast dyes

At the turn of the century, James Morton, a Carlisle weaver, made a trip to London. Standing outside Liberty's window in Regent Street, he examined a display of cotton tapestries he had designed. He was horrified by what he saw. The colours had faded after only 1 week in the shop window.

Immediately, Morton set out to find dyes that were fast to light and began manufacturing them. You can see an advert for some of his dyes in Figure 32.

Morton's success helped to re-establish the reputation of the dye industry in Britain. But cotton presented a problem in other ways. Dyes must be fast to washing and rubbing as well as to light. Many of the dyes which were fast to wool and silk did not bind at all well to cotton.

Figure 32 James Morton built up a range of unfadable dyes in the early 1900s called the Sundour range.

How do dyes stick to fibres?

Protein-based fibres such as wool and silk have free ionisable —COOH and —NH_2 groups on the protein chains which can form electrostatic attractions to parts of the dye molecule. For example, a sulphonate (—SO_3^-) group on a dye molecule can interact with an —NH_3^+ group on a protein chain, as shown in Figure 33.

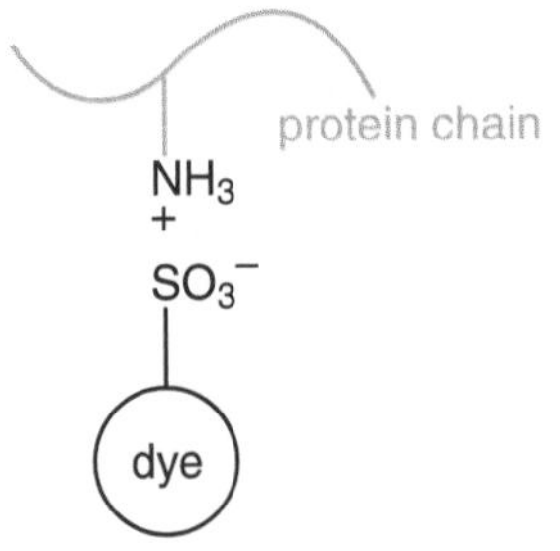

Figure 33 Interaction between a dye molecule and a protein chain.

Cotton, on the other hand, is a cellulose fibre consisting of bundles of polymer chains with no readily active parts. The polymer is a string of glucose units joined as shown in Figure 34.

Dyes such as Alizarin are bound to the fibres using a mordant (look back to Figure 26 on p. 224). Indigo, which is used to dye denim jeans, is a **vat dye**. Here the cotton is soaked in a colourless solution of a reduced form of the dye. This is then oxidised to the blue form of indigo which precipitates in the fibres.

Many azo dyes for cotton are trapped in the fibres because they are insoluble. In contrast, **direct dyes**, such as *Direct Blue 1* (see Figure 35) are applied to the cotton in solution and are held to the fibres by hydrogen bonding and instantaneous dipole–induced dipole forces. Hydrogen bonds are weak compared with covalent bonds and so these dyes are only fast if the molecules are long and straight. They must be able to line up with the cellulose fibres and form several hydrogen bonds.

ASSIGNMENT II

Look at the structure of Direct Blue 1 in Figure 35.

a Which groups of atoms do you think are responsible for the fact that a compound with such large molecules is soluble in water?

b Which groups of atoms would you expect to form hydrogen bonds with cellulose fibres?

c Draw part of the molecule and part of a cellulose fibre (in the style of the lower diagram in Figure 34) to show how you picture these hydrogen bonds.

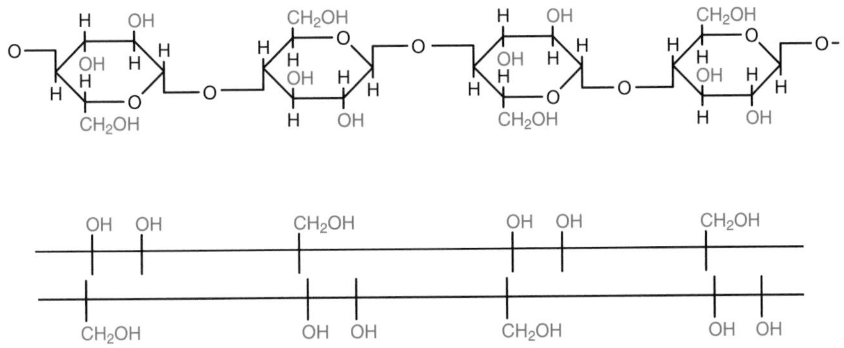

Figure 34 Two ways of depicting a cellulose fibre: in the top diagram the molecule is shown to be a chain of glucose units; in the lower diagram only the reactive —OH groups are shown (in red).

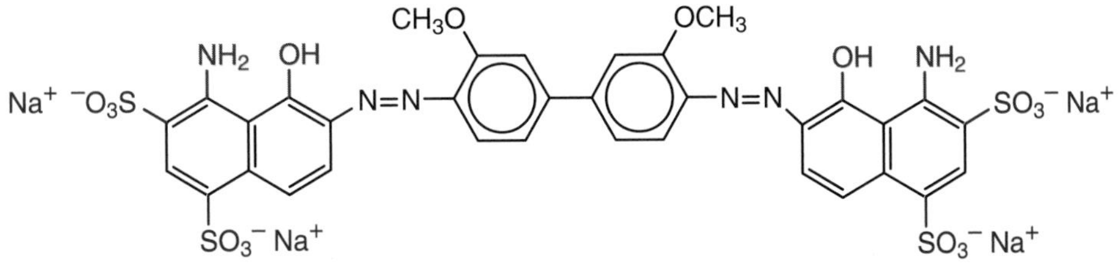

Figure 35 The structure of Direct Blue 1 (CI 24410).

A dyemaker's dream

For many years chemists dreamed of developing dyes which would be held to fibres by strong covalent bonds instead of weak intermolecular forces. They knew that such dyes would be very fast to washing because they would react with the textile materials and become chemically part of the polymer molecules.

The story starts in the early 1950s with a group of chemists working at ICI's research laboratories in Blackley trying to find better dyes for wool.

William Stephen was part of the group. He started with azo dyes and modified the molecules by adding reactive groups, which he hoped might combine with the amino groups of proteins in wool. One idea was to modify an azo dye containing an amino group by reacting it with **trichlorotriazine**, as shown in Figure 36.

Stephen took some samples of his modified dyes to Ian Rattee, the senior technician in the wool-dyeing section of the dyehouse. He hoped that the new dyes would react with wool as shown in Figure 37. However, the results were poor and Rattee was not impressed.

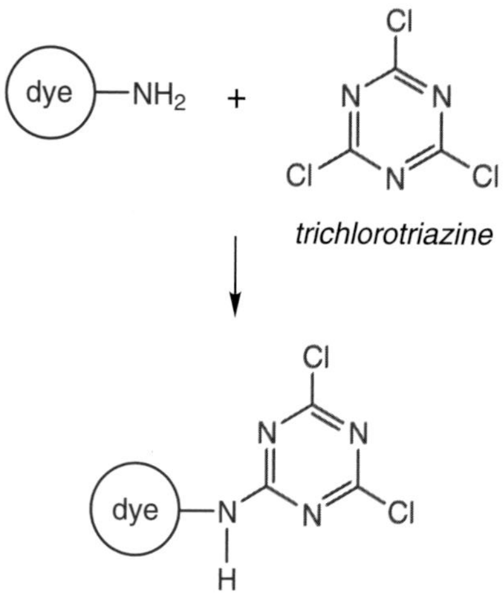

The dream come true

One morning in October 1953, Stephen and Rattee were discussing their experiments with reactive dyes. Stephen pointed out that the reaction of the new dye with the wool would be much more likely to happen under alkaline conditions. Unfortunately, the alkali would damage the wool at the same time.

ASSIGNMENT 12

Look at Figure 37, which shows the reaction Stephen was hoping to bring about between the modified dyes and wool.

a Why are there free —NH$_2$ groups in the protein molecules in wool?

b What is the second product of the proposed reaction?

c Why might this reaction be expected to go better in alkaline solution?

d Suggest an explanation for the fact that wool is damaged by alkali while cotton is not.

Stephen suggested that it might be worth trying the dyes with cotton, because cotton is not damaged by alkali in the way that wool is. The reaction they were hoping to use to bind the dye to the fibres can take place with hydroxyl groups on cotton as well as with amino groups.

Rattee saw the importance of this idea. Dyeing cotton was not part of his responsibility, but that afternoon he took samples of Stephen's dyes and showed that, as predicted, they would react with the hydroxyl groups on cotton when alkali was added.

Figure 36 Building a reactive dye to react with wool.

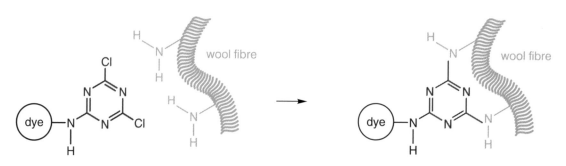

dye molecule becomes covalently
bound to the wool fibre

Figure 37 The planned reaction of the new dye with amino groups in wool fibres.

Figure 38 *The General Dyestuffs Research Laboratory at ICI, Blackley in the early 1950s where work on the first reactive dyes took place.*

The discovery had been made! The idea had at last become a reality – but it was far from certain that it could be made to work in textile mills. A great deal of work lay ahead to develop the innovation, demonstrate its potential and prove that it could be exploited successfully.

ASSIGNMENT 13

Draw a diagram in the style of Figure 37 to show what happens when a fibre reactive dye reacts with cotton in alkaline solution.

The first of the **fibre reactive dyes** are shown in Figure 39. Fibre reactive dyes are now so common that it is hard to recall the feverish excitement and energy released when it was realised that a search which had been going on for around 60 years might be coming to a climax.

In **Activity CD7.1** you can dye samples of cotton cloth with a fibre reactive dye and with a direct dye and compare the fastness to washing.

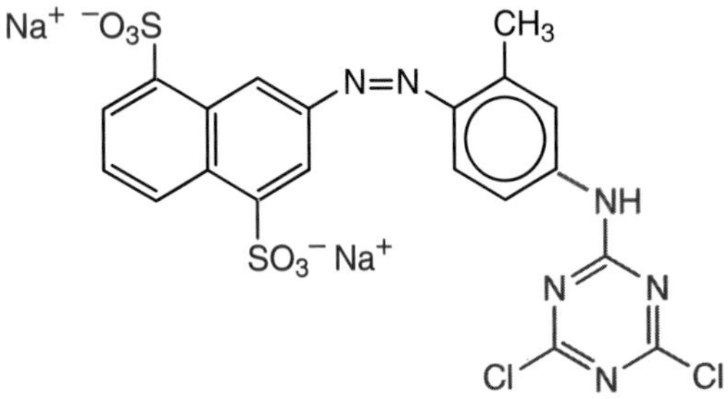

Procion Yellow RS

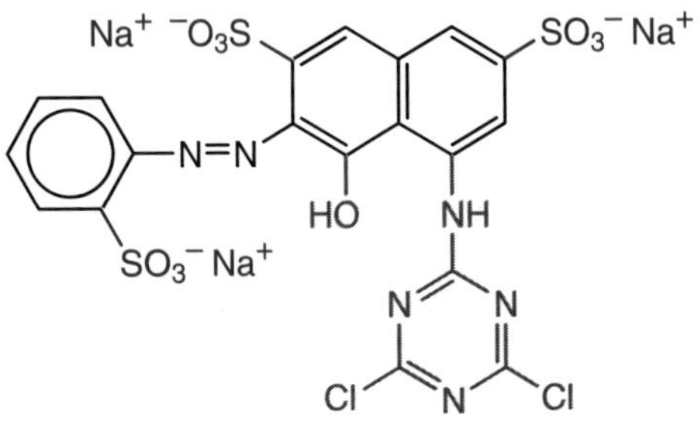

Procion Brilliant Red 2BS

Figure 39 *The first fibre reactive dyes.*

ASSIGNMENT 14

The two dyes in Figure 39 have the same chromophore.

a Copy the structure of one of the dyes and draw a ring round the chromophore.

b Which functional groups in the dye structure you have drawn help to make the dye soluble in water?

c Which part of the molecule you have drawn helps to attach the dye to the cloth? Explain how you decide on your answer.

d What are the differences in the structures of the two dyes which might account for the difference in colour?

You can investigate a mixture of different dyes using paper chromatography in **Activity CD7.2**.

Paradox and problem

The new dyes were fast because of the reaction with the hydroxyl groups in cotton, but this reactivity was itself a problem because there are hydroxyl groups in water too. The reactivity of the dyes was destroyed by hydrolysis.

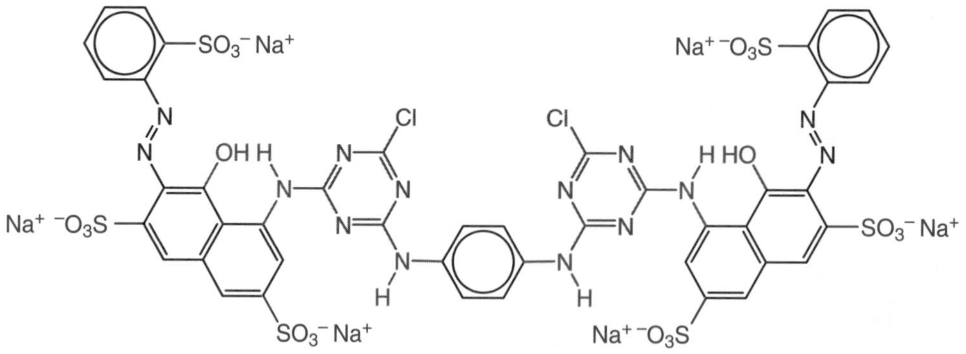

Figure 40 The structure of Reactive Yellow 202 (CI 292775), a modern reactive dye.

This problem had to be overcome because commercial success with the textile industry depended on the possibility of using the dyes when dissolved in water. Stephen developed systems of **buffers** which kept the pH of the solution within strict limits and kept the hydrolysis reaction under control.

The action of buffers is explained in **The Oceans**.

The decision was finally taken to launch the new dyes in March 1956 – exactly 100 years after Perkin discovered his Mauve dye. There was an explosion of activity in all departments – research, patents, production planning, engineering, the dyehouse, costing, sales and publicity. The dream had finally become a reality.

CD8 *High-tech colour*

The work of Stephen and Rattee heralded a new era of bright modern dyes that are fast to light and washing. Other types of reactive dyes soon followed, using different groups to bond the dye to the fibres. The 1980s saw the development of a range of extremely fast, brightly coloured dyes for polyester.

Colour chemists can now produce colours to order for a particular application. It is possible to have the exact shade of the season's fashion colour reproduced perfectly in a range of different fabrics – real colour by design (Figure 41).

One of the most exciting recent developments has been in the modification of dyes for high-tech uses, for example dyes used in high-speed ink jet printers and for electronic photography. In this section, we shall look at these two applications in more detail.

A smudgy problem

Ink jet printers are a boon in many offices. The print-out is clear and sharp like that from a laser printer, but it is obtained at a fraction of the cost. The early machines had one big drawback though. The print was not permanent. A sweaty finger was enough to cause an embarrassing smudge and ruin an immaculate report.

Figure 41 It is even possible to colour leather so that it coordinates with colours of other materials.

How is an ink jet printer different?

In conventional printers, such as typewriters and dot matrix printers, the paper is struck by a key through an inked ribbon. In photocopiers and laser printers, charged ink toner particles are transferred electrostatically to the paper (see **Designer Polymers**).

The ink jet mechanism sprays drops of ink onto the paper from a nozzle in the print head. There are around 120 dots in each centimetre of print.

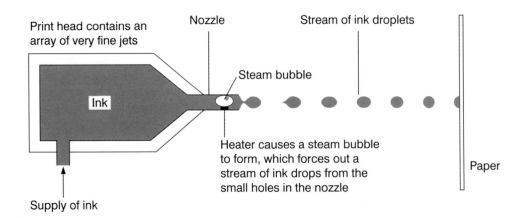

Ink
(pH 7.5–9.0, dye soluble in water)

Paper
(pH 4.5–6.5, dye insoluble in water)

Figure 42 An ink jet printer.

The problem was this. For the method to work, the black dye used in the ink must be soluble in water in the print head, but once on paper it must be insoluble. You can see how one type of ink jet printer works in Figure 42.

Chemists at ICI Colours and Fine Chemicals at Blackley were called in to help sort out the problem. There are about 100 known black dyes to choose from, but none of them met the strict requirements necessary. The ICI chemists decided to concentrate on black food dyes like *Food Black 2,* since these dyes are cheap, non-toxic and light-fast. They then set about adapting the structure to give the required properties.

Their solution to the problem was beautifully simple. They made use of some elementary organic chemistry and the change of pH during the process: most modern papers are acidic whereas the ink solution is slightly alkaline, as shown in Figure 42.

They knew that arene carboxylic acids, such as benzoic acid, are insoluble in neutral or acidic solutions, but dissolve in alkaline solution.

So why not replace some of the sulphonic acid ($-SO_2OH$) groups in the Food Black 2 molecule by carboxylic acid ($-COOH$) groups?

COOH

insoluble in cold water

OH^- ⇌ H^+

COO⁻

soluble in cold water

ASSIGNMENT 15

a Azo food dyes like Food Black 2 are fairly stable. If they break down, they tend to do so at the N=N linkage. It is important that any decomposition products are soluble in water. Explain how the dye is designed to achieve this.

b Why do you think only *some* of the $-SO_2OH$ groups in Food Black 2 were replaced by $-COOH$ groups in the new ink jet dye?

SO₂OH

HOO₂S — N=N —

N⫶N

OH

NH₂

HOO₂S

SO₂OH

Food Black 2

When they did this, the new dye they obtained was a vast improvement, but still not quite perfect. The final refinement was to use the *ammonium salt* of the acid dye in the ink solution, rather than its sodium salt. You may have heated smelling salts, ammonium carbonate crystals, and seen them decompose into gaseous products:

$$(NH_4)_2CO_3(s) \xrightarrow{\text{heat}} 2NH_3(g) + CO_2(g) + H_2O(g)$$
smelling salts

The ammonium salt of the modified dye decomposes in a similar way on heating:

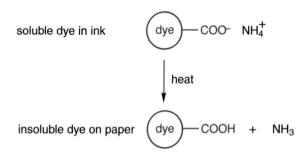

The reaction, together with the change in pH, ensures that conversion to the insoluble carboxylic acid is virtually complete.

One in a million

The challenge here was to find three dyes with the right properties to print photographs taken by an electronic camera.

Electronic cameras have been developed to replace conventional silver halide photography. They contain no film. Instead, the photographs are recorded electronically on a magnetic disc and can be viewed on a television screen. You can then select the pictures you wish to print from the disc.

The printing process works like this. The colours come from just three dyes: a yellow dye, a magenta (or red) dye and a cyan (or blue) dye. These three dyes are dispersed, in sequence, in a thin polyester film which looks like a strip of multicoloured cling film. They then pass in turn beneath a row of thermal heads heated to about 400 °C. These raise the temperature and cause the dyes to diffuse from the ribbon to the white receiver sheet to give the final colour print. The amount of dye transferred depends on the temperature. Figure 43 shows how the printing process works. It's called the **dye diffusion thermal transfer process** (or the **D2T2 process** for short!).

The three dyes had to have some very special properties. Their colours had to blend precisely to cover the whole of the visible colour range – and they had to have a very high thermal stability up to 400 °C. Most organic dyes start to decompose long before this temperature is reached.

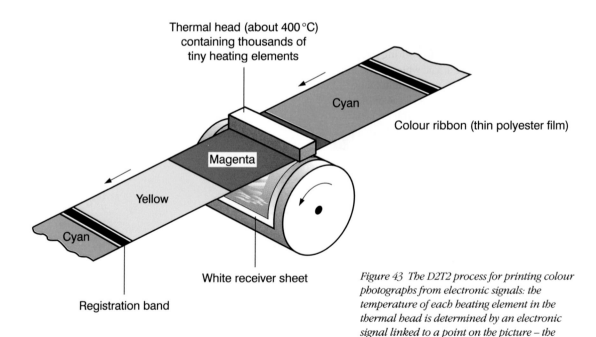

Figure 43 The D2T2 process for printing colour photographs from electronic signals: the temperature of each heating element in the thermal head is determined by an electronic signal linked to a point on the picture – the electronic signal triggers the diffusion of the right amount of each of the dyes.

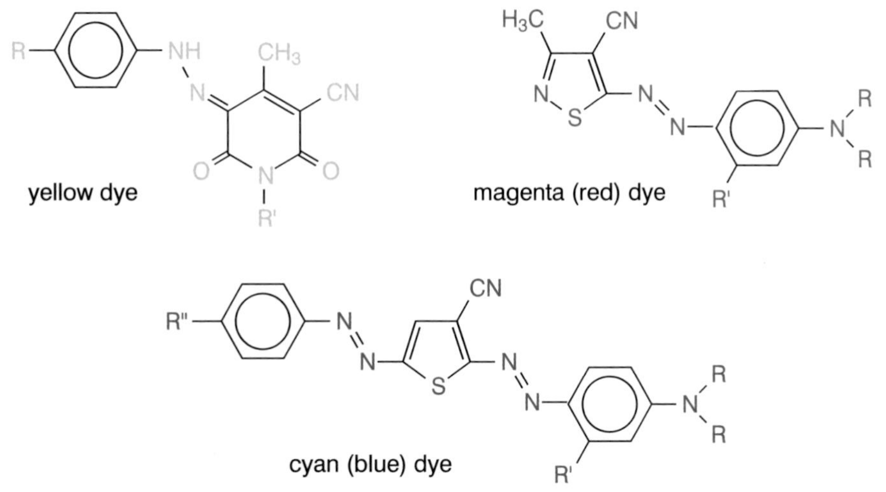

Figure 44 Dyes designed for the D2T2 printing process (R represents an alkyl group).

There are about 1 million known dyes in the world and none would do! Three completely new dyes had to be designed. The colour ribbon and the receiving sheet are both made of polyester, so a natural starting point was to look at existing polyester dyes and try to modify them. Computer models were a great help here: changes can be made to the structure of a dye on the computer screen, and the effect on the properties of the dye seen immediately.

The cyan dye was the most difficult to find. True cyan dyes are scarce. The best ones are the phthalocyanines but these were no good in this case because the particles of the insoluble pigment are too large. The new dyes the researchers came up with are shown in Figure 44 and they all have relatively small molecules. Their structures are rather complex so don't try to remember them!

CD9 *Summary*

Colour by Design is really two connected stories, one about pigments and paints and the other about the development of synthetic dyes. Both stories are a mixture of historical aspects and frontier applications of colour chemistry. They illustrate well the very diverse roles chemists play in society.

The chemists at the National Gallery use a variety of analytical techniques to investigate pigments and paint media. For example, ultraviolet and visible spectroscopy and atomic emission spectroscopy provide information about the chemical composition of pigments, while gas–liquid chromatography is used to determine the medium used to make the paint. Drying oils are unsaturated triesters of propane-1,2,3-triol (glycerol). The drying process involves reaction of the unsaturated chains with oxygen, followed by a polymerisation reaction in which cross-linking takes place between the chains.

The development of synthetic dyes over the last 150 years has been closely linked to the development of organic chemistry. Each contributed to the growth of the other. In this part of the unit you found out about the special structure of benzene and the types of reactions arenes undergo. An understanding of the relationship between the structure of a dye molecule and its function, and of the forces which bind dyes to fibres, is essential when designing new dyes.

Substances appear coloured because they absorb radiation in the visible region of the spectrum. The energy absorbed causes changes in electronic energy and electrons are promoted from the ground state to a higher energy level. Coloured organic compounds often contain unsaturated groups such as $C=O$ or $-N=N-$. These groups are usually part of an extended delocalised electron system called the chromophore. Functional groups, such as $-OH$ and $-NH_2$, are often attached to chromophores to enhance or modify the colour of the molecules. Many coloured inorganic compounds contain transition metals. Here, the radiation absorbed excites a d-electron to a higher energy level.

Activity CD9 will help you to review the important chemical ideas covered in this unit.

THE OCEANS

Why a unit on THE OCEANS?

To many people, the term 'ocean' probably conjures up an image of a seemingly endless expanse of water, of some biological interest but chemically inert.

Yet the oceans are far from being inert. They play an essential part in the cycling of many chemicals (sulphur and nitrogen compounds, for example) throughout the Earth. The oceans absorb and store carbon dioxide, and must be considered together with the atmosphere in any study of the greenhouse effect. The oceans play a further role in controlling our climate through their absorption of solar energy and the consequent production of water vapour and flows of warm water which help drive currents in the air and the seas.

The oceans help to make the Earth hospitable to life and have kept our planet habitable for over 3.5 billion years. Despite their importance, our understanding of ocean processes is far from complete, and they are one of the major sources of uncertainty in scientists' attempts to model future global conditions.

This unit attempts to raise awareness of the importance of the oceans to life on Earth, and to bring out some of the fundamental chemistry which lies behind some ocean processes. Major chemical ideas such as

- molecular-kinetic theory and energy distribution among molecules
- enthalpy changes involving ionic compounds
- the role of entropy changes in determining the feasibility of chemical reactions
- applications of equilibrium to phase changes, solubility and the behaviour of weak acids

are developed, and linked (perhaps unexpectedly) to familiar objects such as shells and rocks, and to the behaviour of water itself.

Overview of chemical principles

In this unit you will learn more about ...

ideas introduced in earlier units in this course

- dissolving and solubility (**From Minerals to Elements**)
- enthalpy changes (**Developing Fuels** and **The Atmosphere**)
- entropy (**Developing Fuels**)
- molecular motion (**The Atmosphere**)
- intermolecular forces (**The Polymer Revolution**)
- chemical equilibrium (**The Atmosphere**, **Engineering Proteins** and **Aspects of Agriculture**)
- interaction of carbon dioxide with water (**The Atmosphere**)
- Group 2 chemistry (**The Elements of Life**)
- acids and bases (**From Minerals to Elements**)

... as well as learning new ideas about

- lattice enthalpies and the Born–Haber cycle
- entropy and the distribution of energy quanta
- entropy changes in a system and its surroundings
- total entropy changes, spontaneity and equilibrium
- solubility products
- weak acids
- the pH scale
- buffer solutions.

01 *The edge of the land*

They that go down to the sea in ships …

In a French restaurant, a plate of shellfish is known as *fruits de mer*, 'fruits of the sea'. For thousands of years mankind has farmed the sea for food, and the easiest creatures to catch must have been the shellfish. But the archaeological records of every seaside culture from Neolithic times onwards contain the remains of marine fish, and the hooks and harpoons used to catch them. Classical literature tells us that the Ancient Greeks devised complicated traps for catching fish and that the Phoenicians and Carthaginians founded many cities around the Mediterranean as fishing settlements – Sidon, for instance, a Phoenician city now in Lebanon, means 'the fishing place'.

Sea creatures provided these ancient cultures with more than just food. The name Phoenicia comes from the Greek word *phoenix* meaning 'purple'. One of the foundations of this country's vast commercial empire, from around 1500 BC to 800 BC, was trade in a brilliant purple dye, extracted from the shells of a species of marine snail. Tyrian Purple, as the dye was called, was used to colour the attire of kings and emperors, and was a symbol of great wealth and power. Tyrian Purple

is chemically similar to indigo, the dye that the Ancient Britons extracted from the woad plant. What a coincidence that the two cultures should have used these two related dyes at about the same time – perhaps blues were in fashion!

Figure 2 The ancient dye, Tyrian Purple, was extracted from the shells of a species of sea snail. The Roman Empire decreed that only members of the ruling family (such as Julius Caesar, above) could wear robes dyed with Tyrian Purple.

Figure 1 Indonesian fishermen; fishing is one of the earliest skills and people have farmed the sea for food for thousands of years.

indigo

Tyrian Purple

(a)

ASSIGNMENT I

Look at the structures of indigo and Tyrian Purple. Both molecules are completely flat because electrons are delocalised around all four rings and across the double bond in the middle.

a What is meant by the term 'delocalised' in this context?

b From what you now know of the structure of Tyrian Purple, explain why the molecule does not exist as optical isomers.

c Tyrian Purple can be formed by substituting bromine atoms into indigo. Describe a general method which is available to chemists for bringing about this kind of reaction.

(b)

Figure 3 The modern kelp industry. (a) Drying seaweed (tangle) before transport to the processing plant. (b) The seaweed processing plant at Girvan in Ayrshire, capable of processing 6 tonnes of dried seaweed per hour to extract the alginate.

Less extensive use has been made of marine plants, but in Japan several species of marine algae (seaweed) are cultivated for food. There are other uses for seaweed: it can be spread on the ground as fertiliser and used as a source of chemicals.

From the early 1700s seaweed, mainly from the Hebrides and Orkney, was used as a source of **alkali**. It took 20 tons of seaweed, collected, transported and burned to produce 1 ton of **kelp**, a brittle, multi-coloured material high in soda ash and potash (sodium carbonate and potassium carbonate) for use in the soap and glass-making industries of Britain.

However, by the 1820s the industry was in trouble. The price fell for a number of reasons: the end of war with France allowed the import of higher quality material, import duty was cut, and a chemical process for making alkali from salt was developed.

Production continued on a smaller scale into the 19th century, with the seaweed used as a source of **iodine**.

Present day uses of seaweed (Figure 3) include the extraction of **alginate**, a thickening or gelling agent used in a wide variety of food products, including pet foods. Other uses range from maintaining the head on a pint of beer, to medical dressings, paper, textiles and pharmaceuticals.

Seaweed is rich in sodium, potassium and iodine, and the marine snail used to make Tyrian Purple is rich in bromine. Both the seaweed and the snail concentrate certain elements found in the sea water they inhabit. Tyrian Purple, for example, contains 38% by mass of bromine, whereas the concentration of this element in sea water is only 0.2%. Bromine is rare in the sea, but even rarer on land; that is why bromine is extracted commercially from the bromide ions in sea water (see **Minerals to Elements** storyline, Section **M1**).

Everyone knows that the sea is salty. Salts are ionic compounds and over 99% of all the dissolved substances in sea water are ionic. Whilst they remain dissolved, of course, the ions are free in solution so – strictly speaking – the sea contains not salts, but a mixture of ions.

Figure 4 shows the abundance of these ions. The proportion of one to another is remarkably constant whichever part of whichever sea you choose to sample.

The composition of sea water has been known for over 100 years, but it is only recently that we have begun to understand in any detail how the sea became salty.

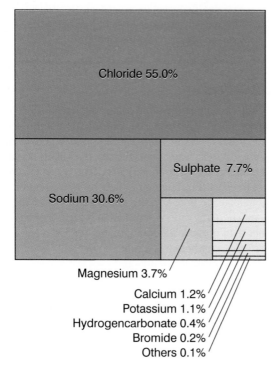

Magnesium 3.7%
Calcium 1.2%
Potassium 1.1%
Hydrogencarbonate 0.4%
Bromide 0.2%
Others 0.1%

Figure 4 Percentages by mass of different ions in sea water.

ASSIGNMENT 2

Use the information in Figure 4 to work out

a the amount in moles of positive charge present in 100 g of sea water

b the amount in moles of negative charge present in 100 g of sea water.

(Remember that 1 mole of doubly charged ions contains 2 moles of charge.)

Comment on the relative magnitudes of the answers you obtain in parts **a** and **b**.

You may find it useful at this point to refer back to the section on the chemistry of ions in solids and solution in **Chemical Ideas 5.1**.

Chemical Ideas 4.5 tells you more about dissolving and solutions and the use of enthalpy cycles.

Chemical Ideas 4.6 introduces you to an important enthalpy cycle for ionic compounds called the Born–Haber cycle.

Why do many ionic compounds dissolve so readily in water? **Activity 01.1** allows you to investigate the relationship between a solvent and the substances it dissolves. In **Activity 01.2** you can investigate changes that occur when an ionic solid dissolves.

You can practise using a Born–Haber cycle in **Activity 01.3**.

Figure 5 The sources of dissolved ions in sea water.

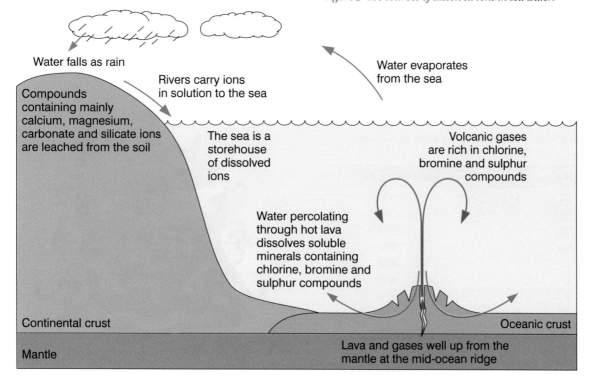

Water falls as rain

Rivers carry ions in solution to the sea

Water evaporates from the sea

Compounds containing mainly calcium, magnesium, carbonate and silicate ions are leached from the soil

The sea is a storehouse of dissolved ions

Volcanic gases are rich in chlorine, bromine and sulphur compounds

Water percolating through hot lava dissolves soluble minerals containing chlorine, bromine and sulphur compounds

Continental crust

Oceanic crust

Mantle

Lava and gases well up from the mantle at the mid-ocean ridge

Salt of the Earth

There is an old Norse myth which tells of a magic salt mill grinding away at the bottom of the ocean, making the sea salt. Strangely enough, this notion is not so far from the truth.

It has been known for a long time that some of the salt comes from the land. Rainwater leaches salts from the soil and rivers wash them into the sea. But the sea contains in abundance some elements that are not found to any great extent in river water – chlorine, bromine and sulphur, for example. The source of these elements remained a mystery until we learned more about the structure of the ocean floors.

Underneath the sediments on the ocean floor are lavas, generated by long, thin, underwater volcanoes called *mid-ocean ridges*. The gases given off from these volcanoes are rich in compounds containing chlorine, bromine and sulphur. Also, as molten lavas meet cold sea water they solidify and shatter. Water streams down through cracks, scouring out soluble minerals. The superheated solutions which re-emerge through *hydrothermal vents* are much richer sources of elements like chlorine, bromine and sulphur than crustal rock. Figure 5 shows the sources of the dissolved ions in sea water.

Scientists think that the water itself escaped from these deep-seated rocks and that sea water was salty right from the beginning. The balance of ions is kept constant by a complicated geochemical cycle that scientists are only just beginning to understand.

ASSIGNMENT 3

Bottled mineral water contains dissolved ions from the soil and rock through which the water has percolated. Most brands of bottled water give an analysis of the contents on their labels. Look at the labels of as many types of mineral water as you can find. Choose the still varieties as the fizzy ones have had carbon dioxide added to them. Compare the contents with the analysis of sea water shown in the table below.

Ion	Contents/mg dm^{-3}
Cl^-	19 000
Na^+	11 000
SO_4^{2-}	2500
Mg^{2+}	1300
Ca^{2+}	400
K^+	400
HCO_3^-	100
Br^-	70
Others	40

a What is the total mass of dissolved solids in 1 dm^3 (1 litre) of
 i sea water
 ii the different sorts of mineral water?

b Compare the analyses for the individual ions in the mineral waters with the figures given for sea water. Comment on any differences. Which of the ions in sea water cannot be accounted for by run-off of water from the land?

Salt sellers

Sea salt has almost the same composition wherever you get it from. However, the concentration of the salt in the water does vary from place to place. Where large volumes of fresh water enter the sea, the salinity (total salt content) is low. So those parts of the sea which are close to estuaries, abundant in icebergs or in areas of high rainfall are not very salty. On the other hand, in places where the rate of evaporation is high, the sea is saltier than normal. Water evaporates quickly in a hot, dry climate or where it is windy.

The two most abundant ions in sea water are sodium (Na^+) and chloride (Cl^-). These ions are also present in the body fluids of all land creatures. An adult human contains about 300 g of dissolved sodium chloride (common salt). Some of this is excreted each day in urine and sweat, and must be replaced if the body is to function normally.

Our need for salt has been recognised for thousands of years and the substance has entered the language, superstition and history of all societies. The word *salary* means, literally, a payment in salt – as was the custom in the Roman legions. In medieval times, important guests sat 'above the salt' at a banquet – and the saying lives on. Spilling salt brings bad luck – a measure of its value hundreds of years ago.

Figure 6 The coat of arms of the Salters' Company: the motto means 'Salt savours all'.

Not only was it important to eat enough salt; salt was also used to preserve food in the days before refrigerators and freezers had been invented.

The obvious place to get salt from was the sea. Along the east coast of Britain around 600 BC was a string of small salt-works. From the remains of their equipment it would seem that these early salters evaporated sea water in shallow pottery dishes set on brick stands over fires.

When the Romans occupied much of Britain, they imported salt from France (Gaul) where it was possible to evaporate the water by the heat of the Sun. After the Romans left, the British salt industry thrived again.

By medieval times, salt was big business. The Guild of Salters controlled the importation of foreign salt into Britain – mainly from the Bay of Biscay – and there were strict rules governing its handling. Salt could only be measured out by officials called 'salt meters' and carried by 'salt porters'. Large profits were made and the investments continue to produce income even today. The modern Salters' Company uses much of this income to support science teaching in schools.

Although most of our salt now comes from underground deposits (see Figure 16 on p. 52), one firm in Essex continues to extract salt from sea water. Essex has a long coastline with numerous shallow inlets from which water evaporates, leaving sea water with a high salinity. First the water is filtered, and heated by natural gas in large stainless steel pans. As the sea water comes to the boil, some of the impurities rise to the surface as a froth and are skimmed off. Then the water is allowed to simmer. Calcium sulphate is one of the first salts to crystallise out and forms a hard scale deposit on the sides of the pans. As the solution becomes more concentrated, crystals of sodium chloride begin to form. This is partly because sodium and chloride ions are most abundant, and partly because the solubility of sodium chloride is relatively low.

After about 15 hours the pile of accumulated crystals reaches the surface of the liquid and heating is stopped. The crystals are raked to the sides of the pan and removed with shovels. It is important to do this before the bitter-tasting magnesium salts crystallise out and spoil the flavour. Luckily, they are present in smaller quantities and are more soluble than sodium chloride so they remain in solution for longer. The early salt makers must have learned by bitter experience not to evaporate sea water to dryness.

ASSIGNMENT 4

Human activities remove only a tiny percentage of salt from the sea. Suggest reasons why, when rivers, volcanic gases and hydrothermal vents continually supply the sea with dissolved ions, the composition of sea water has remained constant for at least the last 200 million years.

… and occupy their business in great waters

The sea is a storehouse for fish and for minerals, and communities which have lived by the sea have always used it as such. Hannah Glasse wrote in 1747, 'Newcastle is a famous place for salted haddocks. They come in barrels and keep a great while'. Haddock were caught off the Northumbrian coast and sea water, boiled over fires of sea coal, provided the salt for preserving. In the winter months, when it was too rough to put to sea, the salted fish provided the community with an assured supply of protein. These people used the coastal waters very much as others used the land. They exploited its resources, but, at the time, their enterprises were small and inefficient enough to avoid over-fishing.

Nowadays, fishing is big business and many traditional fishing stocks have been depleted. Factory ships from Russia and other Eastern European countries travel as far as the waters outside Argentina, accompanied by fuel tankers, salvage tugs, repair boats

Figure 7 Evaporating sea water to make salt at the Maldon Crystal Salt Company in Essex.

Figure 8 Fishing on a large scale; catches must now be regulated to ensure adequate supplies for the future.

Figure 9 *Manganese nodules on a deep ocean floor. The nodules (diameter about 5 cm) are mainly manganese and iron, but also contain traces of nickel, cobalt and copper. They could provide a valuable source of these metals in the future.*

and refrigerator ships that will hold 10 000 tonnes of fish. The pursuit of the whale, for food and for chemicals, has also led to the near extinction of some species.

Sea water provides us with bromine, chlorine and magnesium. More recently, manganese-rich nodules have been collected from the ocean floors (Figure 9).

In future, scientists hope to use the sea as a source of more complex, organic compounds. Many sea creatures, especially those which have no shell to protect them, have evolved poisons which could perhaps be used as medicines. Some of these creatures have been known for thousands of years. For example, the tomb of the Egyptian Pharoah Ti, who lived nearly 5000 years ago, bears a picture of the pufferfish and a hieroglyphic description of its toxicity. In 1968, biochemists determined the nature of *tetrodotoxin*, the virulent poison produced by the pufferfish, and began to investigate its action.

The oceans, however, are not infinitely rich in food and chemicals, nor are they an ever-open bin into which we can throw our waste.

The North Sea, for example, is polluted by industrial effluent, sewage and excess of fertiliser that the major rivers of Northern Europe carry into it. Heavy metal ions poison fish stocks, and the increase in nitrate(V) concentration leads to rapid growth of algae.

02 *Wider still and deeper*

Surveying the seas

In December 1872 a small wooden warship, her guns replaced by scientific equipment, sailed out of Portsmouth. The ship was the HMS *Challenger*. She was fitted out by the Royal Society of London and her crew of 240 men was assigned to investigate 'everything about the sea'.

Figure 10 *The watery planet: a view of the Earth showing the vast expanse of the Pacific Ocean.*

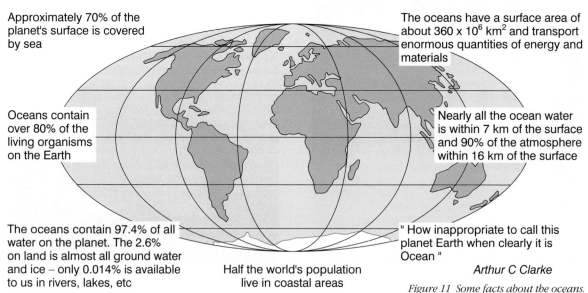

Approximately 70% of the planet's surface is covered by sea

The oceans have a surface area of about 360 x 10^6 km^2 and transport enormous quantities of energy and materials

Oceans contain over 80% of the living organisms on the Earth

Nearly all the ocean water is within 7 km of the surface and 90% of the atmosphere within 16 km of the surface

The oceans contain 97.4% of all water on the planet. The 2.6% on land is almost all ground water and ice – only 0.014% is available to us in rivers, lakes, etc

Half the world's population live in coastal areas

" How inappropriate to call this planet Earth when clearly it is Ocean "

Arthur C Clarke

Figure 11 *Some facts about the oceans.*

Figure 12 HMS Challenger. *Her voyage from 1872 to 1876 was the first systematic study of the oceans.*

Three and a half years later, on 24 May 1876, HMS *Challenger* returned home. During a journey of over 100 000 km the expedition had collected information from most of the oceans, as well as samples of water, sediments and marine life. The voyage of HMS *Challenger* (see Figure 13) was the first systematic study of the oceans, and provided the scientific basis of modern oceanography.

As we have learned more about the seas, we have realised that their role as a storehouse of food and chemicals is almost a sideline compared with the part they play in controlling our climate. Together with the atmosphere, the oceans are at the centre of the system which controls global conditions – the conditions in which we live and under which life has evolved for billions of years.

Perhaps we should have realised much sooner that the oceans have a regulatory role: they are so huge it would be hard to believe they did not influence what happens on Earth. But it is much harder to understand *how* the oceans work. Their vastness makes them difficult to study and it is only recently that our knowledge of them has really begun to grow.

HMS *Challenger* took samples and a range of measurements (such as water depth, the temperature at different depths, and the direction and speed of currents) at 360 sites in 3.5 years.

Even today an ocean survey ship, equipped with much more sensitive equipment, usually only makes measurements at one site per day, travelling 400 km between sites. But accurate information may be of little use if the system you are studying has changed significantly by the time you have finished collecting your data. Some of the events being studied are over in a matter of days or a few weeks. A satellite's greater speed is often a considerable advantage (see Figure 13), even though the measurements it makes may sometimes be less accurate.

Surveys have shown that the depth of the oceans is far from constant. In fact, underwater landscapes are more extreme than those we see above the water (Figure 14). Mount Everest rises 8.85 km above sea level; there are several parts of the ocean which are deeper than 10 km.

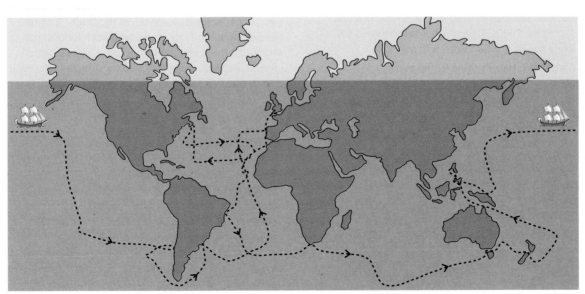

Figure 13 The voyage of HMS Challenger, *which took three and a half years, compared with the area covered by a single satellite over a period of 10 days (shaded grey).*

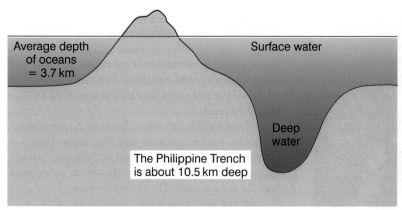

Figure 14 Underwater landscapes are more extreme than those we see above the water.

Making accurate measurements on the oceans can be difficult, and errors can become significant when values are multiplied millions of times to scale them up to ocean-sized quantities. This assignment tells you about one example of such an error.

Fritz Haber was a brilliant German chemist, who developed a process for making ammonia (see **Aspects of Agriculture** storyline, Section **AA3**). Ammonia can be turned into explosives, and Haber's process was used to supply Germany's munitions factories during the First World War. When the war was over, Haber decided that his country's war debts could be paid off by another of his ideas – extracting gold from the sea.

Gold compounds are present in solution in sea water. Their concentration is very low, but Haber calculated that vast quantities of the precious metal could be extracted by special devices fitted to ships. During the 1920s, German ships sailed around the world hoping to return laden with gold.

They didn't. Haber's figure for the gold concentration was hopelessly high, and our estimate has been falling ever since. For example, a survey carried out between 1988 and 1990 set a new maximum level at $1 \times 10^{-11} \, g \, dm^{-3}$; before 1988 it was thought to be $4 \times 10^{-9} \, g \, dm^{-3}$.

a Use the surface area and average depth of the oceans to estimate their volume in km^3 (see Figures 11 and 14).

b Convert your answer to part **a** into dm^3 units. You should now be able to see the 'big number aspect' of calculations on the oceans.

c Estimate the maximum value for the total mass of gold thought to be in the oceans today.

d If all the gold could be extracted from the sea and shared equally among the Earth's population of 6×10^9 people, approximately what would be your share today?

e The price of gold varies from day to day, but it is currently about $10.0 \, £ \, g^{-1}$.

  i How much would your share of the gold be worth?

  ii Explain why you would be unlikely to receive as much as this, and could even make a loss.

Unravelling a complex system

Here is an example of how our understanding of one global process has grown as we have learnt more about the interrelationship of life, the oceans and the atmosphere.

The problem of **acid rain** (or more correctly **acid deposition**) is serious in many parts of the world, and compounds containing oxidised sulphur are among the handful of chemicals involved. In the 1960s, scientists were trying to find out more about acid deposition by measuring the quantities of sulphur which circulated around the land, the oceans and the atmosphere. But their sums didn't add up … there was a missing link in the sulphur cycle (Figure 15).

It is now thought that dimethyl sulphide, $(CH_3)_2S$, produced by seaweeds and other marine algae, is the missing compound in the cycle. Dimethyl sulphide is volatile and quickly finds its way into the atmosphere, where it is partly responsible for the smell of sea air. Once in the atmosphere, it is oxidised to form acidic sulphur compounds.

A North Sea survey showed how significant algal production of dimethyl sulphide can be. Maximum activity occurs in April and May along the coast from Germany to France and probably accounts for about 25% of all acidic pollution over Europe at that time. So we have to accept that some acidic emissions are beyond our control.

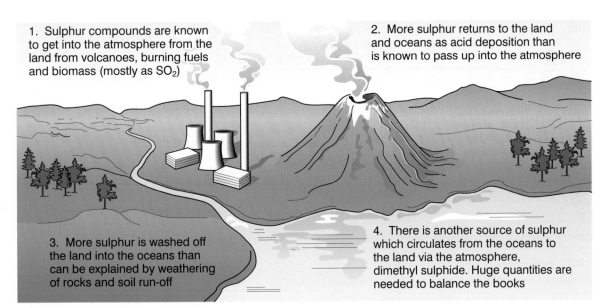

1. Sulphur compounds are known to get into the atmosphere from the land from volcanoes, burning fuels and biomass (mostly as SO_2)

2. More sulphur returns to the land and oceans as acid deposition than is known to pass up into the atmosphere

3. More sulphur is washed off the land into the oceans than can be explained by weathering of rocks and soil run-off

4. There is another source of sulphur which circulates from the oceans to the land via the atmosphere, dimethyl sulphide. Huge quantities are needed to balance the books

Figure 15 The missing link in the sulphur cycle.

ASSIGNMENT 6

a Draw a full structural formula for dimethyl sulphide. What shape would you expect the molecule to adopt?

b Sulphur is in Group 6 of the Periodic Table, along with oxygen.

  i What do we call the series of compounds related to dimethyl sulphide in which the sulphur atom is replaced by an oxygen atom?

  ii Dimethyl sulphide, like its oxygen-containing relative, is volatile. Use your knowledge of intermolecular forces to explain why.

c $(CH_3)_2SO$ and SO_2 are two of the compounds produced in the atmosphere from dimethyl sulphide. Explain why formation of these products corresponds to oxidation of the sulphur.

Figure 16 Microscopic marine algae: some types of marine algae produce dimethyl sulphide which is released into the atmosphere.

Later in this unit you will look at another element, carbon, and the role of the oceans in determining the amount of CO_2 in the atmosphere. But it helps to consider first the part the oceans play in distributing energy around the Earth.

03 *Oceans of energy*

The global central heating system

With the exception of *tidal power*, all our energy comes ultimately from nuclear sources. On Earth, we can mine radioactive compounds and produce from them fuel for *nuclear power stations*. Energy from the nuclear processes which go on in the core of the Earth can be tapped at suitable locations as *geothermal energy*.

But most of our energy comes from outside our planet – from a great nuclear furnace in the Sun. The Sun's rays heat the Earth directly, and through this drive the *winds* and *waves*. *Solar cells* convert sunlight into electricity. *Photosynthesis* uses sunlight to build up fuels: some are used almost immediately, and others have been changed over millions of years into *coal, oil* and *gas*.

When the Sun's rays reach the Earth they can be

- reflected
- absorbed by the atmosphere
- absorbed by the Earth's surface.

Roughly half the energy we receive from the Sun is absorbed by the land and the oceans, and this causes the Earth's surface to warm up (Figure 17). The Earth in turn radiates energy back into space (see **The Atmosphere** storyline, Section **A6**).

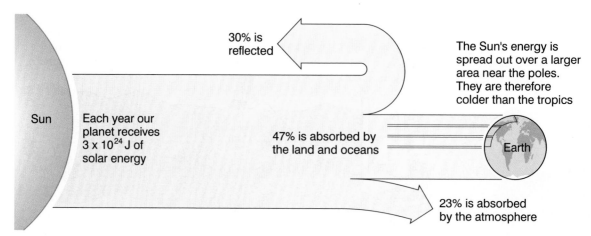

Figure 17 The fate of the Sun's energy reaching the Earth.

If the Earth was a dry lump of rock with no atmosphere, each part of its surface would soon settle down to a situation in which the energy being received from the Sun would, on average, be balanced by energy lost through radiation (see Figure 18). The tropics would be much warmer and the poles even colder than they are – and the Earth would be far less hospitable to life.

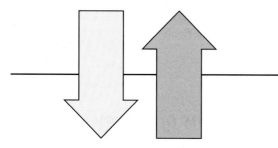

Incoming solar energy at any point is balanced by energy lost to outer space

Figure 18 A planet with neither atmosphere nor oceans.

But the Earth is surrounded by water and gas – both fluids. Temperature differences set up currents in the oceans and atmosphere which spread out the heating effect of the Sun more evenly. Just like warm air from a radiator spreads around a room, currents in the sea and air take thermal energy from the tropics to the colder regions of the Earth.

In fact, the ocean/atmosphere system is even more effective at spreading out energy. Warm water can do more than circulate – it can *evaporate*. Energy is taken in when water evaporates, so the situation must be reversed, and energy must be released, when water condenses. The tropics are cooled by evaporation, and currents in the atmosphere carry the water vapour to colder, high latitude regions where condensation releases energy (Figure 19).

High latitude regions receive more energy than is provided by the Sun alone: they are wetter – but warmer. Figure 21 shows the balance of condensation and evaporation around the world.

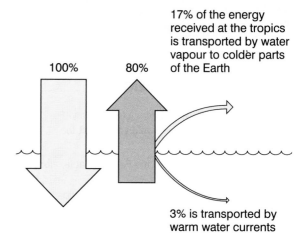

Figure 19 At the tropics of a planet with an atmosphere and oceans, about 20% of the incoming solar energy is transported to colder regions.

In the North Atlantic region, the winds and warm water currents flow from SW to NE. Northern Europe, including the UK, is warmed by energy which has been transported from the tropics and the Caribbean. In winter, as much as 25% of our thermal energy may come this way. Eastern North America does not receive this energy. So winters are much more pleasant in Lisbon, Portugal (latitude 38°N) than in Boston, US (latitude 42°N) (Figure 20).

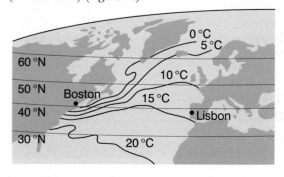

Figure 20 Average surface temperatures in the North Atlantic in February.

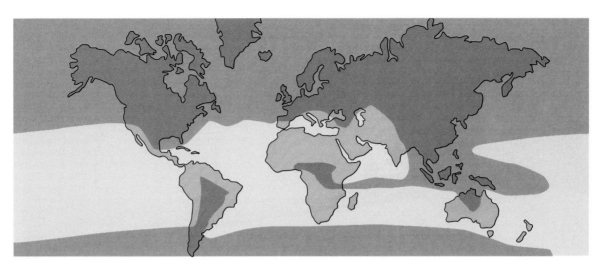

Figure 21 The condensation–evaporation balance of the Earth (dark shaded areas denote regions where condensation exceeds evaporation).

ASSIGNMENT 7

The positions of three Canadian cities are marked on the map below. Explain the patterns in their winter and summer temperatures.

	Victoria	Winnipeg	St. John's
July maximum/°C	17	23	17
January maximum/°C	2	–19	–6

Chemical Ideas 4.4 reminds you of the behaviour of molecules in solids, liquids and gases, and explains it in terms of entropy.

Energy in the clouds

The molecules in liquid water and water vapour differ in one important aspect. In the liquid, attractive forces between the molecules – **intermolecular forces** – hold the molecules onto one another and keep them quite close together. In water vapour, the molecules are much further apart and move about freely.

When water evaporates, changing from liquid to vapour, the intermolecular attractions must be overcome – a process which takes in energy. The enthalpy change of vaporisation, ΔH_{vap}, is our measure of this energy. In the reverse process, condensation, molecules come together again, intermolecular attractions re-form and an equal quantity of energy is released. In other words, condensation is **exothermic** and the enthalpy change is $-\Delta H_{vap}$.

You can remind yourself about hydrogen bonding and the unique properties of water by reading **Chemical Ideas 5.4**.

In **Activity 03.1** you can investigate the enthalpy change of vaporisation of water in more detail.

Figure 22 Energy is taken in from the surroundings when sea water evaporates, and is released when the water vapour condenses in clouds.

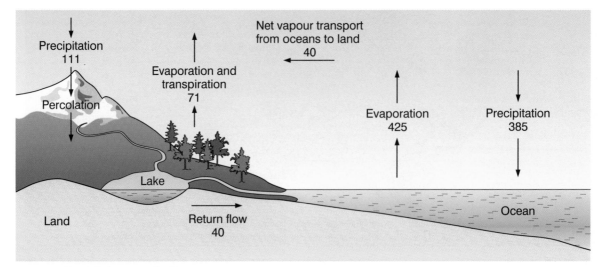

Figure 23 The global water cycle (figures represent quantities moved in 10^{15} kg per year).

Figure 23 illustrates the global water cycle. It summarises the major processes by which water circulates around the world.

Notice that more water evaporates from the oceans than is directly returned to them as precipitation (ie rain and snow). Each year, 40×10^{15} kg of water vapour produced from the sea falls as precipitation over the land. The process makes the land wetter and keeps the rivers flowing. It also makes the land warmer.

So evaporation and condensation of water affect the temperature of different parts of the Earth in two ways:

- by transferring energy from low latitudes to high latitudes
- by warming the land through condensation of water which comes from the oceans.

The UK may seem cold enough and wet enough most of the time, but imagine a world with propanone seas. The enthalpy changes of vaporisation of propanone, water and some other liquids are given in Table 1 (notice that kJ kg^{-1} units are used).

Substance	Formula	ΔH_{vap}/kJ kg^{-1}
water	H_2O	+2260
ethanol	C_2H_5OH	+840
propanone	C_3H_6O	+520
hexane	C_6H_{14}	+330
mercury	Hg	+300

Table 1 Some enthalpy changes of vaporisation.

If our present rain were replaced by the same mass of 'propanone-rain' it would release only about one-quarter of the energy. For enough 'propanone-rain' to fall to keep our temperature the same, there would have to be about four times more rain. Then there would be other problems – an increased fire risk, for example!

Thus, the properties of water make it an ideal liquid for spreading the Sun's energy around the world. Without it, the pattern of evolution and human development would have been very different.

ASSIGNMENT 8

a Use the data in Figure 23 and Table 1 to calculate the thermal energy released each year when water vapour carried from the oceans condenses out over the land.

b The output of a typical power station is about 2000 MW, in other words about 6×10^{16} J per year. Approximately how many power stations would be needed to produce the same energy as that transferred from oceans to land by evaporation and condensation?

Warm water from the west

Western Europe is doubly fortunate with its climate (see Figure 21 on p. 247). Winters are kept mild not only by the precipitation of rain but also by an 'ocean conveyor belt' bringing warm water from the tropics. The 'conveyor belt' is a surface current called the *North Atlantic Drift*. Surface currents are driven by prevailing winds and they alter course when they are deflected by large land masses (Figure 24).

We can find out how good a substance is at storing thermal energy by looking at its *specific heating capacity* (c_p). This is a measure of how much energy we have to put into 1 g of the substance to raise its temperature by 1 K. Looked at the other way, it tells us how much energy we can take out of a substance before it cools down by 1 K.

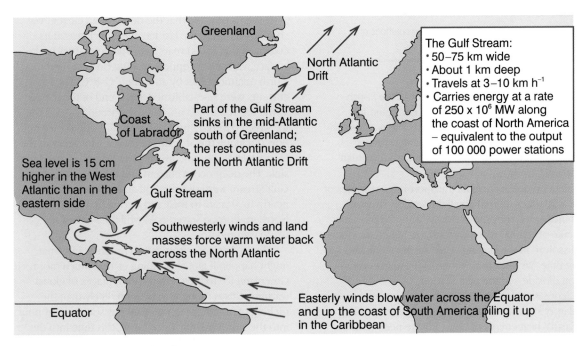

Figure 24 Ocean currents in the Atlantic.

Water has a large specific heating capacity: it will release quite a lot of energy without cooling down too much. Its specific heating capacity is given in Table 2, along with values for some other substances.

Substance	Specific heating capacity/$J\,g^{-1}\,K^{-1}$
water	4.17
ethanol	2.44
hexane	2.27
propanone	2.18
granite	0.82
copper	0.39
mercury	0.14

Table 2 Some specific heating capacities (c_p).

Water is one of the best liquids for transporting energy. A Gulf Stream in a 'propanone sea' would provide us with only half as much warmth.

From Greenland's icy waters

On its way across the North Atlantic, the Gulf Stream meets two currents of cold water: one flows down the eastern side of Greenland, the other flows past the Labrador coast of Northeast Canada (see Figure 25). The currents are fed by melted ice and snow from the Greenland ice-sheet, so their salinity is low. The Gulf Stream has a relatively high salinity because it is a warm current: water evaporates from it leaving the salt behind.

The density of pure water varies with temperature in a complicated way, but as long as we don't go below 4 °C we can say that water is denser when it is colder. Salty water is also denser than pure water. When it meets the East Greenland and Labrador currents, the water in the

Gulf Stream becomes much colder. The cold East Greenland and Labrador currents become saltier. The result of cooling the salty Gulf Stream and making the two cold currents saltier is a body of water which sinks because it is denser than the currents which feed it. The landmass of Greenland deflects the sinking water to the south, to produce a deep water cold current which follows the same route as the Gulf Stream but in the opposite direction (see Figure 25).

Another deep water current is generated in the Antarctic. The temperature is much colder there than in Greenland so this current does not come from melted ice. Instead it results from sea water freezing under the ice shelves. Salt is not taken up into the ice when this happens. It stays behind, so the residual water becomes saltier and sinks.

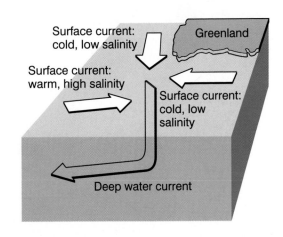

Figure 25 The deep water cold current in the Atlantic Ocean.

Activity O3.2 looks at the effect of freezing solutions of different concentrations.

The two deep water currents meet in the South Atlantic where the geography of the Earth and its rotation cause them to flow eastward. The water slowly rises back to the surface in the Indian and Pacific Oceans, and returns in surface currents to replenish the Atlantic, as shown in Figure 26.

The deep ocean currents transport huge volumes of water – 20 times more than all the world's rivers combined – and they move slowly. Water which sinks may take over 1000 years to resurface. Materials which are dissolved in the sinking water are also removed for a long time. We shall look at one of these materials – carbon dioxide – in the next section.

Scientists are learning more about the ocean circulatory system and the oceans' role in controlling the Earth's heat energy balance. But we need to know much more before these factors can be included with confidence in global climate models – models which explain our present climate and predict how it may change in the future. We know that ocean currents are important but we cannot predict how they may change, and therefore how our climate may be affected, if the Earth warms up.

It seems possible that the deep ocean current was shut down during the last Ice Age, which was at its coldest about 18 000 years ago. The flow of warm surface water also stopped, making lands around the North Atlantic cooler by an extra 6°C–8°C. The currents were re-established at the end of the Ice Age as the Earth warmed up. But the warming caused the ice which covered much of North America to melt and flow out along the St. Lawrence river. The water in the North Atlantic became much less salty and it could not sink. The deep ocean current and therefore the warm Gulf Stream were switched off again. The ice took nearly 1000 years to melt and during that time Northern Europe stayed cold while the rest of the world warmed up.

The ice has gone from North America, but what might happen if global warming caused much more Greenland ice to melt? (In the early stages of global warming this is thought to be more likely than the melting of Antarctic ice.) Low salinity water might pour into the North Atlantic once more, this time into the East Greenland and Labrador currents. The ocean circulatory system might shut down again and Northern Europe might become colder.

It is important to keep saying *might* because, until we have more knowledge, we cannot make accurate predictions. We are however sure that the consequences of global warming would be uneven. So Northern Europe could become colder while the rest of the world was getting hotter.

Figure 26 Deep ocean currents take cold salty water from the Atlantic to the Indian and Pacific Oceans on a giant 'conveyor belt'.

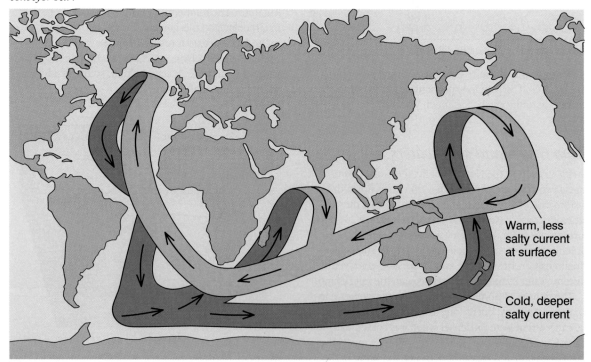

Warm, less salty current at surface

Cold, deeper salty current

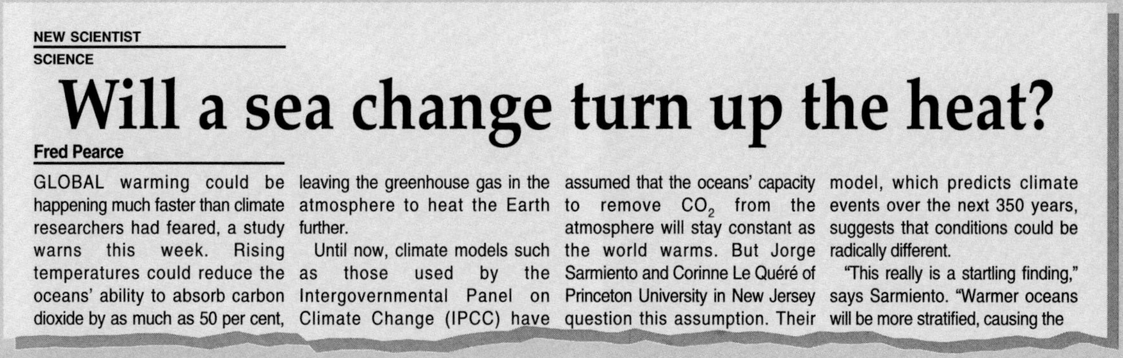

NEW SCIENTIST
SCIENCE

Will a sea change turn up the heat?

Fred Pearce

GLOBAL warming could be happening much faster than climate researchers had feared, a study warns this week. Rising temperatures could reduce the oceans' ability to absorb carbon dioxide by as much as 50 per cent, leaving the greenhouse gas in the atmosphere to heat the Earth further.

Until now, climate models such as those used by the Intergovernmental Panel on Climate Change (IPCC) have assumed that the oceans' capacity to remove CO_2 from the atmosphere will stay constant as the world warms. But Jorge Sarmiento and Corinne Le Quéré of Princeton University in New Jersey question this assumption. Their model, which predicts climate events over the next 350 years, suggests that conditions could be radically different.

"This really is a startling finding," says Sarmiento. "Warmer oceans will be more stratified, causing the

Figure 27 New Scientist article.

This headline in the *New Scientist* introduces another concern about global warming. As the atmospheric temperature increases, so does the temperature of the ocean surface. Less carbon dioxide will dissolve in the warmer water. One estimate is that, unless we reduce the rate at which CO_2 is building up in the atmosphere, up to 50% less gas will dissolve in the oceans, with the consequence that the enhanced greenhouse effect will lead to further global warming.

The child becomes stronger

A phenomenon which has come to be known as **El Niño** (Spanish for 'The Child', a reference to Jesus Christ, because some aspects of El Niño are most obvious around Christmas time) has worldwide repercussions for the weather.

Usually, there is low pressure over northern Australia and Indonesia and higher pressure over the central Pacific Ocean and the coast of South America. Thus winds normally blow along the equator from east to west across the Pacific. The winds carry warm surface water westward and cold deep water, full of nutrients, wells up along the coast of Peru and Equador to replace the water that is pulled out to the west (Figure 28).

However, in some years, the low pressure over northern Australia is replaced by pressure higher than that over the central Pacific. Warm water now flows towards South America and there is no upwelling of the cold deep ocean water and consequently a vastly reduced source of nutrients (Figure 29).

The consequences are profound. The fish stocks, essential for the livelihood of many in Peru and

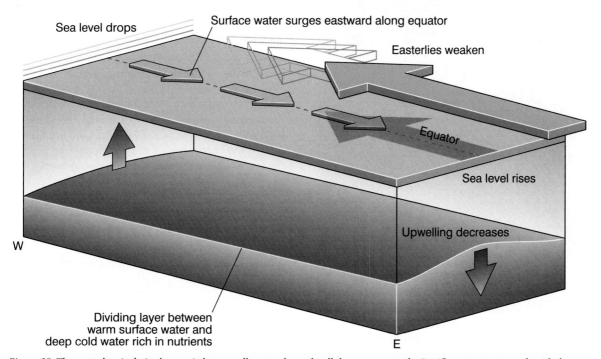

Figure 28 The easterly winds (red arrow) that usually extend nearly all the way across the Pacific retreat eastwards with the onset of El Niño conditions. As a result, along the equator, the slope of the sea surface and the dividing layer between the warm surface water and deep cold water both flatten out.

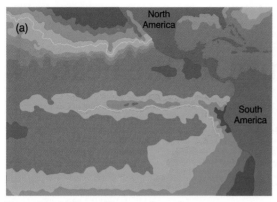

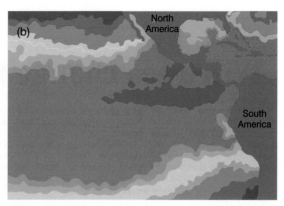

Figure 29 (a) This satellite map shows the temperature pattern of the surface of the sea when the equatorial Pacific is cold. The warmest water is shown dark red, with yellow and green being progressively cooler. (b) This satellite map shows the sea temperature during an El Niño. Note the absence of the nutrient-rich tongue of cold water extending westward from the South American coast which can be seen clearly in (a).

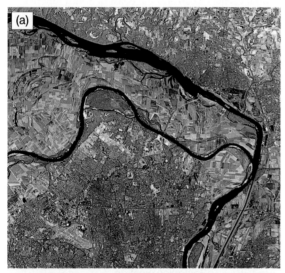

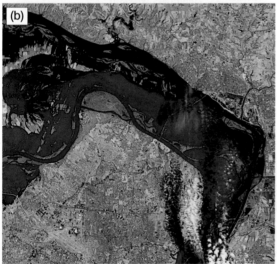

Figure 30 (a) A satellite picture of the Mississippi River (flowing from top left to bottom right) and the Missouri River (flowing from centre left, joining the Mississippi near St. Louis) in July 1988. (b) The flooding of the river in 1993 is thought to be the result of the coupling of the 1992 El Niño with global warming.

Equador, are significantly reduced. Further, tropical rains usually centred over northern Australia and Indonesia move eastwards and the weather all over the world is affected. For example, the El Niño of 1997 was blamed for monsoon rains in the Pacific, droughts and hence forest fires in Indonesia and Australia, flooding in the southern part of the US and winter storms in California. The flooding in the US in 1993 has been blamed on the 1992 El Niño (Figure 30).

The El Niño phenomenon a is well known if not well understood. Some scientists believe that the El Niño is becoming more intense as global warming increases.

04 *A safe place to grow*

Storing carbon dioxide

Imagine you are planning a couple of hours on the beach in the middle of your summer holiday. You have a choice – to take cans of fizzy drink with you, or to leave them in the fridge to drink when you return. From experience you decide to leave the cans in the fridge. You know that more of the fizz stays in a fizzy drink if you keep it cool. If you leave a can standing in the Sun, the gas will come frothing out explosively when you open the top.

The carbon dioxide in the drink is like all gases: its solubility *decreases* as temperature increases. Table 3 shows this for carbon dioxide, oxygen and nitrogen.

Temperature/°C	Solubility/mg per 100 g water		
	CO_2	O_2	N_2
0	338	7.01	2.88
10	235	5.47	2.28
20	173	4.48	1.89
30	131	3.82	1.65
40	105	3.35	1.46
50	86	3.02	1.35

Table 3 Variation of the solubilities of some gases with temperature (at atmospheric pressure).

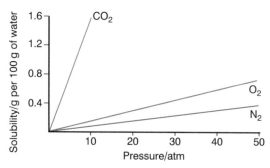

Figure 31 *Variation of the solubilities of some gases with pressure, at 298 K.*

Carbon dioxide is put into fizzy drinks under considerable pressure – about 14 atmospheres in most cases. That's why there is such a 'pop' when you open the can. The high pressure makes more gas dissolve, as Figure 31 shows.

ASSIGNMENT 9

Perrier water is officially described as naturally carbonated, natural mineral water – which means that the fizz comes from carbon dioxide produced naturally underground rather than chemically manufactured gas added from a cylinder. Lots of people prefer the taste of bottled mineral water to their tap water, and many prefer the naturally carbonated drink to the other kind of fizzy water.

a Write an equation for the thermal decomposition of calcium carbonate.

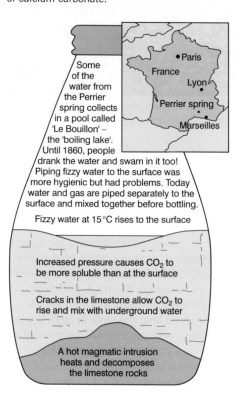

b The pressure below the Perrier spring is 443 kPa (4.4 atm). How many times more soluble is carbon dioxide at this pressure than at atmospheric pressure? (Look at Figure 31 to find the way the solubility of carbon dioxide varies with pressure.)

c Why do you think the Perrier pool was called the 'boiling lake'?

d The Perrier company had problems when they tried piping the carbonated water out of the ground. Describe one difficulty they would have encountered and how it would be avoided by piping the gas and water separately.

A high pressure of gas and a low temperature help to keep your drink fizzy. They also encourage carbon dioxide to dissolve in the oceans, which takes the gas out of the atmosphere and helps to maintain a stable environment on Earth.

You may find it useful here to refer to **The Atmosphere** storyline, Section **A8** and **Chemical Ideas 7.1**, which discuss the dissolving of carbon dioxide in the oceans.

You can read about equilibrium constants in terms of concentrations in **Chemical Ideas 7.2**.

The exchange of carbon dioxide between the atmosphere and oceans takes place quickly – more quickly than if you left some water in a beaker to come to equilibrium. Like so many environmental processes, we cannot explain it in terms of physical factors alone. Uptake of carbon dioxide by the oceans is speeded up by the action of marine life, as shown in Figure 32.

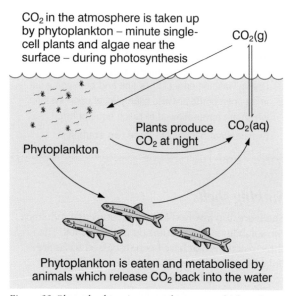

Figure 32 *Phytoplankton increase the rate at which carbon dioxide dissolves in the oceans. They play a key role in removing carbon dioxide from the atmosphere.*

Table 3 and Figure 31 show that carbon dioxide is more soluble in water than oxygen or nitrogen. CO_2 molecules contain polar C=O bonds which can form hydrogen bonds with water molecules.

$$CO_2(g) \rightleftharpoons CO_2(aq) \qquad \text{(reaction 1)}$$

In addition, some of the carbon dioxide molecules react chemically with water and are removed from the equilibrium. More carbon dioxide therefore dissolves to maintain the equilibrium position:

$$CO_2(aq) + H_2O(l) \rightleftharpoons H^+(aq) + HCO_3^-(aq)$$
$$\text{(reaction 2)}$$
$$HCO_3^-(aq) \rightleftharpoons H^+(aq) + CO_3^{2-}(aq) \qquad \text{(reaction 3)}$$

The reaction with water produces a mixture which contains mainly hydrogencarbonate ions (HCO_3^-) and H^+ ions together with some carbonate ions (CO_3^{2-}).

Much of the excess carbon dioxide we release into the atmosphere from the combustion of fuels is absorbed by the oceans. Estimates vary, but it seems likely that 35%–50% is removed this way. The oceans continue to soak up carbon dioxide because surface water, rich in CO_2, is constantly being removed and stored away for hundreds of years in the ocean deeps. The maximum amount of carbon dioxide is removed because this takes place in cold regions where CO_2 is most soluble. So currents, chemistry and marine life together make up a very efficient CO_2 removal system.

_____ ASSIGNMENT 10 _____

The oceans can keep on taking carbon dioxide out of the atmosphere because the dissolving of CO_2 takes place in what chemists call an *open system*. This means that material can enter or leave the system and so prevent equilibrium ever being established.

In this example, $CO_2(aq)$ is removed to the deep oceans so more $CO_2(g)$ dissolves to replace it. Drying washing on a clothes line and beer going flat in an open bottle are also examples of open systems which are never allowed to reach equilibrium.

Explain what happens in these two examples.

Sinking shells

Shells are also involved in the reactions that influence the solubility of carbon dioxide in the oceans. The three reactions that have been discussed so far are linked together – the product of reaction 1 is the reactant in 2, and so on. The three equations can be added together to produce just one equation which shows how the reactants in reaction 1 lead to the products of reaction 3:

$$CO_2(g) + H_2O(l) \rightleftharpoons 2H^+(aq) + CO_3^{2-}(aq)$$
$$\text{(reaction 4)}$$

Remember, the reaction does not happen as simply as this, but the equation should make the next part of the story clearer.

Le Chatelier's principle tells us that any way of removing H^+ or CO_3^{2-} ions from solution will cause more CO_2 to dissolve. Removing H^+ ions by adding a base is one way of doing this. You should already be familiar with this process: carbon dioxide is an acidic gas and it dissolves well in alkaline solution. That's why alkalis like sodium hydroxide are used to absorb CO_2.

Making the sea alkaline is not a very easy way of encouraging the oceans to take up carbon dioxide. But many marine organisms build protective shells composed of insoluble calcium carbonate using CO_3^{2-} ions in the sea water. The shells are often very beautiful. They provide a way of mopping up carbon dioxide – perhaps from someone's factory chimney – and keeping the composition of our atmosphere constant.

_____

At this point you may need to go back to **Chemical Ideas 8.1** to revise ideas about acids, bases and alkalis.

_____

Figure 33 Mussels have protective shells made of calcium carbonate, made from carbon dioxide dissolved in the oceans.

Billions of years ago, the Earth's atmosphere contained very much more carbon dioxide than it does now – probably about 35% carbon dioxide by volume. Once the process of photosynthesis had evolved, marine life had plenty of raw materials to work on in the form of carbon dioxide and water. Shell production flourished. Limestone and chalk rocks are the remains of shells of marine organisms which lived at that time and changed carbon dioxide from the atmosphere into solid calcium carbonate. When you go down into a cave you are quite literally making a journey into the past. When you walk on limestone hills or chalk downland, you are treading on Earth's prehistoric atmosphere.

Calcium carbonate is a good material for shellfish to use for protection at the surface of the oceans. It does not dissolve in sea water. But it does dissolve, slightly,

Figure 34 The chalk cliffs of the Seven Sisters are the legacy of marine organisms which lived billions of years ago.

in pure water. It is an example of a **sparingly soluble solid**. The dissolving of sparingly soluble solids is controlled by equilibria such as

$$CaCO_3(s) \rightleftharpoons Ca^{2+}(aq) + CO_3^{2-}(aq) \quad \text{(reaction 5)}$$

in which the ions in the saturated solution are in dynamic equilibrium with the undissolved solid present.

The position of this equilibrium is determined in the normal way by an equilibrium constant which, because it describes the solubility of a compound, is called a **solubility product** (K_{sp}). The solubility product for reaction 5 is given by

$$K_{sp}(CaCO_3) = [Ca^{2+}(aq)]\,[CO_3^{2-}(aq)]$$

Its value is $5.0 \times 10^{-9}\,mol^2\,dm^{-6}$ at 298 K.

Solubility products are discussed in more detail in **Chemical Ideas 7.7**.

One of two things can happen when Ca^{2+} ions and CO_3^{2-} ions are mixed together in a solution.

- Calcium carbonate may precipitate out of solution. This occurs whenever multiplying the dissolved calcium ion concentration by the dissolved carbonate ion concentration gives a value in excess of K_{sp}. (K_{sp} is the *maximum* value the product of the concentrations of $Ca^{2+}(aq)$ ions and $CO_3^{2-}(aq)$ ions can have in a solution at that temperature.)
- The ions may remain in solution. This will be the case whenever multiplying the concentrations of the two ions gives a number which is smaller than, or equal to, K_{sp}.

Calcium carbonate is a safe material from which to build sea shells because $[Ca^{2+}(aq)]$ and $[CO_3^{2-}(aq)]$ are already high enough at the surface of the sea for the calcium carbonate in the shells to be effectively insoluble. Remember, though, that the shells are in equilibrium with the ions in sea water, and there will be a constant exchange of Ca^{2+} and CO_3^{2-} ions between the two.

ASSIGNMENT II

a Use Hess's law and enthalpy changes of formation in the Data Sheets to calculate the standard enthalpy change for the process

$$CaCO_3(s) \rightarrow Ca^{2+}(aq) + CO_3^{2-}(aq)$$

b Explain in terms of equilibria how calcium carbonate production encourages more carbon dioxide to dissolve from the atmosphere.

But things are different deeper in the ocean where the pressure is higher and the temperature is lower. The value of K_{sp} is greater under these conditions so calcium carbonate is more soluble. There is also a continuous downward drift of material from above. It's like a perpetual snowstorm. In fact the falling material is called *marine snow*. It contains the remains of dead organisms and the waste products from live creatures. Most of the organic material, such as tissue, is consumed or decomposed higher up, but some reaches the deeper water where bacteria break it down to produce carbon dioxide. The shells fall down intact, but then react with the extra carbon dioxide and dissolve. These processes are summarised in Figure 36.

There are no shells on the deep ocean floor: they've all dissolved. The creatures which live there cannot use calcium carbonate for a protective coating.

The calcium carbonate deposits which built up to form our limestone hills could not have formed in deep water. They must have been laid down when our landmass was in shallower seas. The abundance of life suggests also that it was warm, tropical water. Evidence like this helps scientists piece together the distant history of the Earth, and helps us explain how the continents have drifted and how the climate has changed throughout time.

Figure 35 Machair in North Uist in the Outer Hebrides. Machair sand is 80%–90% calcium carbonate from crushed shells. It is driven by waves and wind over grass and peatland, neutralising acid from the peat. Machair is one of the rarest habitats in Europe.

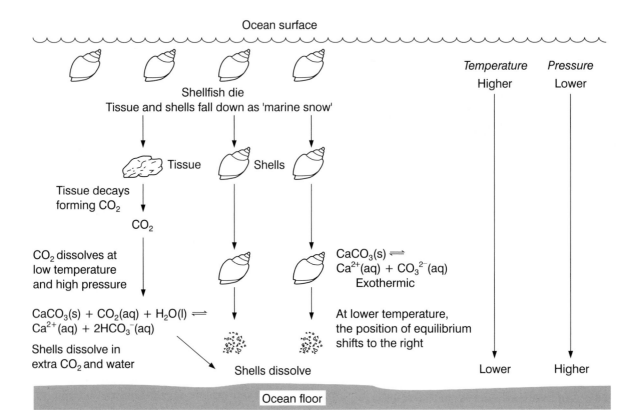

Figure 36 *The dissolving of shells on the deep ocean floor.*

Now might be a good time to revise the properties of s-block elements such as calcium. You can read about these in **Chemical Ideas 11.2**.

Figure 37 *The world's coral reefs, already damaged by record sea temperatures, are threatened by rising carbon dioxide concentrations in the atmosphere and the oceans. Tiny reef-dwelling creatures called coral polyps produce calcium carbonate, but this is more difficult when the concentration of carbon dioxide in the sea water increases.*

ASSIGNMENT 12

Although the dissolving of calcium carbonate

$$CaCO_3(s) \rightarrow Ca^{2+}(aq) + CO_3^{2-}(aq)$$

is a slightly exothermic process, it is accompanied by a large decrease in entropy. We often assume that dissolving is accompanied by an increase in the entropy of the chemicals, but in this case ΔS_{sys} is large and negative. The entropy change which would occur if calcium carbonate dissolved is $-199 \, J \, K^{-1} \, mol^{-1}$. Explain why you think dissolving might lead to such a large *entropy decrease* in this situation.

Getting rid of carbon dioxide

Scientists have predicted that the carbon dioxide concentration in the atmosphere could rise to 550 ppm by 2100. To keep the carbon dioxide concentration stable, even at 550 ppm (it was over 370 ppm in 2000), the emissions of the gas worldwide have to be reduced from about 21 billion tonnes per year to 7 billion tonnes per year by 2100.

There are three ways to reduce the build-up of carbon dioxide in the atmosphere:

- use solar, wind, water and nuclear methods to produce energy instead of using fossil fuels
- use fossil fuels more efficiently
- capture and store carbon dioxide, using a process known as *sequestration*.

The third way is the most revolutionary and four methods are being considered:

- turning CO_2 into useful products (as yet unknown!)
- growing more trees and increasing the organic content of soils
- storing the gas in deep natural trenches on the sea floor, where the pressure will cause it to liquefy
- injecting the gas onto the sea floor – at say 3500 metre depth, it will form a lake of liquid carbon dioxide that should remain undisturbed.

There is a danger that storing the carbon dioxide in the deep ocean in this way *may* disturb the environment. Even so, this may be less damaging than allowing carbon dioxide to be absorbed naturally in the surface waters of the ocean, which is already limiting coral growth.

ASSIGNMENT 13

Stalactites grow slowly. But they are worth waiting for. The growth rate depends on a number of factors, but the average rate is about 1 mm per year. The beautiful stalactite formations you can see in caves today have probably taken several thousands of years to form, and they have been created by the simple shifting back and forth of a chemical equilibrium.

The reaction involved is

$$CaCO_3(s) + CO_2(aq) + H_2O(l) \rightleftharpoons$$
$$Ca^{2+}(aq) + 2HCO_3^-(aq) \text{ (reaction 6)}$$

It's the same reaction as the one responsible for dissolving shells deep under the oceans. Shells also dissolve in the formation of stalactites, but this time deep underground – the limestone rock which is the source of the calcium carbonate is itself composed of the remains of ancient marine organisms.

The same solution dissolves the calcium carbonate in the rocks and in the oceans. It is water with a higher than normal concentration of carbon dioxide which has been produced from decomposition of organic material. 'Marine snow' is broken down in the oceans; in the ground, the decomposer organisms work on the plant and animal remains which fall down into the soil.

So, in the oceans and on land, nature uses the same processes to dissolve and re-form structures using the simple compound, calcium carbonate.

a Use reaction 6 and Le Chatelier's principle to explain why a higher concentration of dissolved carbon dioxide in the ground water causes calcium carbonate to dissolve from limestone.

b Caves and their surrounding rocks are at the same temperature which stays close to 10°C all year round. The carbon dioxide concentration inside a cave is close to the atmospheric value. Explain why calcium carbonate is precipitated when water drips through a cave.

c Suggest why stalactites would be unlikely to form on a planet without life.

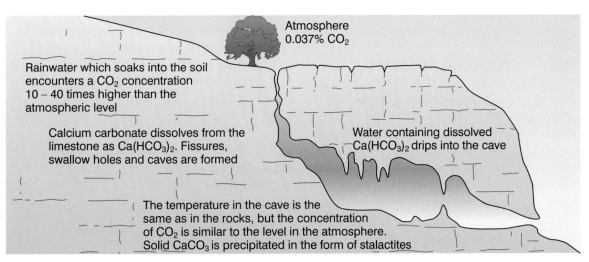

Figure 38 Formation of stalactites.

Figure 39 Stalactites and stalagmites in the cave of the Black Spring, Wales.

Life on Earth

The Earth's early life forms evolved in the oceans – and that's where they stayed throughout most of the Earth's history. The planet was nearly 4 billion years old and life had existed for 3 billion years before life moved onto the land (Table 4).

Deep in the ocean, several kilometres down, there are cracks in the ocean floor called *hydrothermal vents* (see Section **O1**). They let out methane, hydrogen sulphide and clouds of black particles of sulphide minerals, which gives them their name – 'black smokers'. Colonies of tube worms up to 3 m long thrive around the vents. The energy the worms need for survival is provided by colonies of bacteria which live inside them. These bacteria must be very similar to the earliest life forms. They can live without light and oxygen because they gain their energy by using sulphate ions to oxidise the methane and hydrogen sulphide. This is life as it was 3500 million years ago.

Photosynthesis became possible with the evolution of cyanobacteria. They produced oxygen, but it was used up by reducing agents dissolved in the sea water before it could build up in the atmosphere. This was just as well for the cyanobacteria because they cannot

Million years ago		
4600	origin of Earth	first atmosphere lost from hot planet
3800	oldest sedimentary rocks organisms probably present	atmosphere consists of CO_2, NH_3, CH_4 and H_2S
3500	beginning of fossil record bacteria leave fossil remains	life probably uses energy released during fermentation, or from oxidation of organic molecules by SO_4^{2-} ions and NO_3^- ions
2800	first photosynthetic bacteria (cyanobacteria or blue-green bacteria)	oxygen produced for 1 million years it is used up in the oxidation of S^{2-} and Fe^{2+} ions no ozone is able to form in the atmosphere so life on the surface is impossible due to ultraviolet radiation bacteria live in the top 200 m of the ocean, screened from ultraviolet radiation by water
2000	first significant $CaCO_3$ deposits	removal of two greenhouse gases (CO_2 to form $CaCO_3$, and CH_4 by reaction with O_2) leads to major ice age
1800	oxygen begins to build up in the atmosphere	
1500	first green algae – plants with chlorophyll in their cells	
800	first sea animals	1% oxygen in atmosphere formation of ozone begins
400	first land plants	10% oxygen in atmosphere ozone layer protects plant cells from ultraviolet radiation
300	first land animals	21% oxygen in atmosphere

Table 4 The origins of life on Earth.

tolerate oxygen. Sulphate ions (SO_4^{2-}) and nitrate(V) ions (NO_3^-) still had to be used as oxidising agents in respiration. It was only later that organisms could use free oxygen which was dissolved in the sea water or which had built up in the atmosphere.

The build-up of oxygen in the atmosphere is also discussed in **The Atmosphere** storyline, Section **A1**.

In terms of the Earth's history, the ozone layer is a very recent phenomenon. Most early life forms were resistant to ultraviolet radiation. Close relatives of cyanobacteria are alive today (there are plenty of places to hide away from the oxygen – in the mud of salt marshes or in a clump of dead seaweed, for example) and they are still very tolerant of ultraviolet radiation.

The Earth's atmosphere has changed dramatically since the early days of life: reducing agents such as methane and acidic gases such as carbon dioxide have been largely replaced by a neutral, oxidising mixture of nitrogen and oxygen. If life had been forced to evolve on land in contact with the air, the primitive organisms would have become extinct. But they have been protected by their watery environment which has altered remarkably little over billions of years.

Figure 40 Colonies of huge tube worms thrive round hydrothermal vents in the ocean floor.

ASSIGNMENT 14

Molecules such as 2-hydroxypropanoic acid and ethanol were probably present in small quantities in the early oceans. An example of an oxidation using sulphate ions, such as might have been carried out by a primitive bacterium, is shown below.

$$2CH_3CH(OH)COOH(aq) + SO_4^{2-}(aq) \rightarrow$$
$$2CH_3COOH(aq) + S^{2-}(aq) + 2CO_2(g) + 2H_2O(l)$$

a Explain why this equation represents a redox reaction.

b Calculate the enthalpy change produced by the oxidation of 2 moles of 2-hydroxypropanoic acid by sulphate ions. (Assume that the enthalpy changes of formation of the aqueous solutions of the acids are as follows: 2-hydroxypropanoic acid, $\Delta H_f = -694$ kJ mol^{-1}; ethanoic acid, $\Delta H_f = -486$ kJ mol^{-1}. Use the Data Sheets to find the other information you need.)

c i Write an equation for the complete oxidation of 2-hydroxypropanoic acid by oxygen to produce carbon dioxide and water.

  ii Calculate the enthalpy change for the reaction in part **c, i**.

d Use your calculations to explain which is more efficient: the aerobic respiration or anaerobic respiration of 2-hydroxypropanoic acid.

Keeping things steady

The ability of the oceans to withstand external changes has been essential for the unbroken evolution of life. For example, their pH has remained close to 8 for millions of years. Why were the oceans not much more acidic when the atmosphere contained 35% CO_2 – 1000 times its present level?

One reason is that a solution of carbon dioxide in water is a **weak acid**. A weak acid reacts *incompletely* with water. If we represent the acid by the formula HA, we can show the reaction with water by the equation

$$HA + H_2O \rightleftharpoons H_3O^+ + A^-$$

H_3O^+ ions (called **oxonium ions**) make the solution acidic. The position of equilibrium is well over to the left-hand side and only a *fraction* of the acid added to the water reacts to produce oxonium ions. So the solution is not as acidic as it could be if all the acid had reacted.

The equation can be simplified to

$$HA(aq) \rightleftharpoons H^+(aq) + A^-(aq)$$

by leaving out the water, which is present in excess.

For carbon dioxide, only a small proportion of the $CO_2(aq)$ molecules react to form hydrogen ions and hydrogencarbonate ions, as shown in reaction 2 (p. 254).

$$CO_2(aq) + H_2O(l) \rightleftharpoons H^+(aq) + HCO_3^-(aq)$$
$$\text{(reaction 2)}$$

Sometimes, to emphasise the link with the general equation for weak acids, an aqueous solution of carbon dioxide is represented by the formula $H_2CO_3(aq)$. It is as if the reaction

$$CO_2(aq) + H_2O(l) \rightleftharpoons H_2CO_3(aq)$$

had occurred. (For this reason some people use the name 'carbonic acid' for a solution of carbon dioxide.)

Chemical Ideas 8.2 tells you more about solutions of weak acids.

You can compare some of the properties of strong and weak acids in **Activity O4.1**.

If a small amount of alkali is added to a solution of a weak acid such as aqueous carbon dioxide, some of the $H^+(aq)$ ions are removed. But this causes more $CO_2(aq)$ to react with water to restore the position of equilibrium and so replace the $H^+(aq)$ ions. In this way a solution of carbon dioxide can maintain a constant pH even when a small amount of alkali is added.

But what happens if the ocean becomes more acidic? When the proportion of carbon dioxide in the atmosphere, for example, was much higher, the equilibrium in reaction 1 (p. 254)

$$CO_2(g) \rightleftharpoons CO_2(aq) \qquad \text{(reaction 1)}$$

would ensure that the concentration of $CO_2(aq)$ was also higher. This in turn resulted in reaction 2 moving to the right and generating more $H^+(aq)$ ions. You would therefore predict that the pH of the ocean would be lower. But the ocean could resist even this change, as it is an example of a **buffer solution** – a solution that remains within a narrow range of pH despite the addition of acid or alkali.

At this point it would help to read **Chemical Ideas 8.3** which will introduce you to the theory of buffer solutions. You may already have encountered buffer solutions in the **Aspects of Agriculture** storyline, Section **AA2** and in **Activity AA2.6** and in the **Colour by Design** storyline, Section **CD7**.

You can investigate some buffer solutions in **Activity O4.2**.

The commonest type of buffer solution is made up from a weak acid and one of its salts. The weak acid acts as a reservoir of $H^+(aq)$ ions. These can react with any OH^- ions which are added and so prevent the solution becoming more alkaline. The anions from the salt act as bases. They can 'soak up' additions of $H^+(aq)$ ions and keep the solution from becoming acidic.

If we represent the weak acid by HA and the anions in the salt by A^-, simple buffers rely on the shifting back and forth of the equilibrium

$$HA(aq) \rightleftharpoons H^+(aq) + A^-(aq)$$

Figure 41 River waters provide a constant supply of HCO_3^- ions to the oceans from the weathering of limestone rocks.

For the ocean to resist an increase in acidity, there must be a supply of $HCO_3^-(aq)$ ions, corresponding to the $A^-(aq)$ ions from the salt of the weak acid in a normal buffer solution.

There are HCO_3^- ions which flow into the sea in the form of material dissolved in river water. But there are two other reactions which can provide an almost limitless supply of hydrogencarbonate ions. These are reactions 3 and 5 (see pp. 254 and 255)

$$HCO_3^-(aq) \rightleftharpoons H^+(aq) + CO_3^{2-}(aq) \qquad \text{(reaction 3)}$$
$$CaCO_3(s) \rightleftharpoons Ca^{2+}(aq) + CO_3^{2-}(aq) \qquad \text{(reaction 5)}$$

Figure 42 illustrates the processes which prevent the oceans from becoming acidic. The shells, chalk and limestone in the sea are the reservoir of the anions needed to prevent changes in acidity.

What would happen if the carbon dioxide in the atmosphere rose to the high level of several billion years ago? Solid calcium carbonate would dissolve to produce the carbonate and hydrogencarbonate ions needed to remove the extra H^+ ions. Few CO_3^{2-} ions would remain. Most of the carbon would be in the form of HCO_3^- and dissolved CO_2. The sea would be like a mixture of Perrier water and bicarbonate of soda; the shells and white cliffs would disappear!

Limestone deposits could not form while the atmosphere contained 35% CO_2. The earliest sedimentary rocks are *silacaceous* (they consist predominantly of silicon dioxide). The first significant calcium carbonate deposits formed only about 2 billion years ago, over halfway through the Earth's lifetime. Much of the atmospheric carbon dioxide had been used up by then – so much so that its greenhouse effect had diminished and a major ice age had set in.

The carbon dioxide/calcium carbonate system is made up of linked equilibria which respond rapidly to change. In the longer term, ion exchange between H^+ ions in the water and Na^+ or K^+ ions in clay sediments provides another very powerful pH control mechanism.

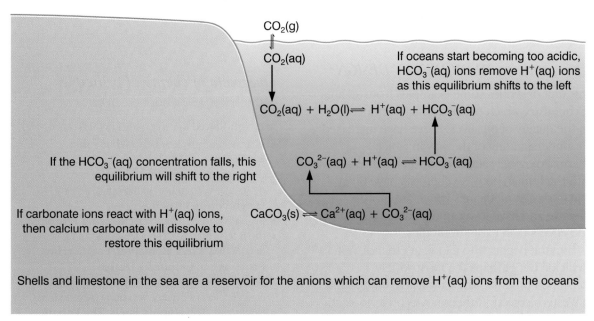

$$CO_2(g)$$

$$CO_2(aq)$$

If oceans start becoming too acidic, $HCO_3^-(aq)$ ions remove $H^+(aq)$ ions as this equilibrium shifts to the left

$$CO_2(aq) + H_2O(l) \rightleftharpoons H^+(aq) + HCO_3^-(aq)$$

If the $HCO_3^-(aq)$ concentration falls, this equilibrium will shift to the right

$$CO_3^{2-}(aq) + H^+(aq) \rightleftharpoons HCO_3^-(aq)$$

If carbonate ions react with $H^+(aq)$ ions, then calcium carbonate will dissolve to restore this equilibrium

$$CaCO_3(s) \rightleftharpoons Ca^{2+}(aq) + CO_3^{2-}(aq)$$

Shells and limestone in the sea are a reservoir for the anions which can remove $H^+(aq)$ ions from the oceans

Figure 42 The buffering action in the oceans – why the oceans do not become more acidic.

This process can only take place at the bottom of the ocean where sea water and sediment are in contact. Deep ocean water circulates slowly, perhaps taking 1000 years to complete one cycle. The ion-exchange equilibria may be important over millions of years, but in the surface water and throughout the oceans on a shorter time scale it is the carbon dioxide/calcium carbonate system which keeps the pH of the ocean stable.

05 Summary

This unit has given you a glimpse of just a few of the processes going on in the oceans. The chemistry followed on from the work you had done previously in **From Minerals to Elements**, and you learned more about solutions, dissolving and enthalpy changes involving ionic compounds.

The oceans play a major role in absorbing and redistributing the energy the Earth receives from the Sun and keeping the planet hospitable to life. Water is an ideal substance to carry out this role because of its unique properties. This led you to revisit ideas on hydrogen bonding and its effect on the properties of water.

In this context you extended your ideas about entropy and saw how entropy changes determine whether or not a process occurs spontaneously.

Some of the materials in the oceans are useful and can be economically extracted. Others, like carbon dioxide, exert a profound influence on the Earth's climate. Life has evolved in the oceans, and these materials have become inextricably linked to the life-cycles of many creatures. Buffering in the oceans, and in the bloodstreams of marine animals, is largely controlled by chemical reactions involving carbon dioxide. But exactly the same systems are found in creatures on land, even humans.

You needed to apply your ideas about acids and bases and chemical equilibria in this setting. Much of this was an extension of earlier work covered in **The Atmosphere**, but you then went on to study solubility products, weak acids, the pH scale and buffer solutions in more detail.

One of the big challenges scientists face is to understand how human activities are likely to affect the global environment. What is becoming clear is that global conditions are determined by the way in which the oceans, the atmosphere, the land and life interact with one another. This is an enormous task involving all branches of science. As this unit has shown, chemistry has a major part to play.

Activity 05 will help you to summarise what you have learned in this unit.

MEDICINES BY DESIGN

Why a unit on MEDICINES BY DESIGN?

This unit describes some examples from an area of chemistry which has had a major influence on the quality of our lives. As we have increased our understanding of the way in which pharmacologically active compounds interact with the human body, so chemists have been able to design medicines which are more effective and have fewer undesirable side-effects than earlier remedies.

The unit begins with a look at ethanol, not a medicine but perhaps the most widely consumed pharmacologically active compound in Western society. It then moves on to look at how two widely different disorders – asthma and heart trouble – can be treated by medicines which selectively activate or deactivate one of the body's nervous pathways. The unit concludes with a study of two medicines which inhibit enzyme action: captopril inhibits the human angiotensin-converting enzyme and penicillin inhibits a bacterial enzyme.

The chemical reactions which are used to synthesise and modify medicines are drawn almost exclusively from the field of organic chemistry, and this unit also serves as a good way of pulling together the organic reactions which you have encountered throughout the course, together with some new work on aldehydes and ketones.

Synthetic chemists use spectroscopic techniques to determine the structure of the compounds they make, and you will use your knowledge of mass spectrometry, infrared, nuclear magnetic resonance, and ultraviolet/visible spectroscopy to follow reactions in this way.

Overview of chemical principles

In this unit you will learn more about …

ideas introduced in earlier units in this course

- pharmaceutical chemistry (**What's in a Medicine?**)
- the importance of molecular shape and molecular recognition in biological activity (**Engineering Proteins**)
- proteins and enzymes (**Engineering Proteins**)
- the interpretation of spectroscopic and g.l.c. data (**What's in a Medicine?**, **Engineering Proteins** and **Colour by Design**)
- reactions of organic functional groups and the classification of organic reactions (several units)
- isomerism (**Developing Fuels**, **The Polymer Revolution** and **Engineering Proteins**)

… as well as learning new ideas about

- the interaction of biologically active molecules with receptor sites
- the role of chemists in designing and making new medicines
- aldehydes and ketones
- the synthesis of organic compounds.

MD1 *Alcohol in the body*

Have you ever taken a *xenobiotic*? No? Well, if you hadn't you wouldn't be alive. Xenobiotics are substances which the body doesn't contain but which can affect it. They can be divided into three types:

- **foods** provide the molecules which give us energy and keep our bodies well maintained
- **drugs** alter the biochemical processes in our bodies, for example changing the way we feel and behave – drugs which lead to an improvement in health are called **medicines**
- **poisons** severely damage our biochemical processes and cause a deterioration in health or even death.

You must have eaten foods, and you would be very unusual if you had not taken at least one medicine. You would have been unwise if you had taken non-medicinal drugs or poisons.

But some substances can behave in more than one way. Their effect depends on the quantity you take and on your state of health. Ethanol is an example; it is a source of energy, but it also affects behaviour, and, in excess, can cause liver damage and even death.

In this unit you will find out how chemists design medicines to perform specific tasks in the body. This involves making new compounds and modifying existing ones. To do this they need to have a 'tool-kit' of reactions which can be used to convert one compound into another. You can read about using organic reactions in this way in **Chemical Ideas 14.1** and **14.2**.

Aldehydes and ketones are important intermediate compounds in many synthetic routes. You can read about their chemistry in **Chemical Ideas 13.7**.

Alcohol as a food

Some people celebrate Christmas with a Christmas pudding ablaze with burning brandy. This is a reminder that energy is released when ethanol reacts with oxygen to produce carbon dioxide and water. Oxidation of ethanol in the body is more controlled and less complete, but it is still highly exothermic.

At the present time, alcoholic drinks contribute about 6% of the total dietary energy intake of people in the UK. In the 17th century they accounted for almost 25% – for children as well as adults. The common drink at that time was beer.

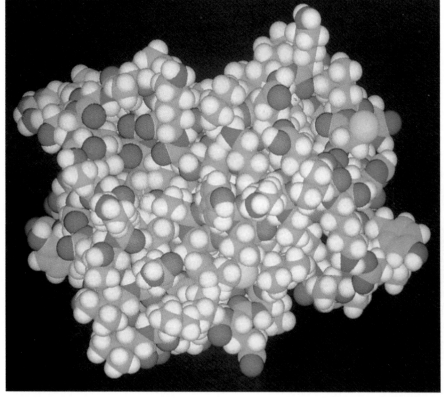

Figure 1 Molecular graphics systems are vital in the design of modern medicines. This shows a computer-generated graphic of interleukin–2, used in the treatment of some cancers.

ASSIGNMENT I

(You will need to refer to the Data Sheets to answer parts of this assignment.)

When ethanol burns, carbon dioxide and water are produced. Under more controlled conditions such as in bacterial metabolism, ethanol can be converted into ethanoic acid.

a Write a balanced equation for the complete combustion of ethanol. Look up the standard enthalpy change for this process.

b i Write a balanced equation for the oxidation of ethanol by oxygen to produce ethanoic acid and water.

ii Using Hess's law and appropriate enthalpy changes of formation, calculate the standard enthalpy change for this reaction.

c Metabolism of ethanol in the human body releases about 770 kJ per mole of ethanol. Suggest what might be likely products of ethanol oxidation in human metabolism.

d Some ethanol is not metabolised. Suggest two ways in which ethanol can be lost from the human body.

This depression of nervous activity has important short-term effects. Drinking alcohol reduces vigilance, slows reaction times and impairs our judgement. It is largely because of these effects that so many laws have been introduced to control the use of alcohol. This is particularly important in relation to driving motor vehicles.

Blood alcohol concentration (BAC) is closely related to the extent of the effects of alcoholic drinks. It is usually defined as

BAC = mg of ethanol per 100 cm^3 of blood

The quantity of alcohol needed to produce a particular BAC varies with age, sex, body weight, how quickly you drink and several other factors. The concentration of ethanol in the blood rises for some time after taking a drink, as the alcohol is absorbed. Then it slowly decreases as the ethanol is excreted or metabolised.

A 'unit' of alcohol is a convenient measure of how much ethanol is contained in a drink. It is an approximate measure, roughly equivalent to

- a half-pint of beer or lager
- a glass of wine
- a single measure of spirits.

Alcohol as a drug

Alcoholic drinks make many people feel better for a short time. They help them to relax or cope with stress. Alcohol can make them feel happier and relieve tension, anxiety or boredom. These effects are all outcomes of a single aspect of the behaviour of ethanol molecules in the body – they depress the activity of the central nervous system. This is explained in more detail in Section **MD2**.

Figure 2 Many road accidents involve either drivers or pedestrians who have drunk too much alcohol.

SAINSBURY'S
CLASSIC SELECTION

CHIANTI CLASSICO

This delicious Chianti Classico comes from the region of Tuscany, famous for its annual 'Palio' horse race. The 187 acre Villa Cerna Estate is located in the heart of the Chianti Classico zone between Florence and Siena, producing wines of finesse and elegance such as this fine example.

STORAGE - It is recommended that this wine be consumed within 2 years of purchase.

STYLE
Rich, full-bodied.

SERVE
At room temperature. Ideal for all red meats and full-flavoured cheeses.

UNITS OF ALCOHOL
This bottle contains 6 glasses.

1.6
units of alcohol per 125ml glass

75cl

CUSTOMER CARELINE
Freephone 0800 636262
INTERNET:
www.sainsburys.co.uk

0114 7056
10005

L9/14OC1043

Figure 3 Some wine bottles now show how many units of alcohol are present in one glass of wine.

"Excuse me Sir, ..."

At present, 80 mg per 100 cm³ of blood is the legal limit for BAC for driving a motor vehicle. When a motorist who is suspected of drink-driving is stopped by the police, a roadside blood test is impractical. A quick BAC estimate is needed in order to decide whether or not to take things further.

Ethanol (the alcohol present in drinks) is almost the only commonly used drug which is sufficiently volatile to pass from the blood to the air in the lungs. This distribution

$$C_2H_5OH(blood) \rightleftharpoons C_2H_5OH(g)$$

is an example of chemical equilibrium, and it is governed by an equilibrium constant (K_c) which has a fixed value at a particular temperature.

In the human body, K_c for this process is 4.35×10^{-4} at body temperature. A measurement of the ethanol concentration in the breath therefore gives an indication of the BAC.

ASSIGNMENT 2

a Write down the expression for K_c for the distribution of ethanol between the blood and the air in the lungs.

b The position of this equilibrium depends on temperature and pressure. These conditions will be the same in all drink-drive suspects. What values will they have?

c Under these conditions, in the human body, K_c for this process has a value of 4.35×10^{-4}. What will be the *breath alcohol concentration* (in mg of ethanol per 100 cm³ of air) which corresponds to the 80 mg per 100 cm³ legal limit for BAC?

One of the earliest successful methods of detection – the 'breathalyser' – is based on a familiar chemical reaction. Orange crystals of potassium dichromate(VI) turn green when they oxidise ethanol to ethanal and ethanoic acid. The extent to which the crystals in a tube change colour is a measure of the BAC.

Figure 4 The 'Lion Alcolmeter' containing a fuel cell has replaced the traditional 'Breathalyser' for roadside tests.

Oxidation of ethanol is a redox reaction and involves electron transfer. An alternative way of finding the ethanol concentration is therefore to measure the voltage of an electrochemical cell which incorporates the reaction. The 'Lion Alcolmeter' is a cell designed to do just this.

An electrochemical cell in which a fuel (such as ethanol) is oxidised is sometimes called a **fuel cell**; the fuel is being oxidised to produce electrical energy rather than being burned to produce heat.

In the 'Alcolmeter' the cell consists of a porous glass plate in which phosphoric acid is absorbed. The plate is sandwiched between coatings of platinum or silver, which act as electrodes.

Oxygen is reduced to water at one electrode and ethanol is oxidised to ethanoic acid at the other. The instrument is calibrated by passing different concentrations of ethanol in air through the instrument and measuring the resulting cell voltage.

The instrument is set so that a green light comes on when there is no alcohol in the breath, an amber light when a small amount is present, amber and red when close to the limit, and red above the limit.

ASSIGNMENT 3

a Write half-equations for the reactions at each electrode of an 'Alcolmeter'. (Remember to include water molecules and/or hydrogen ions where necessary to balance these equations.)

b State which electrode (positive or negative) each of your half-reactions will occur at.

The science behind the 'Alcolmeter' is fairly straightforward, but the technology needed to develop it is more challenging. The instrument must be

- small enough for hand-held use beside the road
- rugged
- safe and easy to use
- reusable with a short turn-round time
- reliable – the consequences of the test can be serious for the person being tested.

Then it has to be marketed and sold.

You can read about alcohols and some of their reactions in **Chemical Ideas 13.2** and **13.4**. The reactions of alcohols are summarised in **Chemical Ideas 14.2**.

Activity MD1.1 investigates the oxidation of alcohols, and the formation and reactions of aldehydes and ketones.

Paul's story

Dr Paul Williams, who worked on the development of the 'Alcolmeter' and is now International Marketing Director for Lion Laboratories Ltd, Barry, South Glamorgan, describes his career.

Figure 5 Paul Williams and the 'Lion Alcolmeter'.

"After A-levels, and a degree in Applied Chemistry, I realised I was interested in forensic chemistry. Luckily for me, a research post was advertised on a project to develop an electrochemically based instrument for the breath alcohol testing of drink-driving suspects.

"I got the job, gained an MSc and PhD while doing it, and at the end of 4 years had developed the 'Alcolmeter' instruments. During this time, I decided that the business life was for me – particularly marketing, where I could use my scientific training to talk to potential customers on their own level.

"So, having developed the 'Alcolmeter', I went out and sold it! There is no better way to learn the trade than to hawk your wares from a company car for a year or so. That set me up to be made Marketing Director for Lion Laboratories.

"Instead of selling, I was now formulating company marketing policy, and getting to talk to 'high fliers' in the alcohol field around the world. The Vauxhall gave way to a Boeing, and home became very often a room in the local Sheraton, Hilton or Intercon. The opportunity to travel to South America, Africa and the Middle East annually, the US at least three times a year and – of course – Europe, meeting police officers and forensic scientists, being able to talk with them and, in many cases, to educate and inform them, are features of the job which I enjoy and find intellectually satisfying.

"Providing expert testimony in court is also a regular part of the job – daunting at first, but easier as it goes along.

"And it was with great pleasure that I became a Fellow of the Royal Society of Chemistry in 1996, based on all this work.

"I think my training did more to help me in my career than simply to provide the chemistry I needed. It improved my memory, and taught me to analyse and evaluate situations logically – which is very useful in the modern commercial world. If you are thinking of 'going commercial' it is probably best not to become too specialised. Keep a broad subject base and make sure your knowledge is practical – that way it is capable of being applied to the job in hand."

Back at the station .....

The techniques used for roadside breath tests give a quick, reliable result but are not used for evidence in court cases as the instrument does not give a print-out at the time of testing. If the roadside test indicates a high level of alcohol, the driver will be taken to a police station for a more accurate determination of BAC.

At the police station, most suspects undergo a second type of breath test, where the ethanol concentration is measured using **infrared spectroscopy** (Figure 6). The driver is asked to breathe continuously into the cell of an i.r. spectrometer and the intensity of radiation absorbed at 2950 cm^{-1} is measured. If the driver is incapable or suffers from a breathing problem such as asthma, blood or urine samples are taken by a doctor. Two samples are taken, one for the forensic service and one for the driver, who can use it to have the analysis checked independently. The sample is analysed in the forensic laboratory using **gas–liquid chromatography (g.l.c.)**.

Figure 6 The Lion Intoxilyzer analyses the ethanol in a suspect's breath by absorption of infrared radiation. An immediate print-out of the result is obtained. This is shown to the driver and can be used as evidence in court cases.

Gas–liquid chromatography is described in **Chemical Ideas 7.6.**

You can read about infrared absorption spectroscopy in **Chemical Ideas 6.4.**

In **Activity MD1.2** you can find out more about the application of g.l.c. to blood alcohol analysis.

ASSIGNMENT 4

The infrared absorption spectrum of ethanol is shown in Figure 7.

The ethanol concentration in the breath of a drink-drive suspect can be determined by measuring the intensity of one of the absorption bands in the ethanol spectrum. The absorption at 2950 cm^{-1} is used in the 'Lion Intoxilyzer 6000 UK' shown in Figure 6.

The suspect's breath will also contain water vapour. This gives rise to strong infrared absorptions centred at about 3800 cm^{-1}, 3600 cm^{-1}, 3200 cm^{-1} and 1600 cm^{-1}.

a Compare the positions of the water absorptions in an infrared spectrum with those of ethanol shown in Figure 7. Suggest why the 2950 cm^{-1} band is chosen for ethanol detection.

b Explain why ethanol and water both give rise to infrared bands in the same region of the spectrum.

c Look at the table of characteristic infrared absorptions in the Data Sheets. What bond is responsible for the 2950 cm^{-1} absorption in ethanol?

d People who suffer from diabetes often produce propanone vapour in their breath. Suggest why infrared breath testing of a person with diabetes might appear to give a positive result even though the person had not been drinking alcohol.

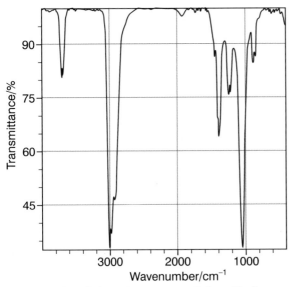

Figure 7 Infrared absorption spectrum of ethanol in the gas phase.

MD2 *The drug action of ethanol*

A touch of nerves

You need to know something about the human nervous system in order to understand how ethanol acts as a drug. The nervous system consists of two parts: the *central nervous system* (the brain and spinal cord) and the *peripheral nervous system* (all other nervous tissues which connect with the organs of the body). A simple overall picture of the nervous system is shown in Figure 8.

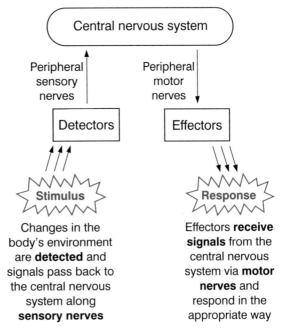

Figure 8 The essential parts of the nervous system.

A more detailed look at the nervous system shows it to be immensely complex. It consists of thousands of millions of nerve cells or *neurons*. These are linked together at connections called *synapses*, as shown in Figure 9.

These interconnecting neurons provide an important system of communication in the body. This is what happens.

An electrical signal called a **nerve impulse** travels along the axon (the long, threadlike extension of the nerve cell) until it reaches a nerve ending. There, the electrical signal causes release of small messenger molecules, called **neurotransmitters**, which carry the message across the synapse (see Figure 10).

Neurotransmitters, released from nerve endings, cross the tiny gap to **receptors** on the dendrites (branches off the main cell body) of another neuron.

Neurotransmitters can either excite or inhibit the electrical behaviour of the next nerve cell.

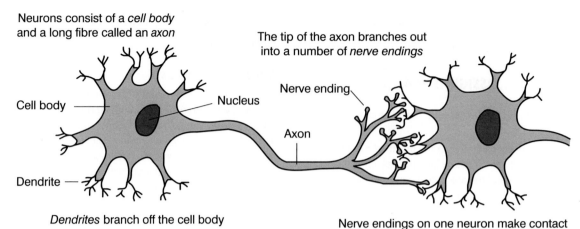

Neurons consist of a *cell body* and a long fibre called an *axon*

The tip of the axon branches out into a number of *nerve endings*

Cell body — Nucleus

Nerve ending

Axon

Dendrite

Dendrites branch off the cell body

Nerve endings on one neuron make contact with dendrites on another neuron at *synapses*

Figure 9   *Connections between nerve cells.*

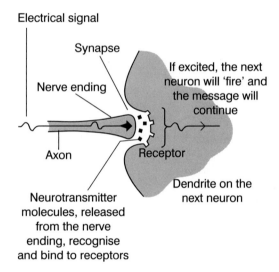

Electrical signal

Synapse

Nerve ending

If excited, the next neuron will 'fire' and the message will continue

Axon

Receptor

Dendrite on the next neuron

Neurotransmitter molecules, released from the nerve ending, recognise and bind to receptors

Figure 10   *Signals pass from neuron to neuron by neurotransmitters released at the synapse.*

Some neuropharmacology

The brain has to process all the information coming in from the sensory nerves, and send out action signals to the appropriate motor nerves. There have to be lots of interconnections between neurons in the brain. At any instant some of these will need exciting (switching on so that they fire nerve impulses more rapidly) and some will need inhibiting (switching off so that they fire nerve impulses more slowly).

Switching on and switching off nerve cells is achieved by movements of ions in and out of the cells. Three ions are important: K^+, Na^+ and Cl^-. There are high concentrations of Na^+ and Cl^- *outside* the nerve cell, but the K^+ concentration is higher *inside*.

When the nerve is at rest, the nerve cell membrane is closed to Na^+ and Cl^- ions, so they can't get in. But K^+ ions can move freely in and out of the nerve through channels in the cell membrane.

Look at Figure 11. Some potassium ions will diffuse out of the cell to try to even out the K^+ concentration, but this leaves behind an excess of negatively charged groups which are attached to the inside of the cell. So the inside of the nerve cell becomes negatively charged. This makes it harder for the positively charged potassium ions to leave. An equilibrium is established in which the cell is at a negative potential of about 60 mV relative to the fluid outside. This is shown in Figure 12.

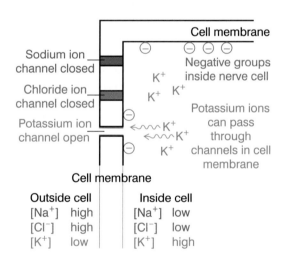

Cell membrane

Sodium ion channel closed

Negative groups inside nerve cell

Chloride ion channel closed

K^+

K^+ K^+

Potassium ion channel open

Potassium ions can pass through channels in cell membrane

K^+

K^+

K^+

Cell membrane

Outside cell	Inside cell
$[Na^+]$ high	$[Na^+]$ low
$[Cl^-]$ high	$[Cl^-]$ low
$[K^+]$ low	$[K^+]$ high

Figure 11   *The processes occurring at a dendrite of a nerve which is at rest.*

This −60 mV is the cell's *resting potential*. The cell 'fires' when neurotransmitters, arriving from an adjacent neuron, cause the sodium ion channels to open. Large numbers of Na^+ ions then flood into the nerve cell. The inside of the cell becomes more positive, possibly to as much as +20 mV (Figure 13). This wave of positive charge is carried along the axon by further movements of Na^+ and K^+ ions, and constitutes the nerve impulse.

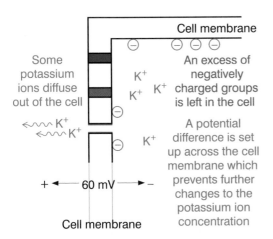

Figure 12 *The establishment of a potential difference across the cell membrane of a nerve cell which is at rest.*

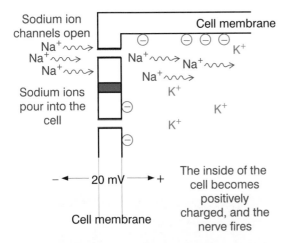

Figure 13 *A nerve cell 'fires' when sodium ions flood in through open sodium ion channels.*

For nerve cells to 'fire' repeatedly, sodium ions are constantly being 'pumped out' of the cell and potassium ions replaced. This involves moving ions against their natural concentration gradients and uses a lot of energy.

Nervous inhibition and GABA

The nerve cell is switched off – it is harder to fire – when its potential is made more negative. This is achieved by opening the chloride ion channels and allowing Cl^- ions to diffuse into the cell.

The compound *GABA* (gamma-aminobutanoic acid) is the neurotransmitter which switches off, or inhibits, nerve cell action in the brain. Its modern, systematic, name is 4-aminobutanoic acid.

$$\overset{\gamma}{H_2N} - CH_2 - \overset{\beta}{CH_2} - \overset{\alpha}{CH_2} - COOH$$

γ–aminobutanoic acid (GABA)

Each ion channel is surrounded by protein molecules embedded in the cell membrane (see Figure 14).

GABA molecules bind to receptor sites on these protein molecules, causing the protein to change shape and open the chloride ion channel. A simple illustration of how this could happen is shown in Figure 15.

Figure 16 illustrates what happens once GABA has bound to the nerve cell, causing the chloride ion channels to open.

Scientists think that one of the effects of drinking alcohol is that ethanol molecules bind to nerve cells near to the GABA receptors. This enhances the effect that GABA has.

Figure 14 *Chloride ion channels in a cell membrane.*

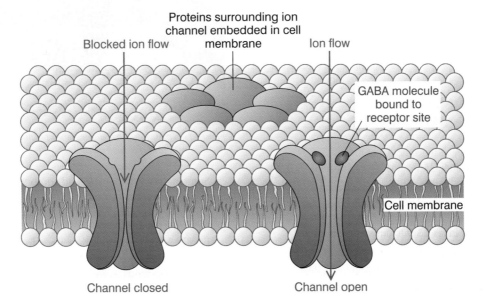

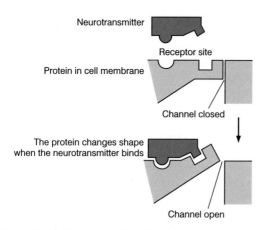

Figure 15 *An illustration of how binding to a receptor can open channels in a cell membrane.*

Neurons are more inhibited if ethanol is present and the action of the nervous system is depressed. Other molecules have a similar effect and some are used medically to treat anxiety, eg benzodiazepines such as *Valium*.

It is very dangerous to drink alcohol if you are taking certain medicines. The benzodiazepines are an example. When ethanol and benzodiazepine molecules are bound to a nerve cell, the effect can be very much greater than when only one of them is present. The inhibition of the nervous system can be severe enough to be fatal.

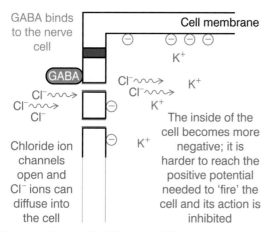

Figure 16 *Nerve cell inhibition by GABA.*

ASSIGNMENT 5

Look carefully at the structure of the neurotransmitter GABA. It is a γ-amino acid.

a In the body, GABA exists as a zwitterion. Draw out the structure of the zwitterion. (You may need to look back to earlier work on amino acids in **Chemical Ideas 13.9** to answer this question.)

b Unlike most α-amino acids, GABA does not exist in D and L isomeric forms. Explain why this is so.

MD3 Medicines that send messages to nerves

Discovering a new medicine

When they start out to design a new medicine, scientists focus their attention on two areas:

- developing a biological understanding of the disease or condition they want to cure
- finding a 'lead compound' (not a compound of the element lead, but a compound which provides a lead – shows some promise and gives some clues).

Sometimes, an idea for a lead compound comes from research into the chemical processes which go on in the body. As you have seen, scientists have some understanding of how the GABA neurotransmitter acts. In this section you will learn how our increased knowledge of another neurotransmitter, *noradrenaline* and its relation, *adrenaline* (a hormone), has helped in the development of medicines to treat asthma and to combat heart disease.

Medicines have also been developed from lead compounds which are the active ingredients in traditional remedies. In **What's in a Medicine?** for example, you learned how aspirin was developed from a natural remedy made from willow bark. In Section **MD4**, you can read about how studies of snake venom led to a medicine for treating high blood pressure.

Figure 17 *Traditional medicines from plants can provide 'lead' compounds for new medicines.*

Some naturally occurring lead compounds have been discovered by accident. A classic example is the discovery of penicillin. You can learn more about this in Section **MD5**.

Finally, some medicines have been found from random screening – looking for activity in a very wide range of compounds, some of which might already be used medically to treat other things.

Despite the increasing success of medicine design based on scientific understanding of the chemical processes which go on in the body, only one in about 10 000 of the compounds which are synthesised survive today's rigorous testing procedures and become commercially available for medical use.

In **Activity MD3.1** you can use your knowledge of organic chemistry to build up a 'toolkit' of organic reactions.

Activity MD3.2 will help you to classify these reactions by the type of mechanism involved.

You can then use these reactions in **Activity MD3.3** to plan the synthesis of some medicines. Also in this activity, you can study spectra to determine the structure of ibuprofen and compounds used in its synthesis from benzene.

You can revise earlier work on the use of spectroscopic techniques in determining the structure of organic compounds by reading **Chemical Ideas 6.4, 6.5, 6.6** and **6.8**.

Adrenaline mimics

Adrenaline and noradrenaline have very similar structures.

HO H
HO — [benzene ring] — *CH(OH)* — CH2 — NH2
HO

noradrenaline

HO H
HO — [benzene ring] — *CH(OH)* — CH2 — NHCH3
HO

adrenaline

Adrenaline is secreted into the bloodstream by the adrenal glands, whereas noradrenaline is produced at the nerve ends. The noradrenaline molecule is the neurotransmitter in many of the synapses where nerves join organs in the body.

Adrenaline and noradrenaline are released when we become nervous, scared or excited; they are known as the chemicals for 'fight, fright and flight'. They prepare the body to cope with a rise in carbon dioxide before this actually happens, and in doing so:

- the blood pressure is increased
- the heart rate is accelerated
- the bronchioles (airways in the lungs) become dilated (widened)
- sweating is increased.

It is the third property, the dilation of the bronchioles, that has led to the development of new medicines to combat asthma.

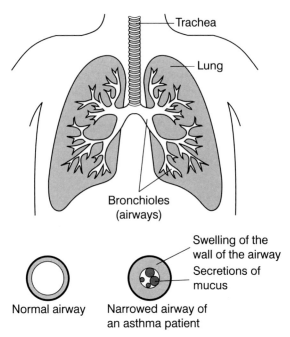

Figure 18 *Asthma patients become short of breath because of narrowing of the airways in the lungs.*

Asthma is a very distressing disease that affects about 5% of the population. The airways in the lungs become narrow and inflamed, and filled with a fluid known as mucus (Figures 18 and 19). This can be very frightening as it makes the sufferer wheeze and become breathless.

One way to relieve an attack of asthma is to widen the narrowed airways, a process known as *bronchodilation* – the very process that adrenaline and noradrenaline are known to promote.

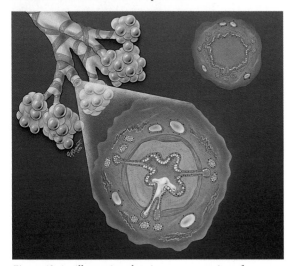

Figure 19 *An illustration showing a cross-section of a bronchiole of an asthmatic compared with that of a healthy person (top right).*

By making modifications to the adrenaline and noradrenaline structures, chemists have been able to design and synthesise molecules which interact more selectively with the receptors.

The compound *isoprenaline* was used for many years. It relieves asthma without causing some of the other effects of adrenaline and noradrenaline. Unfortunately, it still affects the heart rate and blood pressure, sometimes to a dangerous extent.

HO H
HO.
NHCH(CH$_3$)$_2$
HO
isoprenaline

In recent years, isoprenaline has been replaced by *salbutamol* which is more selective and acts only by widening the bronchioles. It binds to receptors on the muscles of the airways, relaxing them and giving relief from breathing difficulties.

HO H
HO
NHC(CH$_3$)$_3$
HO
salbutamol

Neurotransmitters have to be reabsorbed by the nerve cells or changed into deactivated forms and detached from their receptors, otherwise their effect would go on forever. A medicine for treating asthma, however, needs to act for a long time. An advantage of salbutamol is that conversion of one of the phenol groups into a —CH$_2$OH group slows down the rate at which it is broken down in the body, so its effects last longer than isoprenaline.

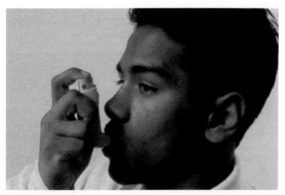

Figure 20 Inhalers containing salbutamol bring quick relief in an asthma attack.

Salbutamol lasts for about 4 hours, which is not long enough to give the sufferer a good night's sleep. This problem has now been solved by making further changes to the structure of the molecule. Although the new compound, *salmeterol*, is not as effective for severe acute attacks, it gives longer-acting protection.

HO H H
HO
N
O
HO
salmeterol

a Name three functional groups which are present in adrenaline.

b What is the difference in molecular structure between adrenaline and noradrenaline?

c In what way are
  i isoprenaline
  ii salbutamol
  different in structure from noradrenaline?

d How would you expect noradrenaline to react with the reagents listed below? Where possible, write structures for the reaction products.
  i FeCl$_3$(aq)   iii HCl(aq)
  ii NaOH(aq)   iv CH$_3$COCl.

e For each of the reagents in part **d**, state whether or not you would expect salbutamol to react in a similar way to noradrenaline.

a The salbutamol molecule is chiral. Draw out the structure of the group of atoms which is responsible for the chirality.

b The enantiomers of salbutamol differ in their pharmacological activity. One isomer is 68 times more active than the other. Suggest a reason for the difference.

Activity **MD3.4** looks in more detail at some of the chemistry and costs involved in producing salbutamol. You can also look at the spectra of some of the compounds involved to determine their structure.

Agonists and antagonists

The structures of adrenaline, noradrenaline, isoprenaline, salbutamol and salmeterol have a common feature. Their activity depends on the presence of the structural fragment

HO H H
N
HO

This is the group of atoms which is involved in binding to the receptor. It's a case of **molecular recognition**. The structural fragment fits precisely into the shape of the receptor, and functional groups on both are correctly positioned to interact.

Molecular recognition was introduced in **Engineering Proteins**, where its importance with regard to DNA, RNA, protein synthesis and enzyme activity was discussed.

The structural fragment shown above is an example of a **pharmacophore** – a group of atoms which confers pharmacological activity on a molecule. The successful medicine, salbutamol, was made by modifying the pharmacophore with a —CH$_2$OH group on the ring and a 2-methylpropyl substituent on the nitrogen atom to make it more selective as a bronchodilator.

Salbutamol is an example of an **agonist** – a molecule which behaves like the body's natural substance in the way it binds to a receptor and produces a response.

In some people the natural response may be a bad thing: for example, an increased heart rate can be dangerous for someone with a heart disorder. What these people need are molecules which compete with the natural, active compound for receptor sites but which have no effect when they are bound.

Molecules like this are called **antagonists**. An antagonist has a structure which is sufficiently like the pharmacophore to allow it to fit the receptor, but it produces no effect because it does not actually possess the pharmacophore. In the case of noradrenaline, if receptors are blocked by antagonists, the neurotransmitter cannot get to the receptor sites to stimulate an increase in heart rate (Figure 21).

Propranolol is a very successful noradrenaline antagonist. It binds to, and blocks, the type of noradrenaline receptors which control the heart muscles. These are called β_1-receptors, so drugs like propranolol are known as *β-blockers*. They are used by people with heart disorders to keep their hearts calm and avoid further trouble.

propranolol

ASSIGNMENT 8

a The structure of the propranolol molecule differs in two important ways from the noradrenaline agonists such as salbutamol. These differences prevent propranolol triggering an electrical impulse when it binds to a receptor.

What are the differences?

b The structures of three medicines are shown below. For each medicine state whether you would expect it to be a noradrenaline agonist, a noradrenaline antagonist, or show no effect related to noradrenaline.

Briefly describe the reasons for your decisions.

i

ii

iii

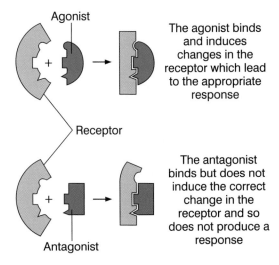

Figure 21 An illustration of an agonist and an antagonist binding to receptors.

The agonist binds and induces changes in the receptor which lead to the appropriate response

The antagonist binds but does not induce the correct change in the receptor and so does not produce a response

Figure 22 Inderal tablets contain the β-blocker, propranolol.

Peter's story

As you will realise by now, devising synthetic routes for the preparation of organic compounds is fundamental to the successful development and production of many medicines. The interview with Peter O'Brien, reported below, provides an insight into the work of an organic synthesis research group. Peter is a lecturer in the Department of Chemistry, University of York, and leader of the research group.

Figure 23 Peter and his research group in the Department of Chemistry at the University of York.

Tell me about your research group.
"The group is made up of two post-doctoral researchers (post-docs). These are people who have successfully completed their PhD and are experienced researchers. There are also four research students studying for a PhD and one for a Masters degree, and myself. The research students are expected to carry out original research and report on it in a thesis (after 3 years for a PhD and after 1 year for a Masters degree). Each thesis is examined and the student interviewed (called a *viva*) by a distinguished chemist from another university along with another member of the York Chemistry Department."

And what is your background?
"I did A levels in chemistry, maths and physics. Then I did my first degree at Cambridge. During two of the summer vacations I worked in two different pharmaceutical companies. I saw at first hand the importance of organic chemistry to industry and to the world at large. In the final year of my undergraduate course I did a research project which involved making new organic molecules. This excited and fascinated me so I stayed on at Cambridge and did a PhD in synthetic chemistry. On completing my PhD, I moved to York as a post-doc and, after three months, was appointed as a lecturer. I have now been a lecturer here for three and a half years."

How do you decide what research problems to tackle?
"It is curiosity driven. From my knowledge of the literature on organic synthesis and certain branches of medicine and the pharmaceutical industry, I identify an aspect of organic synthesis which has not been done before and which could be of interest to the pharmaceutical industry.

"It is actually quite easy to produce new compounds – it is much more challenging and satisfying to develop a more elegant and simple laboratory method to prepare a compound which is already known to be potentially useful. I am particularly interested in problems associated with the synthesis of chiral molecules."

How are the problems broken down for different members of the group?
"Each of the research students must have their own clearly defined project as the award of the higher degree has to be based on their own work. But we try to have people working on related areas and, in particular, on topics related to the work of a more experienced researcher so that they can support each other. I supervise all projects but as people gain experience I become more of a consultant."

What does the actual research involve?
"Organic synthesis involves three main stages. The first is the reaction stage. We carry out reactions at different temperatures, from 120 °C to −78 °C. Often reactions are carried out under nitrogen, which provides an inert atmosphere – this requires special techniques. Sometimes reactions are done in simple round-bottomed flasks and sometimes in more complicated equipment.

"During the second stage the product is purified. This can take a long time. Recrystallisation, distillation and column chromatography are the main techniques used. Proton n.m.r. spectroscopy is the principal method we use of judging the purity of a product. If the spectrum shows peaks that we would not expect from the compound then they are likely to be caused by impurities.

"The third stage is the analysis of the product. This

is where we collect evidence that the compound is what we say it is. For confirmation of the identity of a product we need n.m.r., i.r. and mass spectra. We also need a combustion analysis which provides the percentages of carbon, hydrogen and nitrogen in the compound and so confirms its empirical formula."

What happens next?
"The most satisfying part of the whole process is the publication of our research findings in specialist journals. These publicise our contribution to new knowledge and the research students get to see their names in print."

So how does all of this relate to the work of the pharmaceutical industry?
"Perhaps this is best explained by considering a specific example. Epoxides contain a three-membered ring made up of two carbon atoms and one oxygen. A subset of these compounds are chiral epoxides – they contain a carbon atom bonded to four different groups and therefore they exist in two isomeric forms, called enantiomers. A common problem with chiral compounds is that it is difficult to synthesis one enantiomer rather than a mixture of the two. If you can only make a mixture of the two then you have to use some other reaction to destroy the one you don't want – and so reduce your yield by 50%!

"From the literature I identified a particular enantiomer of an epoxide which is related to an intermediate used in the manufacture of HIV protease inhibitors – these can be used as antiviral compounds in the treatment of people who are HIV-positive. I prepared a research proposal which set out our plans for developing an efficient synthesis of this epoxide enantiomer. A pharmaceutical company, which is interested in the area, liked the proposal and has awarded us a grant to support the research, which I use to buy equipment and chemicals, and to give a scholarship to one of the research students."

MD4 *Enzyme inhibitors as medicines*

Captopril – an ACE inhibitor

Receptors and enzymes are very similar. Both are proteins with precise structures designed to accommodate specific arrangements of atoms on another molecule such as a neurotransmitter or a substrate.

Just as antagonists are used to block receptors, so some medicines work by inhibiting the action of enzymes. A good example is *captopril* – which is widely used to treat high blood pressure (a condition called hypertension). Left untreated, this can lead to serious consequences such as a stroke or a heart attack.

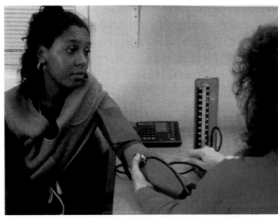

Figure 24 Measuring blood pressure helps to identify people who may be at risk of heart disease.

A small protein built from eight amino acids is known to be a key factor in raising blood pressure. It is called *angiotensin II*. The body creates *angiotensin II* from an inactive 10-amino-acid protein called *angiotensin I*. The enzyme which brings about the loss of the two amino acids in this conversion is called *angiotensin-converting enzyme (ACE)*:

$$\text{angiotensin I} \xrightarrow{\text{ACE}} \text{angiotensin II} + \text{dipeptide}$$

An imbalance in the production of angiotensin II results in high blood pressure, so one method of treatment is to inhibit the enzyme which catalyses its formation. The medicine needs to be an *ACE inhibitor*. Captopril works in this way.

Following the lead

The lead for the development of captopril was the discovery that the venom of the Brazilian Arrowhead Viper brought about its toxic effect by causing a fall in blood pressure.

Figure 25 The venom of the Brazilian Arrowhead Viper provided 'lead' compounds for the development of medicines used to treat high blood pressure.

The venom is a complex mixture, but separation, analysis and further study of the components led to the identification of several small proteins which were powerful ACE inhibitors.

Unfortunately, proteins are difficult to administer by mouth as medicines because they are readily broken down by digestive enzymes present in the stomach.

The snake venom proteins could not therefore be used medically as ACE inhibitors, but their structures gave scientists clues about the enzyme's active site. They learned more by studying the way the enzyme changes angiotensin I into angiotensin II. From this they realised that ACE must be similar to another, better known, enzyme – *carboxypeptidase*.

Combination of these results led to a model in which ACE binds to its substrate (angiotensin I) in three places by three different kinds of interaction:

1. a metal–ligand bond between a Zn^{2+} ion on the enzyme and an atom with a partial negative charge on the substrate

2. a hydrogen bond between an N–H group on the enzyme and a negatively charged atom on the substrate

3. an ionic interaction between a positively charged $-NH_3^+$ group on the enzyme and a negatively charged $-COO^-$ group at one end of the substrate.

These interactions (labelled 1, 2 and 3) between ACE and the three amino acid residues at the —COOH end of angiotensin I are illustrated in Figure 26. Notice too that the enzyme's active site has a precise shape which also fits other groups on the amino acid side chains.

It is essential for the action of ACE, and all enzymes, that the products are bound less strongly than the substrate. This ensures that the products will leave the enzyme and allow more substrate molecules to bind, thus continuing the reaction.

In this case, angiotensin I (the substrate) is bound by three interactions, whereas angiotensin II (one of the products) is only bound by the metal–ligand bond (interaction 1 in Figure 26), and the dipeptide (the other product) is only held by the hydrogen bond and the ionic interaction (interactions 2 and 3 in Figure 26).

A successful ACE inhibitor needs to bind but not react, so it remains attached by all three interactions. The use of computer graphics to study the shape and charge density in the active site played a crucial role in the design of possible compounds.

ASSIGNMENT 9

The primary structure of the angiotensin I substrate is:

Asp Arg Val Tyr Ile His Pro Phe His Leu

The part shown in Figure 26 is the Phe His Leu section at the —COOH end of the chain.

ACE catalyses hydrolysis at the position shown in Figure 26.

Write down the primary structures of the two products of the hydrolysis.

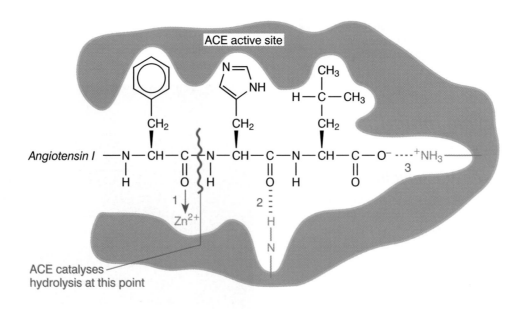

Figure 26 A representation of the structure proposed for the ACE/angiotensin I complex.

Finding the best medicine

All the active proteins which were isolated from the snake venom had proline as the amino acid at the —COOH end of the chain. This was chosen as a good starting point for the development of an ACE inhibitor.

proline

Chemists set about synthesising hundreds of compounds with the general structure

each one with a different R group. They were all capable of interacting at positions 2 and 3 with the ACE enzyme.

The breakthrough came when it was discovered that derivatives of proline which contained an —SH group in the right position to interact with the Zn^{2+} were particularly effective. Table 1 gives an indication of the effectiveness of three derivatives of this kind.

Derivative (Pro = proline residue)	Concentration needed to produce 50% inhibition of ACE activity/10^{-6} mol dm^{-3}
HS—CH_2—$CH(CH_3)$—CO—Pro	0.02
HS—CH_2—CH_2—CO—Pro	0.30
HS—CH_2—CH_2—CH_2—CO—Pro	9.70

Table 1 Effectiveness of various proline derivatives at inhibiting ACE activity.

Notice how the activity depends on the derivative's ability to fit precisely into the enzyme's active site. Small changes to the structure can lead to big changes in activity.

In fact, very small changes may be significant. The orientation of the —CH_3 group in HS—CH_2—$CH(CH_3)$—CO—Pro is particularly important: one isomer is much more active than the one with the —CH_3 group in the other configuration. The more active isomer is called captopril. A space-filling model of captopril is shown in Figure 27.

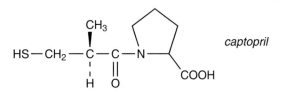

captopril

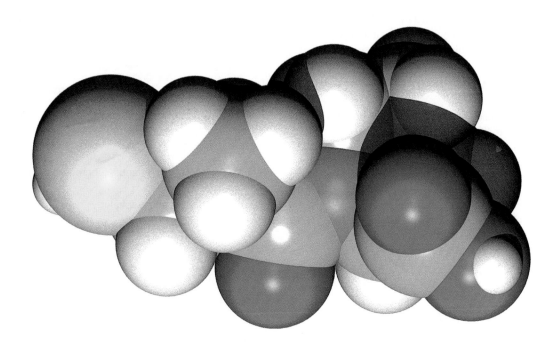

Figure 27 A computer-simulated model of captopril. This is a space-filling model.

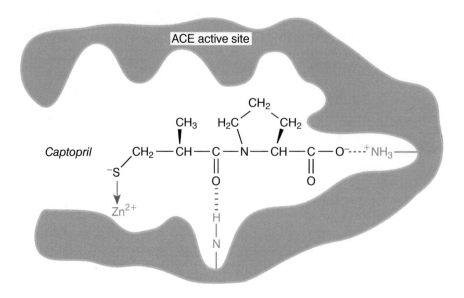

Figure 28 A representation of captopril in the ACE active site (the —SH and —COOH groups on captopril are ionised in the form of the compound which interacts with the enzyme).

Figure 28 shows how captopril is thought to bind to ACE.

Captopril is not a protein and is not broken down by digestive enzymes when it is taken orally. The peptide link which is broken in angiotensin I is not present in captopril. At the corresponding position on the captopril–enzyme complex, there are S–C and C–C bonds which cannot be hydrolysed.

The development of captopril is an excellent example of the logical design of a molecule. The medicine passed the testing procedures and was launched in 1980. It is still in widespread use.

ASSIGNMENT 10

Captopril is a much more active ACE inhibitor than its isomer with the methyl group in the other configuration.

a Discuss what you understand this statement to mean.

b Draw a structure for the other isomer of captopril.

c What other group in captopril is like the methyl group in being able to exist in another configuration in another isomer?

MD5 *Targeting bacteria*

Some medicines inhibit the action of enzymes present in bacteria. As a result, the bacterial cells do not grow and divide, and infection is prevented. Of these medicines, *penicillin* is one of the most widely prescribed.

Types of bacteria

Bacteria are classified according to the chemical composition of their cell walls. The cell wall confers a particular shape to each type (Figure 29). Bacteria multiply by cell division, which occurs approximately every 20 minutes.

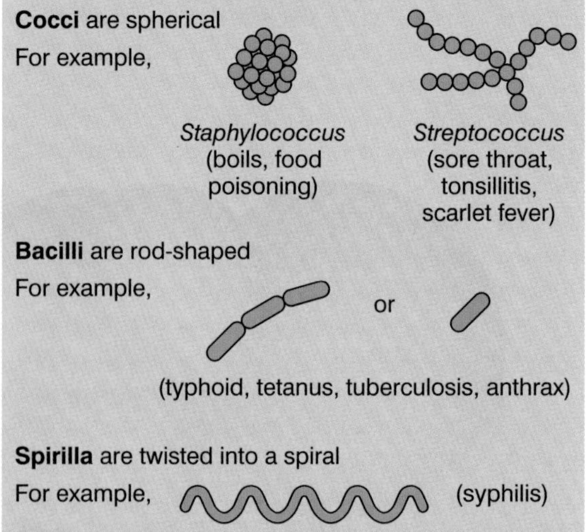

Cocci are spherical
For example,

Staphylococcus (boils, food poisoning)

Streptococcus (sore throat, tonsillitis, scarlet fever)

Bacilli are rod-shaped
For example,

or

(typhoid, tetanus, tuberculosis, anthrax)

Spirilla are twisted into a spiral
For example,

(syphilis)

Figure 29 Types of bacteria and their effect on the body.

Most bacteria are harmless and some are essential to us, such as those in the intestine which aid digestion. A minority, called *pathogens*, can cause infections of, for example, the lungs (pneumonia), brain (meningitis), heart (endocarditis) or bloodstream (septicaemia).

Penicillin first became available during the Second World War. Its use saved the lives of many injured people, who would otherwise have died from bacterial infections. It was hailed as a 'miracle cure'.

The discovery of penicillin occurred as a result of a chance observation. It proved to be an extremely valuable lead compound, and set the scene for the discovery and development of a vast range of **antibiotics**. These are compounds which selectively destroy disease-causing bacteria.

The production of penicillins on a large scale is an early example of biotechnology.

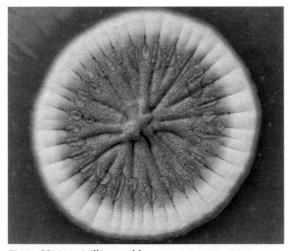

Figure 30 A penicillin mould.

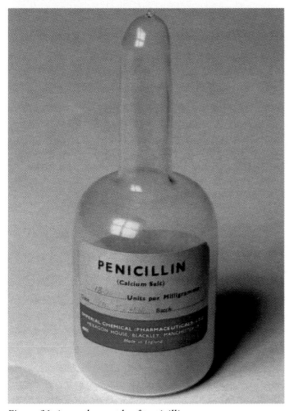

Figure 31 An early sample of penicillin.

The miracle cure

Natural materials have been used in many forms to treat infections in the past. The *Old Testament*, for example, describes the use of fungi and moulds to treat infected wounds.

In 1928, these remedies began to be shown to have a scientific basis, when Alexander Fleming noticed that a mould (*Penicillium notatum*) produced a substance that appeared to inhibit bacterial growth.

Accidentally, one of Fleming's experiments with bacteria had become contaminated with a mould. On return from a holiday, Fleming noticed that bacterial growth was restricted in the areas where the mould had developed. He deduced that the mould had affected the bacteria by producing chemicals which he called **penicillins** after the name of the mould.

Fleming also showed that the culture appeared to be non-toxic to animals. However, he did not go further to see whether penicillin was effective against infections in the animals. This is not surprising as the age of antiobiotics had not begun and penicillin was only considered as an antiseptic for local use.

Interest was revived in May 1940, when Howard Florey and Ernst Chain, in Oxford, showed that penicillin injections in mice were effective against a lethal streptococcal infection. Penicillin was later introduced into the clinic for use in humans, with dramatic consequences.

The antiobiotic era had now dawned. Fleming, Florey and Chain shared a Nobel Prize in Medicine for their work.

During the 1940s, penicillin was extracted in bulk from mould cultures, both in the UK and the US, and it became widely available.

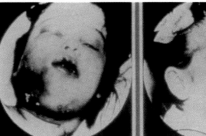

Figure 32 The first child to be treated with penicillin by Dr Wallace Herrell at the Mayo Clinic, US. The photo on the right was taken 4 weeks after treatment.

Getting moulds to do the work

At first, scientists relied entirely on the fermentation of moulds to make penicillins.

The mould *Penicillium notatum* makes several antibiotic compounds as products of its natural metabolism. These were isolated and called, for example, penicillin F, G, K and X.

Researchers found that the mould can be encouraged to produce just one penicillin, and that the type of penicillin produced can be altered, by changing the nutrient on which the mould is grown.

For example, growing the mould with compound I as the nutrient gives just penicillin G. This was the first 'miracle cure'.

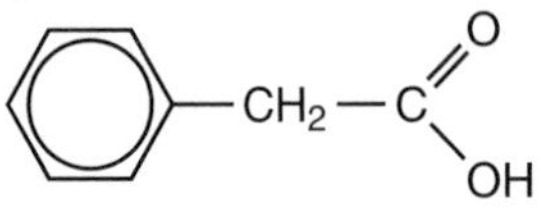

compound I

Unfortunately, penicillin G is only active against a limited variety of bacteria. Also, stomach acidity causes it to lose its activity, so it must be given by injection. Growing the mould with compound II gives the improved penicillin V, which is not so susceptible to attack by acids and can be taken orally.

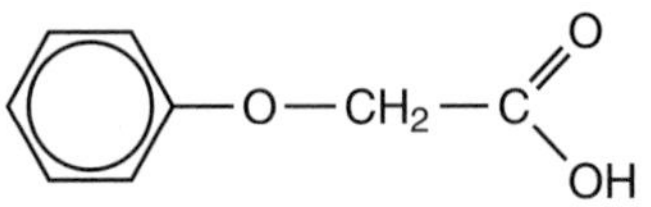

compound II

Improving on nature

When the National Health Service was set up in 1948, Beecham, which become SmithKline Beecham (SKB) in 1989, known then as the 'pills and potions' company, decided to look more closely at the preparation of antibiotic compounds.

A major breakthrough came in the late 1950s. While trying to make a new penicillin, their scientists found the penicillin 'nucleus', the core structural framework of the penicillin molecule. They isolated this compound, called *6-aminopenicillanic acid (6-APA)*, and showed that it has the structure shown below:

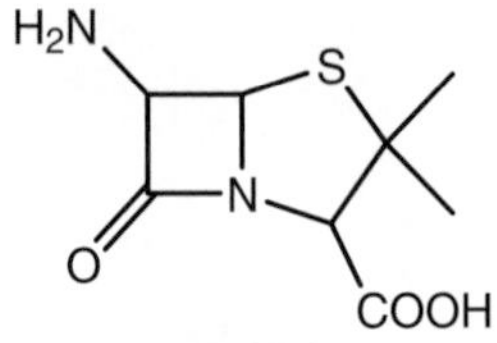

6-APA

This was an important discovery because it helped to make sense of the different types of penicillins. It also enabled scientists to make new penicillins partially by chemical techniques (ie, semi-synthetically), without relying completely on moulds to do the necessary reactions.

6-APA itself has little effect on bacterial growth, but if you add an extra 'side chain' to the amino (—NH_2) group you have a disease-curing penicillin. You can see the structure of a penicillin in Figure 33. The nutrient compounds I and II react with 6-APA to give different side chains.

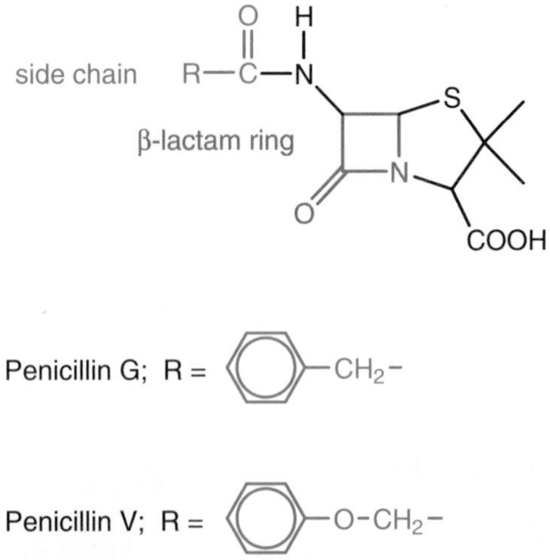

Penicillin G; R =

Penicillin V; R =

Figure 33 Structure of a penicillin – all penicillins have the same basic structure, only the R group varies.

The β-lactam ring

Penicillin contains a fused-ring system containing a nitrogen atom and a sulphur atom. The four-membered ring contains a **cyclic amide** group and is called a β-**lactam ring**.

The simplest β-lactam ring is

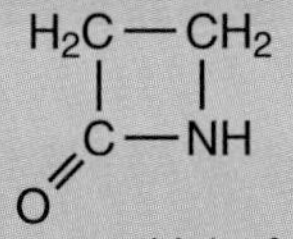

You can think of it as arising from the β-amino acid

$$H_2N - \overset{\beta}{C}H_2 - \overset{\alpha}{C}H_2 - COOH$$

by the —NH_2 group bending round and condensing with the —COOH group on the other end of the molecule.

β-Lactam rings are very sensitive to acids and alkalis. They react readily to give open-chain compounds. This is one of the reasons why it was so difficult to isolate and purify the first penicillins.

Research now focused on chemical reactions which would attach different side chains to 6-APA. An **acylation** reaction is used, as shown in Figure 34.

In **Activity MD5.1** you can make a semi-synthetic penicillin and test its antibacterial activity.

The different penicillins produced have different properties and ranges of antibacterial activity. A doctor decides which of the many penicillins available is the best one to use against the particular infection being treated.

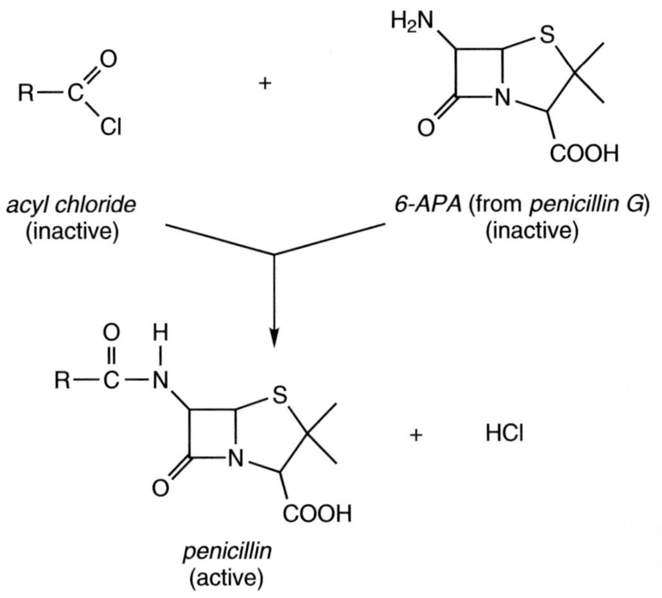

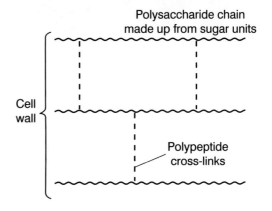

Making a so-called *semi-synthetic* penicillin involves making two biologically inactive molecules and clipping them together to make the active compound

One of the molecules is synthesised in the laboratory. This is the side chain in the form of an acyl chloride. The other is *6-APA*. This is obtained by treating naturally produced *penicillin G* with an enzyme to hydrolyse the amide linkage in the side chain

Figure 34 Preparation of a semi-synthetic penicillin.

ASSIGNMENT II

Use the 'tool-kit' of organic reactions you built up in **Activity MD3.1** to devise a synthesis for phenylpenicillin

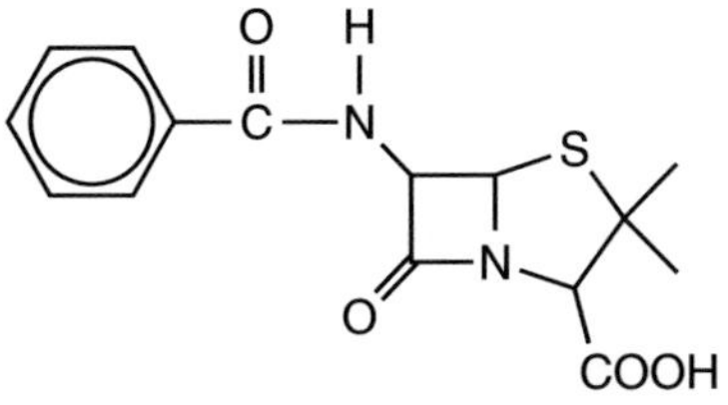

from 6-APA and ⬡—CH₂OH (3 steps).

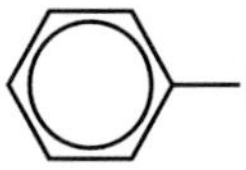

For each step, give the essential conditions and write a balanced equation for the reaction. You will carry out the last step of the synthesis in **Activity MD5.1**.

How penicillin works

Penicillin does not normally attack bacteria that are fully grown or in a resting state. Instead, it stops the growth of new bacteria by inhibiting the action of an enzyme responsible for constructing the cell wall.

A bacterium is protected by a rigid cell wall which is built up from a network of polysaccharide chains joined by polypeptide cross-links (see Figure 35). The shape of the penicillin molecule resembles that of the crucial amino acids used in the cross-linking process. The penicillin reacts with the enzyme which makes the cross-links and inhibits its action. This results in weakening of the cell wall.

The enzyme never satisfactorily completes the wall, and eventually the contents of the cell burst out and the bacterium dies.

The structure of the cell wall in bacteria is unique and different from the cell walls in plants and the cell membranes in animals. One of the amino acids used to make the polypeptide cross-links is D-alanine. Only its isomer, L-alanine, is found in humans. Penicillin is not toxic to humans and animals generally. It selectively targets bacteria.

Penicillin inhibits the action of the enzyme forming the cross-links

Figure 35 A simplified diagram of a bacterial cell wall.

Nature fights back

Sometimes, penicillin-resistant bacteria cause serious outbreaks of infections in hospitals, especially in surgical wards. These bacteria produce an enzyme called *β-lactamase*, which attacks the four-membered β-lactam ring and makes the penicillin inactive.

Chemists have responded by developing new side chains that make the penicillin molecule less susceptible to attack by this enzyme. Compounds produced include methicillin and flucloxacillin.

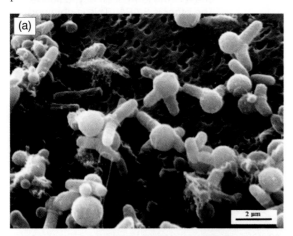

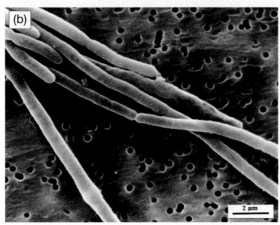

Figure 36 E. coli *being treated with amoxycillin: (a) 35 minutes after the dose and (b) 3 hours after the dose. The cell wall is unable to form normally.*

_____ **ASSIGNMENT 12** _____

New strains of bacteria appear as a result of *mutations*. These are spontaneous changes in the genetic material of the cell. They occur at random and are passed on when cells divide.

a Explain how penicillin-resistant strains of bacteria develop at the expense of susceptible ones.

b When you are prescribed a course of antibiotics, you should always finish the medicine, even if you feel better. Suggest a reason for this.

In **Activity MD5.2** you can make models of some penicillins and investigate how the structure of the side chain affects their antibacterial activity.

A group of scientists working at Beecham Research Laboratories (later SmithKline Beecham) tried an alternative approach to overcome bacterial resistance. In addition to the search for penicillins with new side chains, they reasoned that a compound might be found that would inhibit the action of the β-lactamase enzyme, leaving the penicillin free to carry out its normal antibacterial action.

After investigating many possible compounds, they came up with a natural product called *clavulanic acid*. It has a structure similar to that of a penicillin molecule and is recognised by the β-lactamase enzyme.

clavulanic acid

Clavulanic acid alone has virtually no antibiotic activity, but, when mixed with previously ineffective penicillins, can be used against pathogenic bacteria that produce β-lactamase, with excellent results.

The story of penicillin doesn't finish here. Work continues to produce an ever-widening range of 'magic bullets' – compounds which are targeted at specific bacteria in particular tissues of the body. Because bacteria are constantly changing, these compounds are in constant danger of becoming obsolete. Ways have to be found (like using the β-lactamase inhibitor, clavulanic acid) to prolong their usefulness by overcoming resistance.

_____ **ASSIGNMENT 13** _____

Compare the structure of clavulanic acid with that of the penicillin shown in Figure 32.

a What features, common to both molecules, might the active site of the β-lactamase enzyme recognise?

b Suggest how clavulanic acid might inhibit the action of the β-lactamase enzyme.

MD6 *Summary*

This unit has centred around the search for efficient medicines which are increasingly selective in the way they work. Once scientists understand the way in which a compound interacts with the human body, or with a bacterial cell, they can design new substances which fit selectively onto a receptor site or active site and bring about a desired effect. This means that medicines can be more effective and have fewer side-effects.

The unit began with a study of ethanol and its effect on the body. This allowed you to revisit earlier work on the reactions of alcohols and to learn about the reactions of aldehydes and ketones (formed when alcohols are oxidised). You saw how i.r. spectroscopy and g.l.c. are used to measure alcohol concentrations in the breath.

Central to the storyline was the concept of molecular recognition. The shape and size of a biologically active molecule are crucial to its action. Its structure and precise shape must be known because certain groups may need to be in specific positions to bind the molecule to a receptor site. Computer-generated graphics help chemists to investigate the interactions involved and are the basis of modern drug design.

Synthetic organic chemists play an important role in preparing new compounds and modifying existing ones. So alongside the storyline, the chemical ideas introduced you to organic synthesis and allowed you to revise the various functional groups and organic reactions which you have encountered throughout the course. These were organised into a 'tool-kit' of reactions which you can use to write schemes for converting one organic compound into another.

You also used your knowledge of spectroscopic techniques to determine the structure of organic compounds and to follow the course of a series of reactions.

Activity MD6 will help you to summarise what you have learned in this unit.

Figure 37 A 200 000 litre penicillin fermenter being lifted into position.

VISITING THE CHEMICAL INDUSTRY

Why a unit on the CHEMICAL INDUSTRY?

The chemical industry is a major contributor to our economy. All the principles of chemistry you learn in your chemistry course are applied in one way or another in the chemical industry. A study of the industry enables you to see the practical importance of chemical ideas.

Throughout this course, you are introduced to industrial applications of chemistry as part of the chemical storylines. In this unit, the underlying general principles of industrial chemistry are brought together and reinforced, if possible, by a visit to a particular chemical plant.

Good practical work in the laboratory helps you experience chemistry personally and helps you to understand the underlying chemical ideas. Likewise, visiting a chemical plant will enable you to learn at first hand about the way the chemical industry works and the visit will help you to understand the chemical principles used in the processes you meet.

Overview

In this unit you will have the opportunity to

- have contact with industry and the world of work
- improve your understanding of the relationship between industry and society
- see the industrial and social significance of the chemistry you are studying.

The visit will help you to bring together your ideas on

- chemical manufacturing processes
- how chemical principles can be applied in order to:
  optimise efficiency
  ensure safety
  minimise environmental damage
  minimise economic cost.

VISITING THE CHEMICAL INDUSTRY

VCI1 *Introduction*

The UK chemical industry is a major contributor to both the quality of our life and our national economy. It is the only major sector of UK manufacturing industry which makes a positive contribution to the UK's balance of trade with the rest of the world.

As well as chemists, chemical companies employ people of most other disciplines to fulfil their activities, including accountants, lawyers, linguists, physicists, biologists, mathematicians, computer programmers, chemical engineers and other engineers of many kinds (mechanical, civil, electronic and so on).

The UK chemical industry is the nation's third largest manufacturing industry as measured by the value added, or money generated, in converting basic raw materials to more sophisticated, advanced chemicals and end products. The largest sectors are: food, drink and tobacco; electrical and instrument engineering; and, equal third, chemicals together with paper, printing and publishing. The chemical industry is the largest export earner among the manufacturing industries in the UK and, in global terms, is the fifth largest chemical industry behind the US, Japan, Germany and France in terms of sales (Table 1).

This study of the industry enables you to see the practical importance of some of the chemical ideas you have met or will meet.

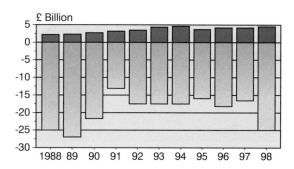

Figure 1 UK trade balance in chemicals and all other manufactured products, 1988–1998 (1 billion = 1×10^9). [UK Chemical Industry, CIA, 1999]

Country	Sales/billion US$
US	391.0
Japan	177.5
Germany	106.6
France*	78.0
UK	52.3
Italy	49.8
Belgium	36.2
Spain	30.4
The Netherlands	29.2
Switzerland	24.7

*Table 1 World chemical sales in 1998 (*excluding synthetic fibres). [Facts and Figures, CEFIC, 1999]*

Figure 2 The steam cracker for ethene production at BP Amoco Grangemouth.

Table 2 shows the ten largest chemical companies in the world. Although only one company has its head office in the UK, most have manufacturing plants here.

Company	Sales/ billion US$	Location of head office
Bayer	31.2	Germany
BASF	30.7	Germany
Merck	26.9	US
Hoechst Group	24.8	Germany
DuPont	24.8	US
Novartis	21.9	Switzerland
Dow	18.4	US
Roche	17.0	Switzerland
ICI	15.5	UK
Rhône-Poulenc	14.7	France

Table 2 The ten largest chemical companies (sales in 1998). [Facts and Figures, CEFIC, 1999]

The chemical business

The chemical industry earns its money by carrying out chemical conversions and selling products either for further chemical reactions or for formulation into final products. These final products cover a very wide range of uses, such as cleaners, paints, inks, medicines, pesticides, polymers, fertilisers, fuels, fuel additives, dyes, and soaps.

ASSIGNMENT I

Consider a typical school day. Make a list of 10 manufactured items that you usually come into contact with during the day between waking up and going to sleep.

Next to each item write down what contribution you think the chemical industry has made to its production. You may prefer to do this with a group of two or three other students.

Now sort your list of items into different categories of chemical industry, for example the polymer industry.

How many items do you have left which have not been dependent on a contribution from the chemical industry?

An important part of this unit is your visit to a chemical plant. **Activity VCI 1** will help you to get the most out of your visit.

VCI 2 *Research and development*

The phrase 'optimum conditions' is much used when discussing industrial processes (as indeed it is in the laboratory).

The phrase 'optimum conditions' is discussed in detail in **Chemical Ideas 15.3**.

When you have prepared a compound from some reactants, either you or your teacher will have decided beforehand what conditions to use. For example:

- what reagents to use
- how much of each reagent
- what temperature
- whether or not to use a catalyst
- how long to allow the reaction to proceed.

You probably did not have time to check whether or not you had the optimum conditions before you did the experiment. However, in industry, much research is carried out to determine optimum conditions. A number of stages can be identified in ironing out difficulties that might occur in transferring the process from the research laboratory to the industrial scale (Figure 3).

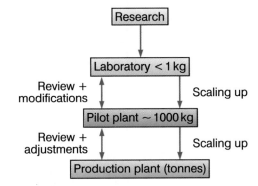

Figure 3 The search for optimum conditions requires several stages before the chemical plant is built.

In the laboratory, small-scale experiments are performed to see whether the manufacture of the compound will be feasible.

Increasingly larger scale tests are carried out to learn as much as possible about the reaction and the effects of changing conditions such as the catalyst or the concentrations of reagents (Figure 4).

Figure 4 Experiments in the laboratory.

At this point, experiments are carried out in a pilot plant, which is a scaled down version of what the full-scale production plant is anticipated to be. Here a significant quantity of product will be made. It is usually necessary to change from glass to steel vessels at this stage and the equipment resembles that used on a chemical plant rather than in the laboratory (Figure 5).

Figure 5 Experiments in a pilot plant.

Finally, the fateful decision (possibly involving the investment of hundreds of millions of pounds) to build the plant is taken (Figure 6).

Figure 6 A full-scale production plant.

At each of these stages, the information obtained is reviewed and modifications are made to the process. When scaling up experiments, industrial chemists must calculate the likely temperature changes as well as the masses of reactants to be used. On a larger scale, a large and rapid temperature rise can cause a reaction to run out of control, causing a fire or even an explosion.

Chemical Ideas 15.3 describes in more detail how the best conditions for the process are considered and decided.

VCI 3 *Building the chemical plant*

The equipment needed to carry out the chemical conversions, the **chemical plant**, is specifically built or adapted to manufacture a required product. Plants are constructed according to the chemical process being operated and their size depends on the demand for the product. The process may be operated in continuous or batch sequences.

Chemical Ideas 15.1 provides details of continuous and batch processes and what materials are used to construct the reaction vessels.

There are many important issues to address in deciding which product to make. For example:

- what is the size of the potential market?
- are there any competitors and, if so, can the company produce the chemical more profitably?
- will the product sell for a price which generates a profit for the company?
- what investment is required to manufacture the product?
- how does the product 'fit in' with the company's other activities?

The plant is the major resource allocated to the manufacture of a product. The economic viability of the product will depend on the returns generated by the capital committed to the plant. The marketing department will generate the necessary information to answer the above questions.

When the first set of questions has been answered, others arise, such as:

- what feedstock should be used?
- does the company have sufficient expertise to develop the process and run the plant?
- how will the plant be made safe for employees and those living nearby?
- how will the environment be protected?
- where should the plant be built?

After answering all of these questions, the company has the information needed to find what are the optimum conditions for making the chemical.

These questions do not simply apply to new plants. Very often, a company will be considering improvements to their existing plants. All the same questions apply and need answering. Further, a company may be manufacturing a chemical and be looking for an alternative process – one for example that is less damaging to the environment, perhaps by using less energy or producing harmless effluent.

The following considerations will be discussed in the rest of this unit:

- the people needed to run a chemical plant
- safety
- environmental issues
- the choice of location
- the choice of feedstock
- making a profit.

How feedstocks are transported is discussed in **Chemical Ideas 15.2**. The feedstocks derived from natural gas and oil are also summarised in this section.

Chemical Ideas 15.5 describes how the costs of a chemical process are calculated.

VCI 4 *People*

The UK chemical industry employs around 230 000 people from many subject backgrounds with many different kinds of expertise and skills. It is necessary, because of the nature of chemical processes, that those directly responsible (chemists, chemical engineers, process operators) are highly qualified and extremely well trained. They work together as a team, with each individual contributing according to his or her different expertise.

The chemical industry is, by and large, capital rather than labour intensive and the relative size of the workforce reflects this. Indeed a large plant, such as one making sulphuric acid, may be operated by only four or five people at a time. Computers and automation are used intensively (Figure 7).

The chemical industry invests heavily in research and development, particularly in pharmaceuticals. Thus, its research and development expenditure was about 7% of sales in 1996, compared to an average of 1.5% for other manufacturing industries.

_____ **ASSIGNMENT 2** _____

The list below shows some of the types of employee, in addition to chemists, chemical engineers and process workers, who might be required in a chemical plant.

Choose *four* from the list and suggest why they might be needed:

accountants   lawyers
linguists   physicists
biologists   computer programmers
engineers
– mechanical   – civil
– electrical   – production
– quantity
biochemists
safety advisors
cleaners
caterers
transport workers
construction workers
fitters
clerical workers
laboratory technicians
market researchers
economists
librarians.

Figure 7 Computers play a major part in the chemical industry, for example in controlling the temperature, pressure and other conditions for all the reactions in the plant.

VCI 5 *Safety matters*

One Saturday afternoon …

"The explosion in the Nypro factory at Flixborough, England, on 1 June 1974 was a milestone in the history of the chemical industry in the UK. The destruction of the plant in one almighty explosion, the death of 28 men on site, and extensive damage and injuries, though no deaths, in the surrounding villages, showed that the hazards of the chemical industry were greater than had been generally believed by the public at large. In response to public concern the government set up not only an inquiry into the immediate causes but also an 'Advisory Committee on Major Hazards' to consider the wider questions raised by the explosion. Their three reports led to far-reaching changes in the procedures for the control of major industrial hazards. The long-term effects of the explosion thus extended far beyond the factory fence."

This paragraph was written in 1994 by Dr Trevor Kletz, one of today's most experienced and distinguished safety advisors in the chemical industry worldwide.

Flixborough is in East Anglia and the plant was concerned with the production of starting materials for the eventual production of nylon. Liquid cyclohexane (a hydrocarbon) was being oxidised with air at about 150 °C and under a pressure of 10 atmospheres. The reaction took place in a series of six reactors, linked to one another.

The cyclohexane is oxidised to a mixture of an alcohol, cyclohexanol, and a ketone, cyclohexanone.

cyclohexane   cyclohexanol   cyclohexanone

Only a small proportion of the cyclohexane is oxidised and the unreacted feedstock is separated from the products and recycled. The product mixture, a mixture of the alcohol and the ketone, is then converted into the starting material for the manufacture of nylon-6.

One of the reactors was developing a crack and was removed for repair. To keep production going, a temporary pipe was connected between the reactors on either side, so that five reactors could be kept in service. This worked well for two weeks until a slight pressure built up, enough to cause the pipe to twist and tear.

The temperature of the liquid cyclohexane, 150 °C, was way above its boiling point. Perhaps as much as 50 tonnes of liquid escaped in a minute and a huge cloud of flammable vapour spread over the plant. It was ignited, perhaps because there was a furnace many yards away, and one of the worse vapour cloud explosions ever was the consequence.

Dr Kletz has this quotation in his book on *Learning from Accidents in the Chemical Industry*:

"It is the success of engineering which holds back the growth of engineering knowledge, and its failures which provide the seeds for its future development."

Following the accident, much more emphasis was placed on reducing the amount of reactant or product that can escape at any one time. It also made chemical engineers and chemists investigate why the crack in the reaction vessel happened at all. It made engineers reconsider the design of temporary pipes. Further, it made engineers consider how the plant should be laid out to prevent loss of life when an accident does occur.

It made everyone in the chemical industry even more safety conscious than they had been and many lessons were learnt from the accident.

… and now

The accident at Flixborough was a defining moment; the realisation that, unless there was a system for safety which was carefully thought out, there could be other accidents which could be as, or even more, serious.

Figure 8(a) The Flixborough works before the disaster. View from the south-east. (b) The works after the explosion, seen from the north. Section 25A is the area where the oxidation of cyclohexane took place. R2525 is the only one of the six reactors remaining.

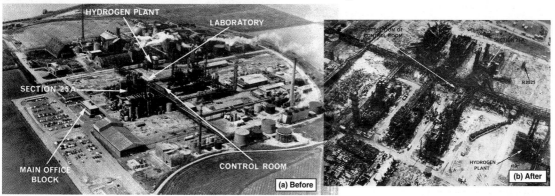

Personal safety is rated very highly and on a typical visit to a production site you may see eye-baths, showers, toxic gas refuges, breathing apparatus, emergency control rooms and, on larger sites, the company's fire brigade, ambulance service and a well-equipped medical centre with its own qualified doctors and nurses.

Some reagents used by chemical companies could become a hazard to people living near the plant in the event of an accident. This particularly applies to poisonous gases or volatile liquids.

Where a plant requires the use of a reagent such as chlorine, the company works with the local authority and emergency services. Emergency procedures are carefully planned and rehearsed at regular intervals.

Leaflets and, possibly, a video will brief people what to do if the plant siren sounds continuously. The siren is tested at monthly intervals and this test ensures everyone recognises the warning. The usual procedure for an emergency is for everyone to go indoors and close all windows and doors until the all clear is given.

Legislation and planning are essential but the key factor is for everyone on the chemical plant to recognise that it is in their interest to work safely. Progress is reflected in the number of days lost in the industry through accidents. This has been reduced by more than 50% over the last 10 years and is now about 1 accident (involving 3 or more days absence from work) every 300 000 hours worked.

Safety legislation

Safety is a major consideration in all operations in the chemical industry. All aspects of safety are affected by national and European Community legislation. In particular, the Health and Safety at Work Act in the UK places responsibility for health and safety with the employer.

ASSIGNMENT 3

a What other legislation do you know about that affects the safe handling of chemicals?

b Does this legislation apply to the work which you do in your study of chemistry?

An analysis of all possible hazards and an examination of safety are applied to any proposed project to build new or modified chemical plant. A Hazard and Operability Study (HAZOP) is a systematic procedure frequently used to carry out such an analysis. Every valve, pipe, vessel, pump, etc, is examined and the risk associated with failure is assessed and minimised by design.

Figure 9 Protective clothing in the chemical industry.

Safety considerations play a crucial part in deciding the plant layout. Legislation requires specific design features to minimise risks of uncontrollable reactions and undesirable emissions. The designs of new plants have to demonstrate that the possible risks have been minimised and that emissions or leaks will be prevented.

On a site, there are people who are responsible for considering specific aspects of safety and for ensuring that every person who works on the site or visits it constantly acts safely and is aware of all possible hazards. Training in the safe operation of the plant is very important.

Figure 10 Safety training at BP Amoco Grangemouth.

VCI 6 *Environmental issues*

The environment is a major factor for chemical companies. It is not acceptable to allow harmful substances to escape into the environment. The resultant pollution which occurs if there is an accidental escape of materials from a plant or disposal of untreated waste products not only damages the environment but also jeopardises the future of the company itself.

Figure 11 Manufacturing chemicals in Widnes, in the 1890s ...

There is an increasing quantity of legislation (local, national, European and international) which regulates the performance of chemical operations with respect to environmental issues.

Waste generation, treatment and disposal are major issues in process development. Minimising waste by getting the right conditions has already been discussed and the adoption of new clean technologies is helping in this area. However, chemical processes will always produce some waste which has to be dealt with.

In the past, waste has been dumped into the nearest convenient place (the atmosphere, old quarry, river, lake, sea). Alternatively, it was contained in purpose-built ponds or tips which have caused problems with toxic materials leaching out into nearby streams and waterways. These methods are no longer acceptable and are becoming illegal unless special permission is given. Figures 11 and 12 show how the chemical industry has changed over the last 100 years.

Figure 12 ... and Widnes now.

Waste must now be treated and can only be disposed of when in a state which is not harmful to the environment. Liquid waste from chemical works has to meet legal requirements on such things as pH and metal ion content before being released into natural waterways or sewage systems. It, therefore, must be treated appropriately, eg to neutralise any acid. Water containing organic waste cannot be discharged into rivers or canals if it would significantly reduce the oxygen content of the water, causing fish to die from lack of oxygen.

Figure 13 To remove waste products from the effluent water, the effluent is treated very carefully before it is discharged into a river, a canal or the sea.

Gases which contain contaminants are purified by bubbling them through neutralising solutions to remove soluble contaminants; particles of dust can be removed by filtration or other methods.

Sulphur dioxide, for example, provides two of the most successful examples of the removal of a potential pollutant – thus improving our environment.

In the manufacture of many metals, for example, copper, lead and zinc, the ores used are sulphides. They are roasted in air to form the oxide as one of the first stages in their manufacture. For example:

$$2PbS(s) + 3O_2(g) \rightarrow 2PbO(s) + 2SO_2(g)$$

The sulphur dioxide is not allowed to escape but is led away from the furnace to a nearby plant which converts it into sulphuric acid by the Contact process.

$$2SO_2(g) + O_2(g) \rightarrow 2SO_3(g)$$
$$SO_3(g) + H_2O(l) \rightarrow H_2SO_4(aq)$$

High grade sulphuric acid is produced.

Sulphur is a contaminant in many coals, so when coal is burnt to generate electricity, large amounts of sulphur dioxide are produced.

Longannet Power Station, the second largest power station in the UK and one of the largest in Europe, burns Scottish coal. The coal contains less than 1% sulphur, which is less than most coals mined in the UK.

Even so, when this coal is burnt, large quantities of sulphur dioxide are put into the atmosphere and there have been many experiments to reduce this amount.

Scientists at ABB Environmental, working with scientists at the power station, have developed a remarkable process in which the flue gases are washed with sea water. Sulphur dioxide dissolves in water to form sulphite ions, $SO_3^{2-}(aq)$ and $H^+(aq)$ ions.

$$SO_2(g) + H_2O(l) \rightleftharpoons SO_3^{2-}(aq) + 2H^+(aq)$$

Sea water is naturally alkaline. The $HCO_3^-(aq)$ ions present react with $H^+(aq)$ ions to form carbon dioxide and water.

$$HCO_3^-(aq) + H^+(aq) \rightleftharpoons CO_2(g) + H_2O(l)$$

Air is then passed through the solution to oxidise the sulphite ions to sulphate ions:

$$2SO_3^{2-}(aq) + O_2(g) \rightarrow 2SO_4^{2-}(aq)$$

At the end of the process, the water, containing a very dilute solution of sulphate ions, is discharged into the nearby river and hence to the sea. The carbon dioxide is discharged into the atmosphere, but, in relative terms, these are small amounts compared to those discharged by combustion of the coal. Discharging carbon dioxide is much more environmentally favourable than discharging sulphur dioxide.

Figure 14 Longannet Power Station.

Other gases which cause great problems in the atmosphere are the oxides of nitrogen, NO and NO_2, known collectively as NO_x. Nitric oxide, NO, is discharged by cars, lorries, factories, power stations, in fact, from anywhere where air is heated to high temperatures and nitrogen and oxygen react:

$$N_2(g) + O_2(g) \rightarrow 2NO(g)$$

Subsequently, the NO is oxidised to nitrogen dioxide, NO_2, in the cool atmosphere.

The oxides are removed catalytically in car exhausts (see **Developing Fuels** storyline, Section **DF6**). However, this process is too expensive for power stations, as very large quantities of gases must be treated.

An ingenious process is being tried at Longannet. Natural gas (CH_4) is fed into the flue gases, which contain nitrogen oxides, above the flames of the burning coal. The methane is oxidised by the oxides of nitrogen to form nitrogen. The overall reaction can be represented as:

$$CH_4(g) + 4NO(g) \rightarrow 2N_2(g) + CO_2(g) + 2H_2O(g)$$

Excess natural gas is then burnt off, producing more heat which is used in the overall production of electricity. The technique, known as *gas reburn*, is being tested on the largest scale ever attempted.

VCI 7 *Location*

The chemical industry in the UK grew up around the source of its major raw materials or near the customer industry which it served or in a location with good transport links.

With increasing specialisation in the UK and a much improved communication network, these factors are no longer the primary considerations in locating manufacturing sites.

New chemical plants are often built near to existing works. This may be because a district specialising in the manufacture of a particular product is more likely to be able to provide the skilled labour needed for new developments. That's why there are specialist steel producers in Sheffield, a traditional centre for steelmaking.

It may be that the feedstock for the new process is already being produced at the existing site. There may be less opposition to building a new plant at an existing works than to the development of a new site.

In addition to these considerations, it may save money if facilities such as a canteen, medical centre and administration block are shared between existing and new plants.

However, many well-established locations are in built-up areas where the increasing concerns of the local population give rise to problems in extending plant or building new facilities.

Increasingly, chemical companies are seeking more remote locations for their operations. Then another problem arises for they are then accused of spoiling the countryside.

The actual site chosen should be served by a good communications network, as building this from scratch would add significantly to the cost of opening the plant. Road, rail and water are all important means of transport, and deep sea access is useful for the import and export of bulk materials, such as oil. Ideally the site would be level, free from danger of subsidence and have scope for further expansion. Waste treatment and disposal are always concerns to be borne in mind.

The story of Paraffin Young

Just west of Edinburgh, about 20 minutes by train or car, lies the site of one of the world's most successful chemical industries, Grangemouth.

Grangemouth looks the perfect location. It is well away from any houses, just in case of serious accidents, yet near enough to cities and towns to allow the many thousands of people who work there to live within a comfortable distance. The site is convenient to a major road network, to allow supplies to be delivered and for its products to be transported away. It is near the edge of the North Sea to allow natural gas and oil to be piped ashore from one of the world's most profitable sources. A triumph of planning, an example for the textbook? Well, the answer is equivocal. It is yes and – alas – no.

Given its proximity to the North Sea oil and gas fields, and that the plant is as modern as anywhere in the world, one might imagine that Grangemouth was chosen as the site at the start of the growth of the chemical industry based on oil – the petrochemical industry. But it can be traced back much further – to 1851, when Queen Victoria had been on the throne for only 14 years and had another 50 to go, when Dr James Young found oil seeping from coal in the area (Figures 15 and 16). He set up a company based on heating the coal gently to collect the crude oil. He then distilled the oil into several fractions: one was 'lighting oil', which we would now call paraffin, another was an oil for lubrication, and another was a solid wax.

Figure 15 Dr James 'Paraffin' Young (1811–1883).

Known as 'Paraffin' Young, his work was copied by many others and within 15 years there were 120 similar companies in the area. However, their market was then taken away by the oil industry in the United States, which had large reserves of oil found in states such as Pennsylvania and Texas.

YOUNG'S PARAFFIN LIGHT & MINERAL OIL COMPANY. LIMITED.
ADDIEWELL WORKS. WEST CALDER. SCOTLAND.

Figure 16 Paraffin' Young's company.

By 1920, only six companies had survived so, to make themselves competitive, they formed a single company 'Scottish Oils'. Eventually this became part of one of the largest oil companies in the world, BP Amoco. It was Scottish Oils that saw the potential of the Grangemouth site as a refinery, distilling crude oil imported from the Middle East (principally Iran, then called Persia) into its various fractions (see **Developing Fuels** storyline, Section **DF3**).

Grangemouth is near to the coast and the smaller tankers could discharge the oil nearby. Later, large tankers had to use Finnart, a port on the west coast of Scotland which has a deep water anchorage, and a pipeline was laid to Grangemouth. The sea, road and rail links ensured that the products could be sent away easily.

In the Second World War, it was impossible to get enough oil from the Middle East, and Grangemouth closed down. It reopened again after the war, but political uncertainty in the Middle East in the 1950s reinforced the obvious conclusion that other sources of oil were needed. Oil and gas is now supplied by the great oilfields of Alaska and the North Sea (Figure 18).

Not only did coal supply Paraffin Young with oil, it was the basis of the British chemical industry until the Second World War. Natural gas and oil began to supersede coal but it wasn't until the late 1950s and 1960s that the huge growth in the petrochemical industry began. With such a good site as Grangemouth, it was natural to build the chemical plants adjacent to the refinery. Similar developments in the UK occurred in Cheshire (Shell) and near Southampton (Esso).

ASSIGNMENT 4

Use a road map (preferably scale 1:250,000) to sketch the area around Grangemouth, showing Edinburgh, Dunfermline, Stirling and Glasgow. Draw in major roads, railways and pipelines.

Use your map, together with information in this section and in **Chemical Ideas 15.2**, to illustrate that the Grangemouth site really does have adequate road, rail and sea connections, and has the ability to manufacture and distribute chemicals.

Figure 17 An aerial view of BP Amoco Grangemouth – the visible emissions are water vapour from cooling towers.

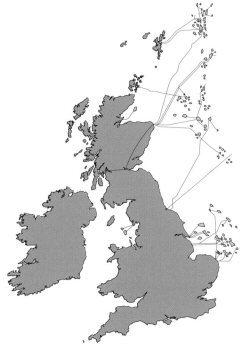

Figure 18 Bringing the oil and gas ashore. The lines show the pipelines from the fields.

Activity VCI 7 is a case study of the birth and development of a chemical manufacturing plant in Cheshire. You will be able to apply some of the ideas you have met in this unit, and revise some basic chemistry in the process.

ASSIGNMENT 5

Consider the company you have visited.

a Why is it on its present site?

b What are the constraints for further growth and employment in the existing area?

c What other locations in the UK or the rest of the world might be appropriate?

d What would the implications of a move be on
  i the company?
  ii the local economy?
  iii the local environment?

VCI 8 *Choosing a feedstock*

The reactants which go into a chemical process are called the **feedstock**.

Feedstock is produced from the raw materials. Sometimes there may be more than one possible raw material to choose from. The choice will normally be decided by cost and suitability for the process.

You can find out more about raw materials and feedstocks in **Chemical Ideas 15.2**.

Ethanoic acid is an interesting example. Over the last 100 years, the feedstock has changed several times. The main source of ethanoic acid used to be from the air oxidation of ethanol (with the ethanol coming from the fermentation of natural products). An example of this is the air oxidation of wine into vinegar. By the 1950s, in the UK, the main feedstock was produced from oil (in particular butane and naphtha).

Butane or naphtha were dissolved in a solvent and heated under pressure at about 200 °C. A range of products was produced including methanoic, ethanoic and propanoic acids. Although the yield of ethanoic acid was small, it was an economic process as the feedstock was relatively cheap. However, it became increasingly difficult to dispose of the by-products.

Now, ethanoic acid is made from methanol and carbon monoxide, using a rhodium/iodine catalyst:

$$CH_3OH(l) + CO(g) \rightarrow CH_3CO_2H(l)$$

It produces a pure product from a relatively cheap feedstock. (The methanol is produced from carbon monoxide and hydrogen.)

ASSIGNMENT 6

There are five ways of manufacturing phenol (C_6H_5OH). Four of them start from benzene (C_6H_6) and one from methylbenzene ($C_6H_5CH_3$).

Methylbenzene is cheaper per tonne than benzene and this would seem to give an advantage for the route starting from methylbenzene.

a Calculate the mass of
  i benzene
  ii methylbenzene
  needed to manufacture 1 tonne of phenol.
  (The relative atomic masses you need are
  A_r: H, 1; A_r: C, 12; A_r: O, 16)

b Explain how your answer to part **a** shows that judging the most economical raw material is not always a straightforward matter of comparing price per tonne.

Changing the feedstock allows for products to be tailor-made. An example is the production of poly(ethene). High density poly(ethene) (hdpe) is produced using a Ziegler–Natta catalyst (for example, $TiCl_4$ and triethylaluminium, see **The Polymer Revolution** storyline, Section **PR3**).

The same reactor can be used to make linear low density poly(ethene)s (lldpe) (see the **Designer Polymers** storyline, Section **DP6**), where the feedstocks are mixtures of ethene and hex-1-ene of varying composition. It is not necessary to shut down the plant when changing from making high density to low density poly(ethene). As the mixture of reactants is changed, so the end product changes and can be collected separately. This is an example of how versatile modern chemical plants are becoming.

Changing the feedstock provides ingenious ways in which flexibility can be incorporated in the chemical industry. For example, both ethene and propene are needed in large quantities to make a variety of polymers, such as poly(ethene), poly(propene), poly(chloroethene) (PVC) and poly(phenylethene) (polystyrene). The demand for each of these can change suddenly and thus a company needs to be able to alter the proportion of ethene to propene that it produces.

Ethene and propene are made by heating alkanes in the presence of steam at about 800 °C–900 °C, a process known as **steam cracking** (see Figure 2). By changing the feedstock as illustrated in Table 3, the proportions of the different products are changed. It is also possible to change the proportion of different products from one feedstock by changing the conditions under which the feedstock is reacted. You can read more about the cracking of alkanes in the **Developing Fuels** storyline, Section **DF4**.

ASSIGNMENT 7

Your company is using naphtha as a feedstock. There is a sudden increase in demand for poly(ethene). Which feedstocks could be used instead of naphtha to help you meet this increase in demand? (Use the data in Table 3.)

Table 3 The percentages of products on cracking various feedstocks.

Product	Feedstock				
	Ethane	Propane	Butanes	Naphtha	Gas oil
hydrogen	5–6	3	1	1	1
methane	10–12	26–28	15–25	13–18	10–12
ethene	75–80	40–45	20–30	25–37	22–26
propene	2–3	14–18	15–25	12–16	14–16
C_4 alkenes	4–6	4–6	15–24	6–10	6–11
others	1	4–6	5–14	22–36	35–44

VCI 9 *Making a profit*

Chemicals have to be sold at a high enough price to make a profit. This profit benefits society in a number of ways:

- in the form of taxes used for public expenditure on, for example, roads, schools or hospitals
- dividends to shareholders who have lent the company money by investing in it
- money for the company to invest in the future, through new ideas, new research, new developments and new plant.

Without this cycle of investment, followed by research and development of new or improved products, the company would gradually cease to be competitive and would eventually go out of business.

It is of paramount importance to determine the optimum conditions for the chemical process to obtain the best yield at minimum cost. The factors which are likely to affect the yield are:

- temperature
- concentration (or pressure for a gas-phase reaction)
- catalyst.

These factors will also affect the rate of reaction. Often a compromise has to be reached between the yield and the rate at which the product is obtained. The Haber process for the manufacture of ammonia from nitrogen and hydrogen is a good illustration of this (see **Aspects of Agriculture** storyline, Section **AA3**).

It would be worthwhile to look again at **Chemical Ideas 15.3** for a discussion on the best conditions for a chemical process.

It is also important to obtain feedstocks as cheaply as possible. This can cause a conflict between a company's interest and what is perceived as a national or social interest. One example, often debated, is whether to use gas or oil to generate electricity in the chemical plant or to use coal, which may be more expensive. Coal has the advantage of being more plentiful in the long term and keeps the coal mining industry alive.

And how should we deal with the countryside? Should we quarry limestone from our beautiful countryside, thereby scarring the landscape and creating traffic, or buy it from abroad? If it can be obtained elsewhere with less damage to the landscape and transported by sea, it may be more expensive but our landscape will be preserved. Costs will rise and we may well price ourselves out of a market, thereby creating unemployment. There are simply no easy answers.

Industry is increasingly trying to save materials and energy, thereby improving the environment and cutting costs. An increasingly important issue is to use any other products of the main reaction (known as co-products) and by-products (from side reactions) as effectively as possible. The overall efficiency of a process is eventually seen in its costs.

You can read more about 'Becoming even more efficient' in **Chemical Ideas 15.4**.

An introduction to the consideration of costs in the chemical industry is given in **Chemical Ideas 15.5**.

VCI 10 *Summary*

In this unit you have brought together and extended many aspects of the chemical industry that you have met in other units of the course.

The unit began by demonstrating the importance of the chemical industry to the many manufactured items we use in our daily lives as well as to our national income.

You then went on to consider the process of research and development of a chemical from the laboratory stage, through pilot plant to the full size plant. These stages are needed in order to establish the optimum conditions for manufacturing the desired product.

To appreciate what is involved in the design of industrial processes you needed to understand and apply ideas about enthalpy changes, factors that affect rates of reactions and factors that affect the positions of equilibria. You also needed to understand the characteristics of batch and continuous processes and the problems posed by scaling-up processes.

When it comes to building the plant, other issues have to be addressed – such as the capital investment required, potential market and competition and likely profit. A new plant is very expensive and mistakes could lead to the collapse of the company.

Other decisions required include:

- what people are required to run the plant
- how the plant will be made safe and friendly to the environment
- where to build the plant
- which feedstock to use.

A visit to a chemical plant allowed you to see how these issues have been addressed in a particular situation.

Activity VCI 10 will help you to summarise what you have learned in this unit.

INDEX

absorption 204

absorption capacity of soils 204

absorption spectra
  of atmospheric gases 84, 85
  infrared 109–10
  of stars 12–13

accidental discoveries: *see* chance discoveries

accidents, chemical plant 289

ACE (angiotensin-converting enzyme) inhibitors 275–8

acetylsalicylic acid 114
  see also aspirin

acid deposition ('acid rain') 57, 244

acids, weak 259

activation enthalpies 70, 160–1

acylation, of 6-APA 280–1

acyl chlorides 124, 143

addition polymers 92, 94, 98

adenine 16, 145, 147

adrenaline 270, 271
  compounds mimicking 271–2

adsorption 204

aerobic respiration 79

aerosol propellants 72, 76

agonists 272–3

agriculture 182–206

agrochemicals 50, 200–5

alanine 142, 281

alcohol (ethanol)
  breath testing for 265–7
  concentration in blood 264–7
  as drug 264–70
  effect on nerve cells 269–70
  as food 21, 263
  units 264

alcohols
  as fuels 21, 31, 32, 39–40
  manufacture of 39, 40

Alcolmeter 265–6

Aldehyde Green (dye) 223

algae, marine 238, 244, 245
  see also seaweed

alginate 238

Alizarin (dye) 224–6, 229

alkanes
  combustion of 20–1
  cracking of 29–30
  isomerisation of 27–8
  in petrol 23, 27, 32, 38
  in crude oil 23
  reforming of 28–9

alkenes
  manufacture of 24, 295
  see also ethene; propene

alkylaluminium catalysts 97, 98, 100

alkynes 102

alloys 165

copper-containing 53, 60, 165

iron-containing 165
  see also steel

polymer alloys 131

aluminium
  alloys containing 165
  in clay minerals 188, 189, 191
  in Earth's crust 184
  in soil 191
  in steelmaking 169
  reforming of 28–9

aluminium oxide catalyst 28

amide group 123

amides 123, 143

amines, reaction, with acids 123
  sequence in proteins 141, 143–50

amino acids 140–3
  coding in mRNA 144, 146
  essential 141
  formation in space 15–16
  general structure 141
  in insulin 140–1, 156–7
  optical isomers 143
  in proteins 144, 147, 153, 277, 282
  representation in peptides 143
  sequence in proteins 141, 143–50

α-amino acids 140, 141

4-aminobutanoic acid: *see* GABA

6-aminopenicillanic acid 280

ammonia
  lost from soil 195
  manufacture of 178, 196–8, 296
  as refrigerant 75

ammonium ions
  oxidation in soil 194
  in soil 184, 190, 193

ammonium nitrate(V) fertiliser 199

amoxycillin, effect on *E. coli* 282

anaemia 4

anaerobic organisms 64, 79

analytical chemists
  pigment identification by 216, 220–1
  steel composition measured by 168

angiotensins 275
  action of ACE 276
  see also ACE inhibitors

aniline: *see* phenylamine

aniline dyes 221–2

antagonists 272–3

Antarctica
  carbon dioxide in ice 87
  deep sea water current 249
  ozone hole 73–4, 75

anthracene, conversion to Alizarin dye 225–6

antibiotics 279
  see also penicillin

anti-codons (in tRNA) 146

anti-knock additives 27, 50

antiseptics 159

Apert, Nicolas 174

aramids 128–130

arenes (aromatic hydrocarbons) 24
  in petrol 38
  see also benzene

argon
  in atmosphere 64, 82
  in steelmaking 170

aromatic polyamides 128

Arrhenius, Svante 82

arsenic(III) oxide 212

asbestos 187

aspartame 141

aspirin 114
  patent for 114
  reason for name 114
  safety aspects 116
  soluble form 115
  synthesis of 113–14
  uses 116–17
  ways of buying 116

asthma 271
  drugs to treat 271–2

atactic polymers 98

Atlantic Ocean, water currents in 249, 250

atmosphere 62–91
  build-up of oxygen 60, 64, 258
  changes over time 60, 64, 254, 258
  composition 64, 82
  structure 63
  as sunscreen 66
  see also stratosphere; troposphere

atmospheric pollution 64
  by primary pollutants 33, 34, 57
  by secondary pollutants 33, 34
  by vehicle emissions 20, 32–3, 35–6, 71
  modelling of 35
  monitoring of 35, 73–5, 84
  simulation experiments 35

atomic emission spectroscopy 168, 215, 220
  see also laser microspectral analysis

atoms, number of: *see* mole

auto-ignition of hydrocarbons 26, 27

azo dyes 227–8, 230

azo pigments 227

azoxystrobin (fungicide) 205–6

azurite (copper mineral) 54

bacilli (bacteria) 278

bacteria
  cell wall structure 281
  denitrification by 194
  digestion of crude oil by 151
  leaching of minerals by 59

CONTRIBUTORS

First Edition

Many people from schools, colleges, universities, industry and the professions have contributed to the Salters Advanced Chemistry course. They include the following:

Central Team

George Burton	Cranleigh School and University of York
Margaret Ferguson (1990–1991)	King Edward VI School, Louth
John Holman (Project Director)	Watford Grammar School and University of York
Gwen Pilling	University of York
David Waddington (Chairman of Steering Committee)	University of York

Project Secretaries

Marilyn Baldwin	University of York
Annie Christie	University of York

Associate Editors

Malcolm Churchill	Wycombe High School
Derek Denby	John Leggott Sixth Form College, Scunthorpe
Frank Harriss	Malvern College
Miranda Stephenson	Chemical Industry Education Centre
Brian Ratcliff	Long Road Sixth Form College, Cambridge
Ashley Wheway	Oakham School

Advisory Committee

Dr Peter Doyle	Zeneca Group
Dr Tony Kirby, FRS	University of Cambridge
Professor The Lord Lewis, FRS (Chairman)	University of Cambridge
Sir Richard Norman, FRS	University of Oxford
Mr John Raffan	University of Cambridge

Steering Committee

Andrew Hunt	Association for Science Education
John Lazonby	University of York
Peter Nicolson	University of York
John Raffan	University of Cambridge
David Waddington (Chairman)	University of York

Sponsors

Many industrial companies have contributed time and expertise to the development of the Salters Advanced Chemistry course. The work has been made possible by generous donations from the following.

The Salters Institute of Industrial Chemistry
The Association of the British Pharmaceutical Industry
BP Chemicals
British Steel
Esso UK
Zeneca Agrochemicals
The Royal Society of Chemistry
Shell UK

Trial schools and colleges

The course underwent full trials in 1990–1992. Both teachers and students involved in the trials contributed essential feedback and comment.

Balby Carr School, Doncaster
Neil Braund and David Hill

Belle Vue Girls' School, Bradford
Ian Metcalf and Lynne Yates

Bushey Hall School, Watford
Neil Bramwell

Franklin Sixth Form College, Grimsby
Adrian Birks and David Whittaker

Greenhead College, Huddersfield
Neil Heeley

Islington Sixth Form Centre, London
Linda Taylor

John Leggott Sixth Form College, Scunthorpe
Derek Denby

King Edward VI College, Totnes
David Waistnidge

Long Road Sixth Form College, Cambridge
Brian Ratcliff

Malvern College
Frank Harriss

Matthew Humberstone School, Cleethorpes
Brian Stewart

Oakham School
Ashley Wheway

Oundle School, Peterborough
Gerald Keeling and Robin Hammond

Reigate College
Mike Shipton and Ann Hubbard

Sir Graham Balfour School, Stafford
Fiona Gerrard, David Clark and Chris Martin

Solihull Sixth Form College
Sue Howes and Bill Russell

Tideway School, Newhaven
Judith Ost

The Trinity School, Nottingham
John Dexter

Tuxford Comprehensive School, Nottinghamshire
Geoff Lloyd

Watford Grammar School
Paul Hayman and David Russell

Wigan and Leigh College
Chris Wright

Wilberforce Sixth Form College, Hull
Graham Harrington and Chris Conoley

Worcester Sixth Form College
Barry Key and Margaret Coles

Wycombe High School, Buckinghamshire
Malcolm Churchill and Veronica Aylward

Assessment advisers

Oxford and Cambridge Schools Examination Board
Howard King
Charles Newbould
John Noel

Revisers

Professor John Holloway	University of Leicester
Dr Ann Hubbard	Reigate College
Dr Richard Jones	University of East Anglia
Mr John Raffan	University of Cambridge

Authors

Susan Adamson	Poynton County High School, Cheshire
Mary Aitken	Redland High School for Girls, Bristol
Terry Allsop	University of Oxford
Eileen Barrett	Royal School of Mines, Imperial College, London
Peter Battye	Garforth Comprehensive School, Leeds
Tim Brosnan	Institute of Education, University of London
George Burton	Cranleigh School and University of York
Ann Daniels	Samuel Ward Upper School, Haverhill
Derek Denby	John Leggott College, Scunthorpe
David Edwards	Huntington School, York
Margaret Ferguson	King Edward VI School, Louth
John Garratt	University of York
Frank Harriss	Malvern College
Paul Hayman	Watford Grammar School
John Holman	Watford Grammar School
Andrew Hunt	Association for Science Education
Roland Jackson	ICI
Glyn James	Christ's Hospital, Horsham
Colin Johnson	Techniquest
Mary Beth Key	University of York
John Lazonby	University of York
Alastair MacGregor	The Leys School, Cambridge
Elizabeth Maltman	St Lawrence College, Thanet

Miranda Stephenson	Chemical Industry Education Centre
Peter Nicolson	University of York
Stuart Nuttall	Bablake School, Coventry
Gwen Pilling	University of York
John Raffan	University of Cambridge
Judith Ramsden	University of York
David Saunders	The National Gallery, London
Mike Shipton	Reigate College
Neil Smith	The Meden School, Mansfield
Kris Stutchbury	Poynton County High School, Cheshire
Tony Travis	Sidney Edelstein Centre, The Hebrew University of Jerusalem
David Waddington	University of York
John Whitmer	Western Washington University, Bellingham, USA
David Whittaker	Franklin Sixth Form College, Grimsby
Michael Withers	Royal Society of Chemistry
Gordon Woods	Monmouth School
Allan Woolley	University College School, London

Expert advisers

Academic consultants

Professor B C Gilbert	University of York
Dr R J Mawby	University of York

Unit advisers

The Elements of Life

Dr R Snaith	University of Cambridge

Developing Fuels

Dr D D Blackmore	Shell, Thornton Research Centre, Chester
Dr A D Barber	Koninklike/Shell Laboratorium, Amsterdam
Dr M du Cane	The College of Petroleum Studies, Oxford
Dr C Hawkins	BP International Ltd, Sunbury Research Centre
Dr K Owens	Automotive Fuels Consultant
Dr J Spaninks	Koninklike/Shell Laboratorium, Amsterdam
Dr J M Vernon	University of York
Professor A Williams	University of Leeds

Minerals to Elements

Dr E Barrett	Royal School of Mines, Imperial College, London
Dr A W L Dudeney	Royal School of Mines, Imperial College, London
Dr G Groslière	Exxon Chemicals

The Atmosphere

Dr C Anastasi	University of York
Dr J C Farman	British Antarctic Survey
Professor M J Pilling	University of Leeds
Dr R Powell	ICI Chemicals and Polymers
Dr J A Pyle	University of Cambridge
Professor F S Rowland	University of California, Irvine

The Polymer Revolution

Dr D Ballard	ICI Chemical and Polymers
Dr D Bott	Courtaulds Research
Dr M Jones	ICI Chemical and Polymers
Dr J Padget	ICI Chemical and Polymers

Dr C B Thomas	University of York
Dr H A Wittcoff	Chem Systems, USA

What's in a Medicine?

Dr M J Davies	University of York
Dr N Broom	SmithKline Beecham
Professor B C Gilbert	University of York
Professor C H Hassall, FRS	Formerly Roche Products Ltd
Dr M Joseph	Zeneca Pharmaceuticals
Dr A J Kirby, FRS	University of Cambridge
Dr J Salmon	Rhône-Poulenc

Engineering Proteins

Professor G G Dodson, FRS	University of York
Professor A R Fersht, FRS	University of Cambridge
Dr R E Hubbard	University of York
Dr Juswinder Singh	University College London
Dr A J Wilkinson	University of York

The Steel Story

Dr F M Garforth	University of York
Dr R J Mawby	University of York
Dr A Nicholson	British Steel
Dr M C Pope	British Steel

Aspects of Agriculture

Dr J D Birchall, FRS	ICI Chemicals & Polymers
Dr M J Bushell	Zeneca Agrochemicals
Dr I M Cunningham	Zeneca Agrochemicals
Dr D L Rowell	University of Reading
Dr A Williams	ICI Fertilisers

Colour by Design

Mr P Gregory	Zeneca Specialties, Blackley
Mr N Hughes	Zeneca Specialties, Blackley
Dr R A Jeffreys	Royal Society of Chemistry
Mr R O Ramwell	Hoechst UK Ltd

The Oceans

Dr C Anastasi	University of York
Professor H Charnock, FRS	University of Southampton
Dr N J P Owen	Plymouth Marine Laboratory
Dr C B Thomas	University of York

Medicines by Design

Dr P Bailey	University of York
Dr N Broom	SmithKline Beecham Pharmaceuticals
Dr W Dawson	Lilly Research
Mr D Madden	National Centre for Biotechnology Education, University of Reading
Dr D Roberts	Zeneca Pharmaceuticals
Dr J Salmon	Rhône-Poulenc

Visiting the Chemical Industry

Dr M J Braithwaite	Allied Colloids
Mr J E Colchester	Zeneca Specialties, Blackley
Dr G C Fettis	University of York

Safety Advisers

Dr P Borrows	ASE Laboratory Safeguards Committee
Mr R Worley	CLEAPSS School Science Service

Second Edition

Central Team

John Lazonby, University of York
Gwen Pilling (Project Director), University of York
David Waddington, University of York

Project Secretary

Sandra Wilmott, University of York

Associate Editors

Derek Denby, John Leggot College, Scunthorpe
John Dexter, The Trinity School, Nottingham
Margaret Ferguson, Lews Castle School, Stornoway
Frank Harriss, Malvern College
Gerald Keeling, Oundle School
Dave Newton, Greenhead College, Huddersfield
Brian Ratcliff, OCR (formerly Long Road Sixth Form College, Cambridge)
Mike Shipton, Oxted School, Surrey (formerly Reigate College)
Terri Vine, Loreto College (formerly Epsom and Ewell School)

Assessment Advisers

John Noel, OCR
Tony Orgee, OCR
Brian Ratcliff, OCR

Expert Advisers

We wish to thank the following for their great help given during the preparation of the Second Edition:
Adrian Abel, Clariant UK Ltd, Pigments and Additives
John Adams, Pfizer, Sandwich
Dr David Alker, Pfizer, Sandwich
Dr Jack Barratt, formerly Department of Chemistry, King's College, London
Dr David Bricknell, CEFIC, Brussels, Belgium
Professor James Carr, Department of Chemistry, University of Nebraska
Dr John Clough, Zeneca Agrochemicals, Bracknell
Louie Cononelos, Kennecott Utah Copper Corporation
Professor Leslie Crombie, FRS, formerly Department of Chemistry, University of Nottingham
Jim Docherty, BP Amoco, Grangemouth
Dr Trevor Dransfield, Department of Chemistry, University of York
Dr Frank Ellis, Glaxo Wellcome, Stevenage
Professor Guy Dodson, FRS, Department of Chemistry, University of York
Dr Paul Fraser, CSIRO Atmospheric Research/CRC for Meteorology, Victoria, Australia
Dr Irene Francois, Chairman, Society for Medicines Research
Dr Duncan Geddes, Royal Brompton and Harefield Hospital Trust
Dr R E Griggs, Zeneca Crop Protection, Haslemere
Dr Frank Hauxell, British Colour Makers Association, Oldham
Dr Adrian Hobson, Knoll Pharmaceuticals, Nottingham

Professor Andrew Holmes, Department of Chemistry, University of Cambridge
Professor Rod Hubbard, Department of Chemistry, University of York
Martin Jones, BP Amoco, Grangemouth
Johnnie Johnson, Institute of Arable Crops Research Station, Harpenden, Rothamsted
Paul Jukes, Elf Atochem
Aoenne Kerkstra, Interconfessioneel Westland College, The Netherlands
Dr Miri Kesner, Weizmann Institute of Science, Rehovot, Israel
Dr Trevor Kletz, Process Safety Consultant, Cheadle
David Knee, Zeneca Agrochemicals, Bracknell
Professor David Knight, Department of Philosophy, University of Durham
Paul Lavell, formerly Brunner Mond
Peter Lawrance, Corus, Grangetown, Middlesbrough
Sandra Marley, Rio Tinto plc, London
Dr Masson, formerly Bradford Grammar School
Dr Jamie MacLeod, York
Ms Norma McGough, British Diabetic Society
Dr Frank McKeever, BP Amoco, Grangemouth
Donald Miller, Scottish Power, Longannet, Alloa
Dr Peter Milner, Smithkline Beecham Pharmaceuticals
Dr Peter O'Brien, Department of Chemistry, University of York
Professor Robin Perutz, Department of Chemistry, University of York
Dr Dick Powell, Department of Chemistry
Bruce Rean, Kennecott Utah Copper Coporation
Dr David Rowell, Department of Soil Science, University of Reading
Dr Wendy Russell, AstraZeneca, Macclesfield
Dr David Saunders, The National Gallery, London
Jane Salter, Fertiliser Manufacturers Association
Dr John Scully, formerly School of Materials, University of Leeds
Michael Stephen, Oundle School
Dr Elizabeth Swinbank, Science Education Group, University of York
Andrew Tait, Bayer AG, Leverkusen, Germany
Dr C B Thomas, University of York
Martin Tims, Esso UK plc, Leatherhead

Dr Jon Warnke, Chem Systems, London
Dr Hazel Wilkinson, Department of Clinical Pathology, York District Hospital, York
Elizabeth Williams, Society of Dyers and Colourists, Bradford
Dr Paul Williams, Lion Laboratories Ltd, Barry
Professor Keith Wilson, Department of Chemistry, University of York

We are also indebted to the following for specific data:
NASA: Total Ozone Mapping Spectrometer (Fig and information: 18a, p.75); The Climate Monitoring and Diagnostic Laboratories of the US National Oceanic and Atmospheric Administration (Fig. 18b, p.75); Observations: CSIRO Atmospheric Research, Australia; University of East Anglia, School of Environmental Sciences, UK. Model predictions: UNEP/WMO, Switzerland (Fig. 21, p.78); Dr C D Keeling and Dr T Whorf, Scripps Institution of Oceanography (Fig. 35, p.86); *Climate Change, the IPCC Scientific Assessments*, eds J T Houghton, G J Jenkins and J J Ephramus, CUP, 1990 (Fig. 36, p.87); The Association of the British Pharmaceutical Industry: *An A to Z of British Medicines Research* (WM8, pp.117–119); The Chemical Industries Association: *The UK Chemical Industry, 1999* (Fig. 1, p.285); CEFIC (The European Chemical Industry Council): Facts and Figures, 1999 (Table 1, p.285; Table 2, p.286); *The Science in the Environment Project*, University of York Science Curriculum Centre (VCI, pp.293–294).

Sponsors

The work on the second edition has been made possible by generous funding from:
The Salters Institute of Industrial Chemistry
Esso UK plc

Guildford College
Learning Resource Centre

Please return on or before the last date shown.
No further issues or renewals if any items are overdue.
"7 Day" loans are **NOT** renewable.

Class: _540 BUR_

Title: _Chemical Storylines_

Author: _BURTON George_